W9-AFD-401

Operator Theory
Advances and Applications
Vol. 98

Editor:
I. Gohberg

New Results in Operator Theory and Its Applications

The Israel M. Glazman Memorial Volume

I. Gohberg
Yu. Lyubich
Editors

Birkhäuser Verlag
Basel · Boston · Berlin

Volume Editorial Office:
Raymond and Beverly Sackler Faculty of Exact Sciences
School of Mathematical Sciences
Tel Aviv University
IL-69978 Tel Aviv

1991 Mathematics Subject Classification 47-06, 34-06, 35-06, 39-06, 58-06

A CIP catalogue record for this book is available from the
Library of Congress, Washington D.C., USA

Deutsche Bibliothek Cataloging-in-Publication Data
New results in operator theory and its applications : the Israel M. Glazman
memorial volume / I. Gohberg ; Yu. Lyubich ed. – Basel ; Boston ; Berlin :
Birkhäuser, 1997
(Operator theory ; Vol. 98)
ISBN 3-7643-5775-4

Printed on acid-free paper produced from chlorine-free pulp. TCF ∞
Cover design: Heinz Hiltbrunner, Basel
Printed in Germany
ISBN 3-7643-5775-4
ISBN 0-8176-5775-4

9 8 7 6 5 4 3 2 1

Table of Contents

Israel Glazman

Operator Theory
Advances and Applications, Vol. 98

ISRAEL GLAZMAN, MATHEMATICIAN AND PERSONALITY

YU. LYUBICH, V.TKACHENKO*

December 21, 1996 would have been the 80-th birthday of Israel Glazman. Born in Odessa, he graduated from Odessa University where his teacher was Mark Krein. After a short period of working in the same University, he was drafted into the army, and returned home to Odessa in November 1940. From July 1941 until the very end of World War II he was an artillery officer at the front lines and took part in the heavy military actions in Stalingrad, Kursk, and Kiev.

In April 1946, I. Glazman came to Kharkov and started to work at the Department of Mathematical Physics, chaired by Naum Akhiezer. In 1955 he became Chairman of this Department.

I. Glazman returned to an active life in mathematics immediately after his demobilization. Very soon he became a distinguished scientist and public figure. He seriously studied Jewish history, and openly protested against the desecration of the Jewish cemetery in Kharkov. He readily helped Jewish young people and never tried to hide his positive feelings towards the State of Israel. Each and every one of these activities were criminal in the eyes of the Soviet authorities. KGB started persecuting and incriminating I. Glazman for Zionist activities. Any attempt of displaying Jewish cultural life, was labeled Zionist activities by the Soviet authorities. The persecution seriously affected his health, and on the tragic day of May 30, 1968, being under this unbearable emotional pressure he took his life.

Glazman's creative work was strongly influenced by the ideas and work of D. Hilbert, R. Courant, H. Weil and M. Krein. In 1949 he took his Candidate degree (Ph.D.) whose contents were later published in [5]. The following year the monograph [1] written jointly with N. Akhiezer was published. It was the first book in the world literature after the pioneering monographs of J. von Neumann [15] and M. Stone [17] on the theory of linear operators in Hilbert space. The book, besides the general theory, contained applications to the spectral theory of differential operators discovered in the forties, partly in Glazman's Thesis. The monograph by Akhiezer and Glazman tremendously enhanced expansion of operator theory ideas into modern mathematics and physics. It is still widely referred to. In 1966 the second edition appeared, and in 1977-78 the broadened two-volume edition [2, 3] was published.

*Gvastela Fellow, partially supported by Israel Science Foundation of the Israel Academy of Science and Humanities and by Israel Ministry of Science

The publications of 1950 brought wide renown to Israel Glazman. N. Dunford and J. Schwartz in their monograph [4] in connection with Glazman's Thesis compared it to the well-known results of K. Kodaira who obtained them independently at the same time. The main result here was a final classification of problems related to the boundary conditions for singular differential operators.

We would like to add that I. Glazman was guided by von Neumann's abstract theory of extension of symmetric operators. It was, possibly, the first nontrivial application of von Neumann's theory. In this context Glazman showed that the defect number m of a differential operator L of order $2n$ on half-axis satisfied the inequalities $n \leq m \leq 2n$, any m from this interval being realizable, which disproves the results due to W. Windau [18] and D. Shin [16], where it was erroneously claimed that for $n > 1$ either $m = n$ or $m = 2n$. I. Glazman investigated the structure of self-adjoint extensions of L. They turned out to be integral operators with Carleman's kernels (in particular, with Hilbert-Schmidt kernels for $m = 2n$).

In the same paper [5] I. Glazman briefly described a new method of investigating the spectrum of differential equations which he later named the *method of splitting*. Using this method, he extended to two-terms operators of order $2n$ Weil's condition $q(x) \to +\infty$ for the discreteness of spectra of operators

$$-\frac{d^2}{dx^2} + q(x). \tag{1}$$

It should be noted that in the forties A. Winter, P. Hartman, C. R. Putman, K. O. Friedrichs, F. Rellich and some other mathematicians[1] obtained a lot of results on a nature of spectrum of operator (1) and its multidimensional analogues. Glazman's method of splitting gave a new general approach to the study of spectra of one-dimensional and multi-dimensional self-adjoint differential operators of second as well as high orders, depending on the behavior of coefficients and the properties of region. I. Glazman summed up his results in his Doctoral[2] Thesis (1958). His monograph [6] related to this field was published in 1963. In that book he wrote: "The striving to find the common foundation for investigation of the spectrum of one-dimensional singular operator of any order, multi-dimensional singular boundary problems and finite-difference operators led to elaboration of operator methods of qualitative spectral analysis and, in particular, the method of splitting. This method permitted to neglect the behavior of coefficients of differential equations outside a neighborhood of singular point and to find out in what way the different singular points or the boundary influence the spectrum. The method of splitting in some sense is the way to localize the singularities."

The essence of the method is completely transparent and, being applied to the simplest operation (1) in $\mathcal{L}^2(0,\infty)$, runs as follows.

Let L be a self-adjoint operator generated in $\mathcal{L}^2(0,\infty)$ by (1) and some boundary condition at $x = 0$. Fix a number $\gamma > 0$ and restrict the functions from domain $\mathcal{D}_L$ of L by the additional conditions $y(\gamma) = y'(\gamma) = 0$. Then the operator L_0 induced by L upon this new variety $\mathcal{D}_{L_0}$ is the orthogonal sum $L_0 = L_1 \oplus L_\gamma$ of some symmetric operators in $\mathcal{L}^2(0,\gamma)$ and $\mathcal{L}^2(\gamma,\infty)$. Extending them to the self-adjoint operators, we obtain the operator $M = L_1 \oplus L_\gamma$,

[1] The detailed list of references is contained in [6]

[2] It is the second degree in fSU, the next after Ph.D.

which is called a *splitting of* L. Now the initial operator L and the splitting M are two finite-dimensional self-adjoint extensions of the same operator L_0. I. Glazman proved that any finite-dimensional self-adjoint extension does not change the continuous spectrum $C(L_0)$. The reason is that for λ to belong to the continuous spectrum $C(L)$ means that there exists an orthogonal and normalized "characteristic" sequence $\{f_n\}_1^\infty$ such that $\lim(L-\lambda)f_n = 0$. Its existence should not depend on the finite-dimensional extension or restrictions of L. It now follows that $C(L) = C(\tilde{L}_1) \cup C(\tilde{L}_\gamma)$ which makes up the principle of splitting. Since $\tilde{L}_1$ is a regular operator, it has the discrete spectrum and hence $C(\tilde{L}_1) = \emptyset$, $C(\tilde{L}_\gamma) = C(L)$.

I. Glazman widely used the splitting method in combination with the method of test functions. The typical fact linking the spectrum of an operator with such test variety follows.

If A is a self-adjoint operator, then $\Delta = (\lambda_0-\delta,\ \lambda_0+\delta)$ contains only a finite number of spectrum points iff the subspace for which $\|(A-\lambda_0)f\| \geq \delta\|f\|$ is of finite codimension.

Using the test function method, I. Glazman investigated the continuous spectrum of an operator $-\Delta + q(P)$ in the entire space $\mathbf{R}^n$ or in a domain $G \subset \mathbf{R}^n$. He proved, for example, that if a domain D contains balls of whatever great radii, and for any $\eta > 0$ the integral $\int_{M_\eta} |q(P)|^2 dv$, with $M_\eta = \{x \in G\ :\ |q(P)| > \eta\}$, is finite, then $C(L) = (0, \infty)$.

In [6] the principle of splitting was expanded to the partial differential operators. The splitting itself may be carried out as described before. Namely, let Γ be an arbitrary bounded closed surface dividing a domain G into two parts: G_1 and G_2. Any self-adjoint extension $\tilde{L}$ of the operator $L = -\Delta + q(P)$, defined on the functions from $\mathcal{D}_L$ vanishing in a neighborhood of Γ, is called a splitting of L. In contrast with the one-dimensional case, this extension is infinite-dimensional. If now $\lambda \in C(L)$, then using the Green formula, one finds that a characteristic sequence $\{f_n\}_1^\infty$ for the operator $L_\gamma = -\Delta + q(P) - \gamma$ in G converges uniformly on compact sets of G. Starting from the sequence $\varphi = f_{2n} - f_{2n-1}$ it is possible to construct a characteristic sequence for the same operator in G_2. It yields the identity $C(L) = C(L_\Gamma)$. The same argument shows that if $\lim_{|x|\to\infty} q(x) = 0$, then $C(L) = [0, \infty)$.

I. Glazman also applied the method of splitting to the study of the oscillation properties of solutions of the equation

$$(-1)^n y^{(2n)} + q(x)y = \lambda y. \tag{2}$$

For $n = 1$ this equation is called oscillatory, if its solution has an infinite number of zeros. It was known to H. Weil that Eq.(2) for $n = 1$ is oscillatory iff there is an infinite number of spectrum points of

$$L = (-1)^n \frac{d^{2n}}{dx^{2n}} + q(x)$$

to the left of λ. I. Glazman introduced a notion of oscillating equation for any $n \geq 2$. Equation (2) is called *oscillatory* if for any t there exists its nontrivial solution with more than one zero of multiplicity n to the right of t. Here we cite one result on the oscillatority of the operator L due to I.Glazman.

Let $\alpha_n = (2n-1)!!2^{-n}$. If $q(x) \geq -\alpha_n^2 x^{-2n}$, then L is non-oscillatory, and, if $q(x) \leq -(\alpha_n^2 + \delta)x^{-2n}$ for some $\delta > 0$, then it is oscillatory. For $n = 1$ this statement coincides with the well-known theorem of Knezer.

In 1957 I. Glazman returned to the extension theory of ordinary differential operators, this time, J-symmetric operators.

Let J be a conjugation in a complex Hilbert space H, i.e., J is anti-linear, continuous, and J^2 is the unit operator. The most typical example is the complex conjugation in $\mathcal{L}^2$. A linear operator A in H is called *J-symmetric*, if $(Af, Jg) = (f, JAg)$. This condition is satisfied for an ordinary differential operator

$$L = \sum_{k=0}^{n}(-1)^k \frac{d^k}{dx^k} p_k(x) \frac{d^k}{dx^k} \tag{3}$$

with complex coefficients. Any dissipative J-symmetric operator A with the dense domain may be extended to J-self-adjoint dissipative operator. In [7] I. Glazman proved that for the operator L given by formula (3) with $\mathrm{Im} p_k \geq 0$ (implying its dissipativity) the number m of linearly independent solutions of equation $Ly = \lambda y$, $y \in \mathcal{L}^2(0,\infty)$, does not depend on λ, $\mathrm{Im}\lambda < 0$, and satisfies the inequality $m \geq n$. For $m < 2n$ the resolvent R_λ of J-symmetric dissipative extension of L is an integral operator with Carleman's kernel, and for $m = 2n$ it is a Hilbert-Schmidt integral operator. Thus I. Glazman successfully expanded the main results of the extension theory of symmetric differential operators to J-symmetric dissipative operators.

It should be noted that the criterion of beauty in mathematics was of exceptional importance for Israel Glazman. It is easy to surmise that it was intrinsically connected with his outstanding musical talent. He had studied the art of violin playing under P.S. Stolyarskii, at the famous Odessa Musical School which gave David Oistrakh and Boris Goldstein to the world. Glazman was the favorite student of Stolyarskii who presented him with the thing he cherished most of all, his violin.

After demobilization in 1946, it took I. Glazman some time to make his final choice between mathematics and music. When, finally, he became a professional mathematician, his profound musical talent revealed itself in his works. For illustration we will mention a remarkable mathematical miniature created by Glazman in connection with a theorem of Mukminov [14]. In [9] he proved that if A is a bounded dissipative operator in a Hilbert space, $\{\lambda_n\}_1^\infty$ is a subset of the eigenvalue set such that

$$\sum_{m \neq n} \frac{\mathrm{Im}\lambda_m \mathrm{Im}\lambda_n}{|\lambda_m - \lambda_n|^2} < \infty,$$

then the system of corresponding normalized eigenvectors is the Bary basis in its linear hull. Glazman's elegant arguments require merely a few lines.

In the sixties I. Glazman organized a Department of Functional Analysis and Applied Mathematics at the Institute of Low Temperatures. Outstanding young mathematicians, V. P. Gurarii, V. Matsaev, V. D. Milman, came to work there. At this period of time I. Glazman switched to applied mathematics. He was interested in general problems of minimization, engineering problems of vibrations of mechanical systems, systems of control and many other fields. Among his works on these subjects we mention [10] and [11] where the minimization algorithms were suggested for a smooth functional in R^n with a finite number of critical points. These algorithms do not need *a priori* bounds for the derivatives.

Teaching mathematics was of great importance in Glazman's life. His lectures combined simplicity and clarity with depth and content. The manner of presentation was brilliant, often paradoxical. His response to a question was instant, and his connection with his audience was incessant. In Glazman's opinion the student's ability to solve problems was of paramount importance. This pedagogical principle was completely developed in the book [12] written jointly with Yu. Lyubich. It is an introduction to a linear functional analysis on a finite-dimensional level, containing more then 2400 statements without proofs. A possibility to prove them is provided by a special arrangement of this material. In [13] P. Halmos evaluated book [12] as "a beautiful and exciting contribution to the problem literature".

All people who were fortunate enough to have had contact and worked with Israel Glazman will forever remember this outstanding man, his exceptional personal charm, brilliant mathematical gift, his wit and vivacity.

References

1. N. I. Akhiezer, I. M. Glazman, *Theory of Linear Operators in Hilbert Space.* Gostechisdat, Moscow, 1950.

2. N. I. Akhiezer, I. M. Glazman, *Theory of Linear Operators in Hilbert Space.*Nauka, Moscow, 1966 (English translation: N.Y: F. Ungar Publishing Co., 1961 - 1963).

3. N. I. Akhiezer, I. M. Glazman, *Theory of Linear Operators in Hilbert Space.* Vyshcha Schkola, Kharkov, **I**, 1977; **II**, 1978 (English translation: Boston, Pitman Adv. Publ. Progr., 1981).

4. N. Dunford, J. T. Schwartz, *Linear Operators.* **II**, New York, Interscience Publ., 1963.

5. I. M. Glazman, To a theory of singular differential operators. Uspechi Mat. Nauk, **5**, 6, 1960.

6. I. M. Glazman, *Direct Methods of Qualitative Spectral Analysis of Singular Differential Operators.* Nauka, Moscow, 1963.

7. I. M. Glazman, On spectral nature of multidimensional boundary problems. Doklady Ak. Nauk, **82**, 2, 1952.

8. I. M. Glazman, On an analogue of extension theory of Hermitian operators and non-symmetric one-dimensional boundary problem on a semi-axis. Doklady Ak. Nauk, **13**, 3, 1957.

9. I. M. Glazman, On decomposability in a system of eigenvectors of dissipative operators. Uspechi Mat. Nauk, **13**, 3, 1958.

10. I. M. Glazman, On gradient relaxation for non-quadratic functionals. Doklady Ak. Nauk, **154**, 5, 1964.

11. I. M. Glazman, Relaxation on surfaces with saddle-points. Doklady Ak. Nauk, **161**, 4, 1965.

12. I. M. Glazman, Yu. I. Lyubich, *Finite-Dimensional Linear Analysis.* Nauka, Moscow, 1969 (English translation: MIT Press, Cambridge Mass.-London, 1974).

13. P. R. Halmos, The heart of Mathematics. Amer. Math. Monthly, **87**, 1980.

14. B. R. Mukminov, On expansion in series of eigenfunctions of dissipative kernels. Doklady Ak. Nauk, **99**, 4, 1954.

15. J. von Neumann, *Mathematische Grundlagen der Quantenmechanik*, Berlin, Springer, 1932.

16. D. Shin, On quasi-differential operators in a Hilbert space. Math. Sbornik, **13,** 1943.

17. M. H. Stone, *Linear Transformation in Hilbert Space and their Applications to Analysis,* New York, Amer. Math. Soc. Colloq. Publications, **15**, 1932.

18. W. Windau, Uber lineare Differentialgleichungen vierter Ordnung mit Singularitäten und die zugehörigen Darstellung willkürliche Funktionen. Math. Ann., **83**, 1921.

Yu.Lyubich, Department of Mathematics,
Technion – Israel Institute of Technology, Haifa 32000, Israel

V.Tkachenko, Department of Mathematics and Computer Science,
Ben-Gurion University of the Negev, Beer–Sheva 84105, Israel

AMS classification # 01A70

Operator Theory
Advances and Applications, Vol. 98

PRINCIPLE OF WEAKLY CONTRACTIVE MAPS IN HILBERT SPACES

YA. I. ALBER and S. GUERRE–DELABRIERE

We introduce a class of contractive maps on closed convex sets of Hilbert spaces, called weakly contractive maps, which contains the class of strongly contractive maps and which is contained in the class of nonexpansive maps. We prove the existence of fixed points for the weakly contractive maps which are a priori degenerate in general case. We establish then the convergence in norm of classical iterative sequences to fixed points of these maps, give estimates of the convergence rate and prove the stability of the convergence with respect to some perturbations of these maps. Our results extend Banach principle previously known for strongly contractive map only.

1 Introduction

Let B be a real Banach space with B^* its dual space, J be a duality mapping and $< w, v >$ be a dual product in B, i.e., a pairing between $w \in B^*$ and $v \in B$ ($< y, x >$ is an inner product in Hilbert space H, if we identify H and H^*). Let A be an operator acting from convex closed set $\Omega \subset X$ into itself and strongly contractive, i.e., there exists a constant $0 < q < 1$ such that for all x and y in Ω,

$$||Ax - Ay|| \leq q||x - y||. \tag{1.1}$$

Then the well-known Banach principle asserts that a fixed point x^* of the map A exists in Ω (i.e. $Ax^* = x^*$), and it is unique. The approximating sequence defined by

$$x_{n+1} = Ax_n, \quad n = 1, 2, ... \tag{1.2}$$

converges strongly to x^*. The convergence rate is exponential, and the convergence is stable with respect to perturbations of the operator A. In this case, the inequality

$$< x - Ax, J(x - x^*) > \geq q||x - x^*||^2$$

holds and the fixed point x^* is called nondegenerate.

As $q \to 1$, the above estimates (for convergence rate, stability, etc.) become worse, and they do not remain true in general if $q = 1$. In this case the operator A is called nonexpansive:

$$||Ax - Ay|| \leq ||x - y||. \tag{1.3}$$

We say that A is properly nonexpansive if it is nonexpansive and inequality (1.3) admits no refinement. In this case, the inequality

$$< x - Ax, J(x - x^*) > \geq 0$$

holds, the fixed point x^* is degenerate and a problem of its finding is a priori unstable. The existence of x^* is asserted on bounded closed convex sets Ω of strictly convex Banach spaces and convergence of the sequence (1.2) can be only weak (see [7, 9, 10]), meanwhile the Cesaro means of the iterative process (1.2) converge weakly on unbounded sets (see [5, 7]). The strong convergence of iterations is only secured by applying some regularizing procedure.

The condition (1.1) on A can be written as:

$$||Ax - Ay|| \leq ||x - y|| - a||x - y||,$$

where $a = 1 - q$. The functional $a||x - y||$ characterizes "a measure of deviation" of the strongly contractive map from nonexpansive one. We will now consider more general situation(cf. [11], p.52).

Definition 1.1 *A map A will be called weakly contractive on a closed convex set Ω in the Banach space B if there exists a continuous and nondecreasing function $\psi(t)$ defined on R^+ such that ψ is positive on $R^+ \setminus \{0\}$, $\psi(0) = 0$, $\lim_{t \to +\infty} \psi(t) = +\infty$ and $\forall x, y \in \Omega$,*

$$||Ax - Ay|| \leq ||x - y|| - \psi(||x - y||). \tag{1.4}$$

Remark 1.2 *If in addition, Ω is a bounded set then the hypothesis $\lim_{t \to +\infty} \psi(t) = +\infty$ is not necessary.*

Remark 1.3 *It follows from Definition 1.1 that $\psi(t) \leq t$ for all $t \in R^+$.*

A simple example of the inequality (1.2) is the following:

$$||Ax - Ay|| \leq ||x - y|| - a(||x - y||^\alpha \wedge ||x - y||^\beta),$$

where $\alpha > 1$ and $\beta \leq 1$

The paper is organized as follows. In the Section 2, we are going to prove the existence of a unique fixed point for weakly contractive map.

In the Section 3, we get an analogue of the Banach principle, called the principle of weakly contractive maps: the iterations (1.2) converge in norm to the unique fixed point x^* of A. Moreover, we give estimates of general type for the convergence rate and establish stability of the convergence with respect to some perturbations of the operator A. In this case, the inequality

$$< x - Ax, J(x - x^*) > \geq \psi(||x - x^*||)$$

holds and x^* is uniformly degenerate.

In the Section 4, we prove the convergence in norm for another iterative process, namely

$$x_{n+1} = x_n - \omega_n(x_n - Ax_n), \quad n = 1, 2, ..., \tag{1.5}$$

where ω_n is a sequence of positive numbers which converges to 0, $0 \leq \omega_n \leq 1$. In this case we also obtain estimates of general and power type for the convergence rate and stability.

For simplicity, we formulate our results in Hilbert spaces. Nevertheless, all of them are valid at least in uniformly convex and uniformly smooth Banach spaces. The proofs use a general theory of nonlinear recursive inequalities [1, 2].

2 Existence of a fixed point for weakly contractive maps

Theorem 2.1 *If A is a weakly contractive map on $\Omega \subset H$ then it has a unique fixed point $x^* \in \Omega$.*

Proof. Let $x_0 \in \Omega$ be given and let $\lambda_n < 1$ be a sequence of positive numbers converging to 1. For each $n \geq 1$, let A_n be a map from Ω to Ω defined for all $x \in \Omega$ by:

$$A_n x = (1 - \lambda_n)x_0 + \lambda_n Ax.$$

Then for any $x, y \in \Omega$, $||A_n x - A_n y|| \leq \lambda_n ||x - y||$ and A_n is strongly contractive for all $n = 1, 2, ...$ By the Banach principle, A_n has a unique fixed point x_n.

Let us prove firstly that the sequence x_n is bounded. Indeed, the equality

$$x_n = (1 - \lambda_n)x_0 + \lambda_n Ax_n$$

implies

$$\frac{\lambda_n - 1}{\lambda_n}(x_n - x_0) = x_n - Ax_n. \tag{2.1}$$

Thus taking the inner product $< . , . >$ of both terms of (2.1) with $x_n - x_0$ one obtains

$$0 \geq \frac{\lambda_n - 1}{\lambda_n}||x_n - x_0||^2 = < x_n - Ax_n, x_n - x_0 > .$$

But, by definition,

$$< (x_n - Ax_n) - (x_0 - Ax_0), x_n - x_0 > \geq \psi(||x_n - x_0||)||x_n - x_0||.$$

So,

$$||x_0 - Ax_0||||x_n - x_0|| \geq$$

$$< x_n - Ax_n, x_n - x_0 > + ||x_0 - Ax_0||||x_n - x_0|| \geq \psi(||x_n - x_0||)||x_n - x_0||.$$

Dividing by $||x_n - x_0||$, we obtain that

$$\psi(||x_n - x_0||) \leq ||x_0 - Ax_0||.$$

By the properties of ψ, this implies that the sequence x_n is bounded.

Taking up again the equality

$$\lambda_n(x_n - Ax_n) = (1-\lambda_n)(x_0 - x_n)$$

we get then that

$$||x_n - Ax_n|| \leq \frac{1-\lambda_n}{\lambda_n}||x_n - x_0||,$$

which proves that the sequence $||x_n - Ax_n||$ converges to 0. As well known [10], this implies that A has a fixed point.

The uniqueness of the fixed point is obvious. ■

3 The principle of weakly contractive maps for the iterative process (1.2)

Theorem 3.1 *Let A be a weakly contractive map on $\Omega \subset H$. Then the iterative process (1.2) starting at $x_1 \in \Omega$, converges strongly to the fixed point x^* and the following estimate holds:*

$$||x_n - x^*|| \leq \Phi^{-1}\Big(\Phi(||x_1 - x^*||) - (n-1)\Big),$$

where Φ is defined by $\Phi(t) = \int \frac{dt}{\psi(t)}$ and Φ^{-1} is its inverse function.

Proof. For all $n \in I\!N$, we can write:

$$||x_{n+1} - x^*|| = ||Ax_n - Ax^*|| \leq ||x_n - x^*|| - \psi(||x_n - x^*||).$$

The sequence of positive numbers $\{\lambda_n\}$ defined by $\lambda_n = ||x_n - x^*||$ satisfies the following inequality

$$\lambda_{n+1} \leq \lambda_n - \psi(\lambda_n).$$

This implies, in particular, that the sequence $\{\lambda_n\}$ is nonincreasing. Thus it converges to some λ such that $\psi(\lambda) \leq 0$. By the hypothesis on ψ, this implies that $\lambda = 0$. Further, we can apply the result of [2], which gives the estimate: for all $n \geq 1$,

$$\lambda_n \leq \Phi^{-1}\Big(\Phi(\lambda_1) - (n-1)\Big). \tag{3.1}$$

Indeed, by definition of Φ, $\forall n \geq 1$, there exists $\bar{\lambda}_n \in [\lambda_{n+1}, \lambda_n]$ such that:

$$\Phi(\lambda_n) - \Phi(\lambda_{n+1}) = \int_{\lambda_{n+1}}^{\lambda_n} \frac{dt}{\psi(t)} = \frac{\lambda_n - \lambda_{n+1}}{\psi(\bar{\lambda}_n)} \geq \frac{\lambda_n - \lambda_{n+1}}{\psi(\lambda_n)} \geq 1.$$

This implies that for all $n \geq 1$,

$$\Phi(\lambda_{n+1}) \leq \Phi(\lambda_n) - 1 \leq ... \leq \Phi(\lambda_1) - n.$$

Hence (3.1) holds.

Since $\psi(t)$ is strictly positive as $t > 0$, Φ and Φ^{-1} are strictly increasing. Thus the function $\chi(t) = \Phi^{-1}\Big(\Phi(\lambda_1) - (t-1)\Big)$ is strictly decreasing.

Since $\psi(t) \leq t$, we have that

$$\lim_{\epsilon\to 0} \Phi(1) - \Phi(\epsilon) = \lim_{\epsilon\to 0} \int_\epsilon^1 \frac{dt}{\psi(t)} \geq \lim_{\epsilon\to 0} \int_\epsilon^1 \frac{dt}{t} = +\infty.$$

Thus

$$\lim_{\epsilon\to 0} \Phi(\epsilon) = -\infty.$$

Taking inverse functions one obtains that

$$\lim_{s\to -\infty} \Phi^{-1}(s) = \lim_{t\to +\infty} \chi(t) = 0$$

Thus $\lambda_n = ||x_n - x^*||$ tends to 0 as $n \to +\infty$ and this completes the proof of the theorem. ■

Remark 3.2 *$\Phi(t)$ means the antiderivative with the null constant.*

Remark 3.3 *If in Definition 1.1, (1.4) is replaced by the inequality*

$$||Ax - Ay|| \leq ||x - y|| - a\psi(||x - y||),$$

where a is some positive constant then

$$||x_n - x^*|| \leq \Phi^{-1}\Big(\Phi(||x_1 - x^*||) - a(n-1)\Big).$$

Examples of functions Φ and Φ^{-1} for $\psi(t) = t^p$, $p > 1$, $\psi(t) = c(e^t - 1)$, $t \geq 0$, $0 < c < 1$, $\psi(t) = t/(t+1)$, $t \geq 0$, are given in [2].

Let us consider two stability theorems for the convergence of these approximations: we study the perturbed iterative method

$$y_{n+1} = A_n y_n, \quad n = 1, 2, ..., \tag{3.2}$$

where maps A_n are closely related to A, in two different ways.

Theorem 3.4 *Let Ω be a closed convex set in the Hilbert space H and let A be a weakly contractive map from Ω to Ω. Suppose that there exist sequences of positive numbers $h_n, \beta_n, \delta_n, \mu_n$ and ν_n converging to 0 as $n \to \infty$, a finite positive function $g(t)$ defined on R^+, a sequence of positive functions $\psi_n(t)$ on R^+ and a sequence of maps A_n from Ω to Ω such that for all $x, y \in \Omega$, $n \geq 1$ and $t \geq 0$,*

$$||A_n x - A_n y|| \leq (1 + \beta_n)||x - y|| - \psi_n(||x - y||) + \mu_n, \tag{3.3}$$

$$||A_n x - Ax|| \leq h_n g(||x||) + \delta_n, \tag{3.4}$$

$$|\psi_n(t) - \psi(t)| \leq \nu_n. \tag{3.5}$$

If $\sum_{n=1}^{+\infty} \beta_n < +\infty$ then the iterative process (3.2) starting at arbitrary $y_1 \in \Omega$ converges in norm to the fixed point x^* of A.

Proof. For all $n \geq 1$, one has

$$||y_{n+1} - x^*|| = ||A_n y_n - Ax^*|| \leq ||A_n y_n - A_n x^*|| + ||A_n x^* - Ax^*||$$
$$\leq (1+\beta_n)||y_n - x^*|| - \psi_n(||y_n - x^*||) + \mu_n + h_n g(||x^*||) + \delta_n$$
$$\leq (1+\beta_n)||y_n - x^*|| - \psi(||y_n - x^*||) + \gamma_n,$$

where $\gamma_n = h_n g(||x^*||) + \delta_n + \mu_n + \nu_n$.

The sequence of positive numbers $\{\lambda_n\}_{n\geq 1}$ defined by

$$\lambda_n = ||y_n - x^*||,$$

satisfies:

$$\lambda_{n+1} \leq (1+\beta_n)\lambda_n - \psi(\lambda_n) + \gamma_n.$$

This inequality is a particular case of a more general inequality which was studied in [2]:

$$\lambda_{n+1} \leq (1+\beta_n)\lambda_n - \alpha_n\psi(\lambda_n) + \gamma_n, \tag{3.6}$$

where $\sum_{n=1}^{\infty} \alpha_n = +\infty$ and $\dfrac{\gamma_n}{\alpha_n} \to 0$. For each $n \geq 1$ there are two possibilities: either

$$\psi(\lambda_n) \leq (\frac{1}{\mathcal{A}_n} + \frac{\gamma_n}{\alpha_n}), \tag{3.7}$$

where $\mathcal{A}_m = \sum_1^m \alpha_i$, or

$$\psi(\lambda_n) > (\frac{1}{\mathcal{A}_n} + \frac{\gamma_n}{\alpha_n}). \tag{3.8}$$

Let us introduce two sets: the set $I^1 = \{n_l : l = 1, 2, ...\} \subset I\!N$, such that the inequality (3.7) is satisfied for all $n = n_l$ and the complementary set $I^2 = I\!N \setminus I^1$, such that the opposite inequality (3.8) holds for all $n \in I^2$. We prove that I^1 is unbounded. Suppose that I^1 is bounded. Then there exists a number N_0 such that

$$\psi(\lambda_n) > (\frac{1}{\mathcal{A}_n} + \frac{\gamma_n}{\alpha_n}), \quad \forall n \geq N_0,$$

and

$$\lambda_{n+1} \leq (1+\beta_n)\lambda_n - \frac{\alpha_n}{\mathcal{A}_n}, \quad \forall n \geq N_0.$$

Make the change of variables: $\lambda_1 = \mu_1$ and $\lambda_n = \mu_n \prod_{i=1}^{n-1}(1+\beta_i)$. Then

$$\mu_{n+1} \leq \mu_n - \frac{1}{\prod_{i=1}^{n}(1+\beta_i)}\frac{\alpha_n}{\mathcal{A}_n}, \quad \forall n \geq N_0. \tag{3.9}$$

Since the infinite product $\prod_{i=1}^{\infty}(1+\beta_i)$ converges by the hypothesis $\sum_{i=1}^{\infty}\beta_i < +\infty$, (3.9) implies that

$$\mu_{n+1} \leq \mu_n - \frac{1}{\bar{C}}\frac{\alpha_n}{\mathcal{A}_n}, \quad \forall n \geq N_0,$$

where $\prod_{i=1}^{\infty}(1+\beta_i) \leq \bar{C}$. Then μ_n satisfies the following inequalities:

$$\mu_{n+1} \leq \mu_{N_0} - \frac{1}{\bar{C}} \sum_{i=N_0}^{n} \frac{\alpha_i}{\mathcal{A}_i}, \quad \forall n \geq N_0.$$

By the same hypothesis, the series $\sum_{n=1}^{\infty} \alpha_n$ does not converge and by Abel-Dini's theorem, the series $\sum_{n=1}^{\infty} \frac{\alpha_n}{\mathcal{A}_n}$ does not converge either. Since $\mu_n \geq 0$ for all $n \geq 1$, we get a contradiction.

Thus I^1 is unbounded and there exists a subsequence $\{\psi(\mu_{n_l})\}_{l\geq 1}$ of $\{\psi(\mu_n)\}_{n\geq 1}$ which converges to 0. By the properties of ψ, the same subsequence $\{\mu_{n_l}\}_{l\geq 1}$ of $\{\mu_n\}_{n\geq 1}$ converges also to 0.

Let $[n_l, n_{l+1}] = [n_l, n_l+1, n_l+2, ..., n_{l+1}]$. Since $\lambda_{n_l} \leq \bar{C}\mu_{n_l}$ and

$$\lambda_{n_l+1} \leq (1+\beta_{n_l})\lambda_{n_l} - \alpha_n\psi(\lambda_{n_l}) + \gamma_{n_l},$$

we obtain that

$$\lambda_{n_l+1} \leq (1+\beta_{n_l})\bar{C}\mu_{n_l} + \gamma_{n_l},$$

and for all $n \in [n_l+2, n_{l+1})$

$$\lambda_n \leq (1+\beta_{n_l+1})\lambda_{n_l+1}.$$

Then the full sequence $\{\lambda_n\}_{n\geq 1}$ converges to 0. Therefore, $\{y_n\}_{n\geq 1}$ converges in norm to x^*. ■

Remark 3.5 *In this theorem, (3.4) can be replaced by the weaker condition:*

$$||A_n x^* - Ax^*|| \leq \delta_n. \quad ■$$

Assumptions (3.3) and (3.5) can be omitted if y_n are uniformly bounded or A_n approximate the map A uniformly. In fact, the following statement is valid.

Theorem 3.6 *Let Ω be a closed convex set in the Hilbert space H and let A be a weakly contractive map from Ω to Ω. Suppose that there exist sequences of positive numbers, $\{h_n\}_{n\geq 1}$ and $\{\delta_n\}_{n\geq 1}$, a finite positive function $g(t)$ defined on R^+ and a sequence of maps $\{A_n\}_{n\geq 1}$ from Ω to Ω such that, for all $x \in \Omega$ and for all $n \geq 1$ the inequality (3.4) holds. Then,*
1. If the iterative sequence (3.2) starting at $y_1 \in \Omega$ is bounded, then it converges in norm to the fixed point x^* of A as $\{h_n\}$ and $\{\delta_n\}$ tend to 0.
2. If $h_n = 0$ for all $n \geq 1$, strong convergence of $\{y_n\}$ to x^* takes place if $\{\delta_n\}$ tends to 0.

Proof. For all $n \geq 1$, we can write:

$$||y_{n+1} - x^*|| = ||A_n y_n - Ax^*|| \leq ||Ay_n - Ax^*|| + ||A_n y_n - Ay_n||$$

$$\leq ||y_n - x^*|| - \psi(||y_n - x^*||) + (h_n g(||y_n||) + \delta_n)$$

We define for all $n \geq 1$, $\gamma_n = h_n g(||y_n||) + \delta_n$. Then by the hypothesis, the sequence $\{\gamma_n\}_{n\geq 1}$ converges to 0.

As before, if $\lambda_n = ||y_n - x^*||$, one obtains that this sequence satisfies the inequalities for all $n \geq 1$:

$$\lambda_{n+1} \leq \lambda_n - \psi(\lambda_n) + \gamma_n$$

This is a particular case of the inequalities (3.6) of Theorem 3.4 and in the same way we find that the sequence $\{\lambda_n\}_{n\geq 1}$ converges to 0. Therefore, again $\{y_n\}_{n\geq 1}$ converges in norm to x^*. ■

Example. The map $Ax = \sin x$ from $[0, 1]$ to $[0, 1]$ is weakly contractive. Indeed, for all $0 < y < x < 1$, we have:

$$0 \leq \sin x = x - \frac{x^3}{3!} + \frac{x^5}{5!} + \dots$$

and

$$0 \leq \sin y = y - \frac{y^3}{3!} + \frac{y^5}{5!} + \dots .$$

The series

$$\sin x - \sin y = (x - y) - \frac{1}{3!}(x^3 - y^3) + \frac{1}{5!}(x^5 - y^5) + \dots$$

is of Leibnitz type, because for all $n \geq 1$:

$$\frac{x^{2n+1}}{(2n+1)!} - \frac{y^{2n+1}}{(2n+1)!} \geq \frac{x^{2n+3}}{(2n+3)!} - \frac{y^{2n+3}}{(2n+3)!}.$$

Hence,

$$\sin x - \sin y \leq (x - y) - \frac{1}{3!}(x^3 - y^3) + \frac{1}{5!}(x^5 - y^5).$$

For any positive a and b $(a > b)$

$$a^r - b^r \leq r a^{r-1}(a - b), \quad r > 1.$$

In particular,

$$x^5 - y^5 \leq 5x^4(x - y).$$

Therefore we can write

$$5(x^3 - y^3) = 5(x - y)(x^2 + xy + y^2) \geq 5x^2(x - y) \geq$$

$$5x^4(x - y) \geq x^5 - y^5.$$

This gives

$$\frac{x^5 - y^5}{5!} \leq \frac{x^3 - y^3}{4!},$$

and

$$\sin x - \sin y \leq (x - y) - \frac{1}{8}(x^3 - y^3).$$

Since

$$x^3 - y^3 \geq (x - y)^3,$$

we have

$$|\sin x - \sin y| \leq |x - y| - \frac{1}{8}|x - y|^3.$$

In this example $\psi(t) = t^3/8, \Phi(t) = -4t^{-2}$, $\Phi^{-1}(z) = (-z/4)^{-\frac{1}{2}}$ and fixed point $x^* = 0$ is degenerate of the order 2, because $F(x^*) = 0, F'(x^*) = 0, F''(x^*) = 0, F'''(x^*) \neq 0$, where $F(x) = x - Ax = x - \sin x$. Theorems 3.1 and 3.6 imply that the iterative sequence $x_{n+1} = \sin x_n$ strongly converges to 0 with the estimate (see also the Remark 3.3), for all $x_1 \in [0,1]$,

$$x_n \le \frac{1}{\sqrt{x_1^{-2} + \frac{n-1}{4}}}, \quad n = 1, 2, \dots,$$

and the convergence is stable. We find again a well known estimate (cf. [8]). ■

4 The principle of weakly contractive maps for the iterative process (4.1)

Now we will consider the iterative process

$$x_{n+1} = (1 - \omega_n)x_n + \omega_n A x_n, \quad n = 1, 2, \dots . \tag{4.1}$$

The following convergence theorem is true.

Theorem 4.1 *Let $\{\omega_n\}$ be a sequence of positive numbers such that $\sum_1^\infty \omega_n = \infty$, $\sum_1^\infty {\omega_n}^2 < \infty$, and $\omega_n \le \omega \le 1$ for all $n \ge 1$. Let Ω be a closed convex set in Hilbert space H and let A be a weakly contractive map from Ω to Ω. Then the iterative sequence (4.1) starting at $x_1 \in \Omega$ converges in norm to the fixed point x^* of the map A and the following estimate holds:*

$$||x_n - x^*||^2 \le \bar{C}\Phi^{-1}\Big(\Phi(||x_1 - x^*||^2) - \frac{2\sum_1^{n-1}\omega_i}{\sqrt{\bar{C}}(1 + 4\omega^2)}\Big), \tag{4.2}$$

where Φ is defined by $\Phi(t) = \displaystyle\int \frac{dt}{\psi_1(t)}$, $\psi_1(t) = \psi(t^{\frac{1}{2}})t^{\frac{1}{2}}, \Phi^{-1}$ is its inverse function and $\bar{C} \ge \Pi(1 + 4{\omega_i}^2)$

Proof. Let us note that $\psi_1(t)$ is strictly increasing because $\psi(t)$ is nondecreasing. Therefore $\psi_1^{-1}(z)$ exists. For all $n \ge 1$, we can write:

$$||x_{n+1} - x^*||^2 = ||x_n - x^* - \omega_n(x_n - Ax_n)||^2$$

$$\begin{aligned} &\le ||x_n - x^*||^2 - 2\omega_n < (x_n - Ax_n) - (Ax^* - x^*), x_n - x^* > + \omega_n^2||x_n - Ax_n||^2 \\ &\le ||x_n - x^*||^2 - 2\omega_n\psi(||x_n - x^*||)||x_n - x^*|| + 4\omega_n^2||x_n - x^*||^2 \\ &= (1 + 4\omega_n^2)||x_n - x^*||^2 - 2\omega_n\psi(||x_n - x^*||)||x_n - x^*||. \end{aligned} \tag{4.3}$$

Here we have used the estimate:

$$||x_n - Ax_n|| = ||(x_n - Ax_n) - (x^* - Ax^*)||$$

$$\leq ||x_n - x^*|| + ||Ax_n - Ax^*|| \leq 2||x_n - x^*||.$$

Let us define a sequence of positive numbers $(\lambda_n)_{n\geq 1}$ by $\lambda_n = ||x_n - x^*||^2$. This sequence has the following property: for all $n \geq 1$,

$$\lambda_{n+1} \leq (1+4\omega_n^2)\lambda_n - 2\omega_n\psi(\lambda_n^{1/2})\lambda_n^{1/2} \leq (1+4\omega_n^2)\lambda_n - 2\omega_n\psi_1(\lambda_n). \tag{4.4}$$

As before, let

$$I^1 = \{n|\ n \geq 1,\ \psi_1(\lambda_n) \leq \frac{1}{\sum_{i=1}^n \omega_i}\}$$

and

$$I^2 = \{n|\ n \geq 1,\ \psi_1(\lambda_n) > \frac{1}{\sum_{i=1}^n \omega_i}\} = I\!N \setminus I^1.$$

Let us show that I^1 is unbounded. If this is not true then there exists $N_0 \geq 1$ such that for all $n \geq N_0$, $\psi_1(\lambda_n) > \dfrac{1}{\sum_{i=1}^n \omega_i}$. Then for $n \geq N_0$,

$$\lambda_{n+1} \leq (1+4\omega_n^2)\lambda_n - \frac{2\omega_n}{\sum_{i=1}^n \omega_i} \leq \lambda_{N_0}\prod_{j=N_0}^{n}(1+4\omega_j^2) - 2\sum_{j=N_0}^{n}(\frac{\omega_j}{\sum_{i=1}^j \omega_i})$$

$$\leq \lambda_{N_0}\prod_{j=N_0}^{\infty}(1+4\omega_j^2) - 2\sum_{j=N_0}^{n}(\frac{\omega_j}{\sum_{i=1}^j \omega_i})$$

Since the series $\sum_{j=N_0}^{\infty}(\frac{\omega_j}{\sum_{i=1}^j \omega_i})$ does not converge and $\lambda_n \geq 0$ for all $n \geq 1$, a contradiction is obtained. Hence, I^1 is unbounded and there exists a subsequence $\{\psi_1(\lambda_{n_l})\}_{l\geq 1}$ of $\{\psi_1(\lambda_n)\}_{n\geq 1}$ which converges to 0. By the properties of ψ_1, the subsequence $\{\lambda_{n_l}\}_{l\geq 1}$ of $\{\lambda_n\}_{n\geq 1}$ also converges to 0.

The inequalities $\lambda_{n+1} \leq (1+4\omega_n^2)\lambda_n - \omega_n\psi_1(\lambda_n)$ imply, in particular, that the sequence $\{\lambda_n\}_{n\geq 1}$ has the following property: for $n \in [n_l+1, n_{l+1}]$,

$$\lambda_n \leq \lambda_{n_l}\prod_{i=n_l}^{n-1}(1+4\omega_i^2) \leq \bar{C}\lambda_{n_l}$$

So the whole sequence $\{\lambda_n\}_{n\geq 1}$ converges to 0.

Show now (4.2) holds. First of all, establish for all $\alpha \in [0,1]$ the inequality

$$\alpha^{\frac{1}{2}}\psi_1(t) \geq \psi_1(\alpha t). \tag{4.5}$$

In fact, by hypothesis, $\psi(\alpha t) \leq \psi(t), \forall \alpha \in [0,1]$. Then

$$\psi_1(\alpha t) = \psi(\alpha^{\frac{1}{2}}t^{\frac{1}{2}})\alpha^{\frac{1}{2}}t^{\frac{1}{2}} \leq \psi(t^{\frac{1}{2}})\alpha^{\frac{1}{2}}t^{\frac{1}{2}} = \psi_1(t)\alpha^{\frac{1}{2}}, \tag{4.6}$$

i.e. (4.5) is true.

Since $(\prod_{i=1}^{\infty}(1+4\omega_i^2))^{-1} \leq 1$, the change of variables in (4.4) $\lambda_1 = \mu_1$ and $\lambda_n = \mu_n\prod_{i=1}^{n-1}(1+4\omega_n^2)$ gives the following: for all $n \geq 1$,

$$\mu_{n+1} \leq \mu_n - \frac{2\omega_n}{\prod_{i=1}^n(1+4\omega_i^2)}\psi_1(\mu_n\prod_{i=1}^{n-1}(1+4\omega_i^2))$$

$$\leq \mu_n - \frac{2\omega_n}{(1+4\omega_n^2)\sqrt{\prod_{i=1}^{n-1}(1+4\omega_i^2)}}\psi_1(\mu_n) \leq \mu_n - \frac{2\omega_n}{\sqrt{\bar{C}(1+4\omega_n^2)}}\psi_1(\mu_n). \tag{4.7}$$

The same argument as in Theorem 3.1, gives easily: for all $n \geq 1$,

$$\mu_n \leq \Phi^{-1}\Big(\Phi(\mu_1) - \frac{2\sum_{i=1}^{n-1}\omega_i}{\sqrt{\bar{C}(1+4\omega_n^2)}}\Big),$$

that is

$$\lambda_n \leq \chi(n),\ \ \chi(n) = \bar{C}\Phi^{-1}\Big(\Phi(\lambda_1) - \frac{2\sum_{i=1}^{n-1}\omega_i}{\sqrt{\bar{C}(1+4\omega^2)}}\Big).$$

Thus λ_n converges to 0 as $n \to +\infty$ and this completes the proof. ■

The estimate of the convergence rate can be improved if the function $\psi_1(t)$ is assumed to have a stronger property.

Theorem 4.2 *If in addition to the hypotheses of Theorem 4.1, $\psi_1(t)$ is convex, or more generally, $\alpha\psi_1(t) \geq \psi_1(\alpha t)$ for all $0 \leq \alpha \leq 1$, then*

$$||x_n - x^*||^2 \leq \bar{C}\Phi^{-1}\Big(\Phi(||x_1 - x^*||^2) - \frac{2\sum_1^{n-1}\omega_i}{1+4\omega^2}\Big). \tag{4.8}$$

Proof. Note that the inequality $\alpha\psi_1(t) \geq \psi_1(\alpha t)$ follows from convexity of the function $\psi_1(t)$. Then (4.7) can be rewritten as follows:

$$\mu_{n+1} \leq \mu_n - \frac{2\omega_n}{\prod_{i=1}^{n}(1+4\omega_i^2)}\psi_1(\mu_n\prod_{i=1}^{n-1}(1+4\omega_i^2)) \leq$$

$$\mu_n - \frac{2\omega_n}{1+4\omega_n^2}\psi_1(\mu_n).$$

Like in Theorem 4.1, one obtains from this the estimate (4.8). ■

We formulate now the convergence theorem without assumption $\sum_1^\infty {\omega_n}^2 < \infty$.

Theorem 4.3 *Let $\{\omega_n\}$ be a sequence of positive numbers such that $\sum_1^\infty \omega_n = \infty$, $0 < \omega_n \leq 1$, $\lim_{n\to\infty}\omega_n = 0$. Let Ω be a closed convex set in Hilbert space H and let A be a weakly contractive map from Ω to Ω. Then the iterative sequence (4.1) starting at $x_1 \in \Omega$ converges in norm to the fixed point x^* of the map A.*

Proof. (4.3) is rewritten as:

$$||x_{n+1} - x^*||^2 \leq ||x_n - x^*||^2 - 2\omega_n\psi_1(||x_n - x^*||^2) + \omega_n^2||x_n - Ax_n||^2.$$

We can estimate $||x_n - Ax_n||$ in the last term, namely:

$$||x_{n+1} - Ax_{n+1}|| = ||(1-\omega_n)x_n + \omega_n Ax_n - Ax_n + Ax_n - Ax_{n+1}||$$

$$\leq (1-\omega_n)||x_n - Ax_n|| + ||Ax_n - Ax_{n+1}||$$

$$\leq (1-\omega_n)||x_n - Ax_n|| + ||x_n - x_{n+1}||$$
$$= (1-\omega_n)||x_n - Ax_n|| + \omega_n||x_n - Ax_n|| = ||x_n - Ax_n||.$$

Thus always

$$||x_n - Ax_n|| \leq ||x_1 - Ax_1|| = K.$$

This implies

$$||x_{n+1} - x^*||^2 \leq ||x_n - x^*||^2 - 2\omega_n\psi_1(||x_n - x^*||^2) + K^2\omega_n^2,$$

or with $\lambda_n = ||x_n - x^*||^2$,

$$\lambda_{n+1} \leq \lambda_n - 2\omega_n\psi_1(\lambda_n) + K^2\omega_n^2.$$

The statement of Theorem follows now from [2] (see also [4]). ∎

One can get estimates of the convergence rate for the approximating sequence (4.1) with general function $\psi_1(t)$ in the hypotheses of last theorem. However, such estimates have sufficiently complicated analytical description. Therefore we consider here the important particular cases, when

$$\psi_1(t) = t^p,\ p > 1,\ \omega_n = \omega n^{-s},\ 0 < s \leq 1,\ \omega = const. > 0. \tag{4.9}$$

Denote:

$$u(n) = \left(C^{1-p} + 2\omega(p-1)\frac{c_0 - 1}{c_0}(\ln n - \ln 2)\right)^{-\frac{1}{p-1}},\quad C = max\{Q, C_1\},$$

$$Q = \left(||x_1 - x^*||^{2(1-p)} + 2\omega(p-1)\frac{c_0-1}{c_0}\ln 2\right)^{-\frac{1}{p-1}},$$

$$C_1 = (2^{-1}\omega K^2 c_0)^{\frac{1}{p}} + \omega^2 K^2,\quad c_0 > 1,\quad K = ||x_1 - Ax_1||.$$

Theorem 4.4 *Let $\{\omega_n\}$ be a sequence (4.9) with $s = 1$. Let Ω be a closed convex set in Hilbert space H and let A be a weakly contractive map from Ω to Ω. Then the iterative sequence (4.1) starting at $x_1 \in \Omega$ converges in norm to the fixed point x^* of the map A and the following estimate holds: if a constant $c_0 > 1$ satisfies the condition*

$$2\omega C_1^{p-1}\frac{c_0-1}{c_0} \leq \frac{1}{p},$$

then

$$||x_n - x^*||^2 \leq u(n),\ n = 2, 3, \ldots.$$

Denote in the next theorem:

$$v(n) = \left((2^{-1}\omega K^2 c_0)^{\frac{1}{p}} + \omega^2 K^2\right)(n-1)^{-\frac{s}{p}},\quad C = max\{Q, C_1\},$$

$$u(n) = \left(C^{1-p} + 2\omega(p-1)\frac{c_0-1}{c_0}\frac{p-1}{1-s}(n^{1-s} - 2^{1-s})\right)^{-\frac{1}{p-1}},$$

$$Q = \left(||x_1 - x^*||^{2(1-p)} + 2\omega(p-1)\frac{c_0-1}{c_0}\frac{p-1}{1-s}(2^{1-s} - 1)\right)^{-\frac{1}{p-1}},$$

$$C_1 = (2^{-1}\omega K^2 c_0)^{\frac{1}{p}} + \omega^2 K^2,\quad c_0 > 1,\quad K = ||x_1 - Ax_1||.$$

Theorem 4.5 *Let $\{\omega_n\}$ be a sequence (4.9). Let Ω be a closed convex set in Hilbert space H and let A be a weakly contractive map from Ω to Ω. Then the iterative sequence (4.1) starting at $x_1 \in \Omega$ converges in norm to the fixed point x^* of the map A and the following statements hold:*

1. Suppose that $0 < s < 1$,

$$\frac{s}{p} \geq \frac{1-s}{p-1},$$

and a constant $c_0 > 1$ satisfies the condition

$$2\omega C_1^{p-1} \frac{c_0 - 1}{c_0} \leq \frac{s}{p}.$$

Then

$$||x_n - x^*||^2 \leq u(n), \quad n = 2, 3, \dots .$$

2. Suppose that $0 < s < 1$ and

$$\frac{s}{p} < \frac{1-s}{p-1}.$$

(i) If $Q \leq C_1$ and a constant $c_0 > 1$ satisfies the inequality

$$2\omega C_1^{p-1} \frac{c_0 - 1}{c_0} \leq \frac{2^s s}{p},$$

then

$$||x_n - x^*||^2 \leq v(n), \; n = 2, 3 \dots .$$

(ii) In all remaining cases the estimates

$$||x_n - x^*||^2 \leq u(n), \quad 2 \leq n \leq \bar{\xi},$$

$$||x_n - x^*||^2 \leq v(n), \; n > \bar{\xi},$$

are satisfied, where $\bar{\xi}$ is a unique root of the equation $u(\xi) = v(\xi)$ in the interval $[2, \infty)$.

The proofs of Theorem 4.4 and 4.5 follow from the general theory of nonlinear recursive inequalities [1, 2]. We emphasize that the asymptotic estimates are:

$$||x_n - x^*||^2 = O\Big((\ln n)^{-\frac{1}{p-1}}\Big), \; if \; s = 1,$$

and

$$||x_n - x^*||^2 = O(n^{-\frac{1-s}{p-1}}), \; if \; 0 < s < 1. \blacksquare$$

We study further the stability of iterative process (4.1) with the perturbed map A_n, i.e. consider a convergence of the sequence:

$$y_{n+1} = (1 - \omega_n) y_n + \omega_n A_n y_n, \quad n = 1, 2, \dots , \tag{4.10}$$

where $\{\omega_n\}_{n \geq 1}$ is a sequence of positive numbers such that $0 \leq \omega_n \leq 1$, $\sum_1^\infty \omega_n = +\infty$ and $\sum_1^\infty \omega_n^2 < +\infty$.

Theorem 4.6 *Suppose that there exist sequences of positive numbers $\{h_n\}_{n\geq 1}$ and $\{\delta_n\}_{n\geq 1}$ converging to 0, a finite positive function $g(t)$ defined on R^+ and a sequence of maps $\{A_n\}_{n\geq 1}$ from Ω to Ω such that, for all $x \in \Omega$ and for all $n \geq 1$, the inequality (3.4) is carried out. If either $h_n = 0$ for all $n \geq 1$, or the iterative sequence (4.10) starting at $y_1 \in \Omega$ is bounded, then (4.10) converges in norm to the fixed point x^* of the map A.*

Proof. Indeed,

$$||y_{n+1} - x^*||^2 = ||y_n - x^* - \omega_n(y_n - A_n y_n)||^2$$

$$= ||y_n - x^*||^2 - 2\omega_n < y_n - A_n y_n, y_n - x^* > +\omega_n^2 ||y_n - A_n y_n||^2$$

$$= ||y_n - x^*||^2 - 2\omega_n < y_n - Ay_n - A_n y_n + Ay_n, y_n - x^* > +\omega_n^2 ||y_n - A_n y_n||^2$$

$$\leq ||y_n - x^*||^2 - 2\omega_n < y_n - Ay_n, y_n - x^* > +2\omega_n ||Ay_n - A_n y_n|| ||y_n - x^*|| + \omega_n^2 ||y_n - A_n y_n||^2$$

$$\leq ||y_n - x^*||^2 - 2\omega_n \psi_1(||y_n - x^*||^2) + 2\omega_n ||Ay_n - A_n y_n|| ||y_n - x^*|| + \omega_n^2 ||y_n - A_n y_n||^2$$

$$\leq ||y_n - x^*||^2 - 2\omega_n \psi_1(||y_n - x^*||^2)$$

$$+2\omega_n(h_n g(||y_n||) + \delta_n)||y_n - x^*|| + \omega_n^2 ||y_n - A_n y_n||^2.$$

Furthermore:

$$||y_n - A_n y_n||^2 \leq 2(||y_n - Ay_n||^2 + ||Ay_n - A_n y_n||^2) \leq 2(4||y_n - x^*||^2 + (h_n g(||y_n||) + \delta_n)^2).$$

Let for all $n \geq 1$

$$\gamma_n = 2\omega_n(h_n g(||y_n||) + \delta_n)||y_n - x^*|| + 2\omega_n^2(h_n g(||y_n||) + \delta_n)^2 + 8\omega_n^2 ||y_n - x^*||^2.$$

Then by the hypothesis of the theorem, the sequence $\{\gamma_n\}_{n\geq 1}$ converges to 0.

Then, taking up again the initial inequality, we get:

$$||y_{n+1} - x^*||^2 \leq ||y_n - x^*||^2 - 2\omega_n \psi_1(||y_n - x^*||^2) + \gamma_n.$$

Define $\lambda_n = ||y_n - x^*||^2$. Then the sequence $\{\lambda_n\}_{n\geq 1}$ satisfies the following inequality

$$\lambda_{n+1} \leq \lambda_n - 2\omega_n \psi_1(\lambda_n) + \gamma_n.$$

This inequality is exactly the same as in Theorem 3.6: in the same way, we can get that $\{\lambda_n\}_{n\geq 1}$ converges to 0. Therefore, again, $\{y_n\}_{n\geq 1}$ converges in norm to x^*. ∎

Remark 4.7 *The hypothesis in Theorem 4.6 that the approximating sequence is bounded, can be removed. For this, we have to modify the approximating sequence and to define it by:*

$$y_{n+1} = (1 - \omega_{n+\sigma})y_n + \omega_{n+\sigma} A_n y_n,$$

where σ is a sufficiently large integer. The proof of this result is longer and more complicated.

Definition 4.8 *A map A from a closed convex set Ω into itself is said to be weakly subcontractive if there exists fixed point $x^* \in \Omega$ for A and a continuous and nondecreasing function $\psi(t)$ defined on R^+ such that ψ is positive on $R^+ - \{0\}$, $\psi(0) = 0$, $\lim_{t\to+\infty} \psi(t) = +\infty$ and for all $x \in \Omega$,*

$$||Ax - x^*|| \leq ||x - x^*|| - \psi(||x - x^*||).$$

It is clear from the proofs, that the above convergence and stability theorems remain true if A is only weakly sub-contractive. For Example of Section 3 we have

$$|\sin x| \leq |x| - \frac{|x|^3}{3!}.$$

Then the iterative sequence $x_{n+1} = \sin x_n$ converges to 0 with the estimate

$$x_n \leq \frac{1}{\sqrt{x_1^{-2} + \frac{2n}{7}}}$$

for all $n \geq 1$.

References

[1] Ya. I. Alber, On the solution of equations and variational inequalities with maximal monotone operators, *Soviet Math. Dokl.*, 20 (1979), 871-876.

[2] Ya. I. Alber, Recurrence relations and variational inequalities, *Soviet Math. Dokl.*, 27 (1983), 511-517.

[3] Ya. I. Alber, Metric and generalized projection operators in Banach spaces: properties and applications, in "Theory and Applications of Nonlinear Operators of Accretive and Monotone Types" (A. Kartsatos, Ed.), pp. 15–50, Marcel Dekker Inc., 1966.

[4] Ya. Alber and S. Reich, An iterative method for solving a class of nonlinear operator equations in Banach spaces. *Panamerican Math. J.*, 4 (1994), 39-54.

[5] J.B. Baillon, R.E. Bruck and S. Reich, On the asymptotic behavior of nonexpansive mappings and semigroupes in Banach spaces. *Houston J. of Math.*, 4 (1978), 1-9.

[6] B. F. Browder, Nonlinear operators and nonlinear equations of evolution in Banach spaces. Proceedings of Symp. in Pure Math. Vol XVIII, Part 2, A.M.S. 1976.

[7] R.E. Bruck, On the convex approximation property and the asymptotic behavior of nonlinear contractions in Banach spaces. *Israel J. of Math.*, 38 (1981), 304-314.

[8] A. Chamber-Loir, Etude du sinus itere. *Revue de Math. Spe* , 1993, 457-460.

[9] K. Goebel and W. A. Kirk, Topics in metric fixed point theory. Cambridge Studies in Advanced Maths 28, Cambridge Univ. Press (1990).

[10] K. Goebel and S. Reich, Uniform convexity, hyperbolic geometry and nonexpansive mappings. Pure and Applied Maths 83, Marcel Dekker (1984).

[11] M.A Krasnosel'skii, G.M. Vainikko, P.P. Zabreiko, Ya.B. Rutitskii, V.Ya. Stetsenko, Approximate solution of operator equations. Wolters-Noordhoff Publishing, Groningen, 1972.

Ya. I. Alber, Technion-Israel Institute of Technology, Haifa, Israel,
S. Guerre-Delabriere, University Paris VI, Paris, France.

AMS (MOS) 1991 Mathematics Subject Classification: 47H10, 47H15, 47H09, 47H17, 65J15.

Operator Theory
Advances and Applications, Vol. 98

POTENTIALS ASSOCIATED TO RATIONAL WEIGHTS

D. ALPAY and I. GOHBERG

Dedicated to the memory of I.M. Glazman

In this paper we consider the problem of building the spectral function of a canonical differential equation when the potential is given. We restrict ourselves to the case where the spectral function is rational. An algorithm is proposed which allows the construction of the spectral function from the values of the potential and of a number of its derivatives at the origin. The approach is based on the solution of the partial realization problem for systems.

1 Introduction

Let H denote a differential operator of the form

$$(Hf)(t) = -iJ\frac{\mathrm{d}\, f}{\mathrm{d}\, t}(t) - V(t)f(t), \quad t \geq 0, \tag{1.1}$$

where

$$J = \begin{pmatrix} I_m & 0 \\ 0 & -I_m \end{pmatrix} \quad \text{and } V(t) = \begin{pmatrix} 0 & k(t) \\ k(t)^* & 0 \end{pmatrix}. \tag{1.2}$$

Here, $k(t)$ is a $\mathbb{C}^{m\times m}$–valued function with entries in $L_1(0,\infty)$. It is sometimes called the potential of the differential operator, or the local reflexivity coefficient function (see [5] for this latter interpretation). Associated to the operator H are a number of functions which play an important role, in particular the scattering function and the spectral function. To define the scattering function, consider for real λ the $\mathbb{C}^{2m\times m}$–valued solution of the equation

$$-iJ\frac{\mathrm{d}}{\mathrm{d}\, t}X(t,\lambda) - V(t)X(t,\lambda) = \lambda X(t,\lambda),$$

subject to the boundary conditions

$$\begin{aligned}(I_m \ \ -I_m)X(0.\lambda) &= 0,\\ (I_m \ 0)X(t,\lambda) &= e^{-i\lambda t}I_m + o(1) \ \ (t\to\infty).\end{aligned}$$

Such a solution exists and is unique (see [11], [6]). It has the property that

$$(0 \ I_m)X(t,\lambda) = S(\lambda)e^{i\lambda t} + o(1) \ \ (t\to\infty).$$

The function S is called the scattering matrix; to give its properties, we first recall that the Wiener algebra $\mathcal{W}^{m\times m}$ consists of the matrix-valued functions of the form

$$Z(\lambda) = D - \int_{-\infty}^{\infty} z(t)e^{i\lambda t}dt \tag{1.3}$$

where $D \in \mathbb{C}^{m\times m}$ and $z \in L_1^{m\times m}(\mathbb{R})$. Note that $D = \lim_{\lambda\to\pm\infty} Z(\lambda)$; we will use the notation $D = Z(\infty)$. The subalgebra $\mathcal{W}_-^{m\times m}$ consists of the elements of the form (1.3) for which the support of z is in $\mathbb{R}_-$. Similarly, $\mathcal{W}_+^{m\times m}$ consists of the elements of the form (1.3) for which the support of z is in $\mathbb{R}_+$.

The scattering function S has the following properties: it takes unitary values, belongs to $\mathcal{W}^{m\times m}$ and $S(\infty) = I_m$, and it admits a Wiener–Hopf factorization: it can be written as

$$S(\lambda) = S_-(\lambda)S_+(\lambda), \tag{1.4}$$

where S_- and its inverse are in $\mathcal{W}_-^{m\times m}$ and S_+ and its inverse are in $\mathcal{W}_+^{m\times m}$.

We now turn to the spectral function. The operator H defined by (1.1) is selfadjoint when restricted to the space D_H of $\mathbb{C}^{2m}$–valued functions f which are absolutely continuous and which satisfy the initial value $(I_m \ \ -I_m)f(0) = 0$. Let W be a $\mathbb{C}^{m\times m}$–valued function which is continuous on the real line and for which $W(\lambda) > 0$ for all real λ. It is called a spectral function for the operator H if there is a unitary mapping $U : L_2^{2m}(0,\infty) \to L_2^m(W)$ such that $(UHf)(\lambda) = \lambda(Uf)(\lambda)$ for $f \in D_H$, where $L_2^m(W)$ is the Hilbert space of $\mathbb{C}^m$–valued measurable functions g such that $\int_{-\infty}^{\infty} g(t)^*W(t)g(t)dt < \infty$. If S given by (1.4) is the scattering function of the operator (1.1), then the function

$$W(\lambda) = S_-(\lambda)^{-1}S_-(\lambda)^{-*} \tag{1.5}$$

is a spectral function of H, and the map U is given in terms of the continuous orthogonal polynomials of M.G. Krein (see [8], [6]). We will call this function the spectral function of the operator H; it is uniquely determined from the scattering function S and the condition $W(\infty) = I_m$. Let $W \in \mathcal{W}^{m\times m}$, with $W(\infty) = I_m$. The function W admits Wiener–Hopf factorizations $W = W_+W_+^* = W_-W_-^*$, where W_- and its inverse are in $\mathcal{W}_+^{m\times m}$ and W_- and its inverse are in $\mathcal{W}_-^{m\times m}$. The function W is the spectral function of the differential operator (1.1) with scattering matrix–function $S = W_-^{-1}W_+$.

The direct scattering problem consists in computing the scattering function S or the associated spectral function W defined by (1.5) from the function V. The inverse scattering problem is the other way around and consists in reconstructing the function V from the spectral function W, and it was solved in [9], [10], [11]. In [2] was presented an explicit solution of the inverse scattering problem when the scattering function S (or equivalently the spectral function W) is rational; in order to formulate it, we first recall a number of facts from the theory of realization of matrix–valued rational functions. Any $\mathbb{C}^{m\times m}$–valued rational function W, analytic on the real line and at infinity with $W(\infty) = I_m$, can be written as

$$W(\lambda) = I_m + C(\lambda I_n - A)^{-1}B, \tag{1.6}$$

where $(A, B, C) \in \mathbb{C}^{n\times n} \times \mathbb{C}^{n\times m} \times \mathbb{C}^{m\times n}$. In this paper, we denote by $\mathbb{C}^{m\times n}$ the space of m–rows and n–columns matrices with complex entries, and $\mathbb{C}^m$ is short for $\mathbb{C}^{m\times 1}$; the identity matrix of $\mathbb{C}^{m\times m}$ is denoted by I_m, or simply by I. The adjoint of a matrix A is denoted by A^*. Such an expression (1.6) is called a realization of W. The realization is called minimal if the number n in (1.6) is as small as possible and the minimal such n is called the McMillan degree of W. Two minimal realizations of W are similar: namely, if $W(\lambda) = I_m + C_i(\lambda I_n - A_i)^{-1}B_i$, $i = 1, 2$ are two minimal realizations of W, there exists a (uniquely defined and invertible) matrix $S \in \mathbb{C}^{n\times n}$ such that

$$A_2 = SA_1S^{-1} \quad B_2 = SB_1 \quad C_2 = C_1S^{-1}. \tag{1.7}$$

For these facts and more information on the theory of realization of matrix–valued functions, we refer to [3] and [12].

Take W to be rational of the form (1.6), analytic on the real line, and suppose that $W(\lambda) > 0$ for all $\lambda \in \mathbb{R}$, and $W(\infty) = I_m$. Then, W is of the form (1.3) with

$$k(t) = \begin{cases} iCe^{-iuA}(I-P)B & \text{if } u > 0 \\ -iCe^{-iuA}PB & \text{if } u \leq 0 \end{cases}$$

where P denotes the Riesz projection corresponding to the eigenvalues of A in the open upper half–plane (see [4, p.287]). It is the spectral function of a differential operator of the form (1.1). The potential is expressed as follows: assume that the realization (1.6) of W is minimal and denote by P the Riesz projection corresponding to the eigenvalues of A in the open upper half–plane $\mathbb{C}_+$. Then, with $A^\times = A - BC$,

$$k(t) = 2C\left(Pe^{-2itA^\times}|_{\mathrm{Im}\,P}\right)^{-1}PB. \tag{1.8}$$

It is more convenient to express some formulas in terms of the function $\kappa(t)$ introduced by

$$\kappa(t) = C\left(Pe^{tA^\times}|_{\mathrm{Im}\,P}\right)^{-1}PB. \tag{1.9}$$

Then, $k(t) = 2\kappa(-2it)$ and for $\ell \geq 0$,

$$k^{(\ell)}(0) = 2(-2i)^\ell\kappa^{(\ell)}(0). \tag{1.10}$$

Let $X(t) = Pe^{tA^\times}|_{\mathrm{Im}\,P}$. Then, $X'(t) = PA^\times e^{tA^\times}|_{\mathrm{Im}\,P}$. Furthermore,

$$(X^{-1})'(t) = -X^{-1}(t)X'(t)X^{-1}(t).$$

Using these formulas it is easily computed that

$$
\begin{aligned}
\kappa(0) &= CPB,\\
\kappa^{(1)}(0) &= -CPAB + (CPB)^2,\\
\kappa^{(2)}(0) &= CPA^2B - (CPAB)(CPB) - (CBP)(CPAB) +\\
&\quad +2(CPB)^3 - (CPB)(CB)(CPB),\\
\kappa^{(3)}(0) &= -CPA^3B + (CPA^2B)(CBP) + (CPB)(CPA^2B)\\
&\quad -(CPB)(CAB)(CPB) + 6(CPB)^4 + (CPAB)^2\\
&\quad -3(CPAB)(CPB)^2 - 3(CPB)^2(CPAB)\\
&\quad +2(CPB)(CB)(CPAB) + 2(CPAB)(CB)(CPB)\\
&\quad -3(CPB)(CB)(CPB)^2 - 3(CPB)^2(CB)(CPB)\\
&\quad +(CPB)(CB)^2(CPB).
\end{aligned}
$$

Further, in Proposition 4.1, it is proved that, for any $\ell \geq 0$, the following equality holds:

$$CA^{\ell}B = CA^{\ell}PB + (CA^{\ell}PB)^*. \tag{1.11}$$

Using this equality, we can invert the previous equations and obtain the following equalities (in fact, we use (1.11) only starting with the equality for CA^2PB):

$$
\begin{aligned}
CPB &= \kappa(0),\\
CAPB &= -\kappa'(0) + (CPB)^2\\
&= -\kappa'(0) + \kappa(0)^2,\\
CA^2PB &= \kappa''(0) + (CPAB)(CPB) + (CPB)(CPAB) - 2(CPB)^3 + (CPB)(CB)(CPB)\\
&= \kappa''(0) + \kappa(0)^3 - \kappa(0)\kappa^{(1)}(0)\\
&\quad -\kappa^{(1)}(0)\kappa(0) + \kappa(0)\kappa(0)^*\kappa(0)\\
CA^3PB &= -\kappa^{(3)}(0) + (CPA^2B)(CPB) + +(CPB)(CPA^2B)\\
&\quad -(CPB)(CAB)(CPB) + 6(CPB)^4 + (CPAB)^2\\
&\quad -3(CPAB)(CPB)^2 - 3(CPB)^2(CPAB)\\
&\quad +2(CPB)(CB)(CPAB) + 2(CPAB)(CB)(CPB)\\
&\quad +(CPB)(CB)^2(CPB)\\
&\quad -3(CPB)(CB)(CPB)^2 - 3(CPB)^2(CB)(CPB)\\
&= -\kappa^{(3)}(0) + \kappa(0)\kappa^{(2)}(0) + \kappa^{(1)}(0)^2\\
&\quad +\kappa^{(2)}(0)\kappa(0) - \kappa(0)^2\kappa^{(1)}(0)\\
&\quad -\kappa(0)\kappa^{(1)}(0)\kappa(0) - \kappa^{(1)}\kappa(0)^2\\
&\quad -2\kappa(0)\kappa(0)^*\kappa^{(1)}(0) - 2\kappa^{(1)}\kappa(0)^*\kappa(0)\\
&\quad +\kappa(0)\kappa^{(1)}(0)\kappa(0) + \kappa(0)^4 + \kappa(0)^2\kappa(0)^*\kappa(0)\\
&\quad +\kappa(0)\kappa(0)^*\kappa(0)^*.
\end{aligned}
$$

We show in Theorem 3.1 that these computations can be extended and that for every $\ell \geq 1$, the matrix $CA^{\ell}PB$ is a noncommutative polynomial function of the matrices

$$\kappa(0), \ldots, \kappa^{(\ell)}(0)$$

and of their conjugates. This remark can be used in order to solve the direct spectral problem and to calculate the spectral function from the values of the reflection coefficient function $k(t)$ and of a number of its derivatives at the origin, in the case when it is known that the spectral function is rational. This is based on the results of partial realization from system theory, which allow to reconstruct C, A and B and hence the function $I_m + C(\lambda I - A)^{-1}B$ from a finite number of matrices $CA^\ell B$. We especially use results from [7]. In Theorem 3.2 and Theorem 3.3 is presented an algorithmic way of computing the formulas for the matrix $CA^\ell B$ in terms of the $\kappa(0), \ldots, \kappa^{(\ell-1)}(0)$ and of their conjugates. Finally, in Theorem 3.5 the functions which arise as potentials of differential operators of the form (1.1), i.e. which can be written as (1.8) are characterized. In proving these theorems, we will need some results on reconstructing a matrix–valued rational function W analytic at infinity (and with $W(\infty) = I_m$) from part of its Laurent expansion at infinity

$$W(\lambda) = I_m + \sum_{j=1}^{\infty} \frac{W_j}{\lambda^j}.$$

The paper consists of five sections; this introduction is the first. In section 2, we review results on the partial realization problem. In section 3 we present the main results of the paper. The proofs of these results are presented in section 5; preliminary computations are first presented in section 4.

It is a pleasure to thank M.A. Kaashoek for his comments on the paper, and in particular for shortening the proofs of the propositions in Section 4. We also thank J.W. Helton, S. Kojcinovic and M. Stankus for their help in the computation of the formulas appearing in the preceding page using their NCAlgebra program.

2 Partial realization for rational matrix functions

This section contains the main results about the partial realization problem, which will be used later for computing the spectral function of the operator (1.1) from the values of $k(0), \ldots, k^{(\ell-1)}(0)$ and of their adjoints. Recall that the triple $(A, B, C) \in \mathbb{C}^{t \times t} \times \mathbb{C}^{t \times q} \times \mathbb{C}^{p \times t}$ is called a realization of the matrices $W_1, \ldots, W_r \in \mathbb{C}^{p \times q}$ if $W_j = CA^{j-1}B$ for $j = 1, \ldots, r$. The realization is called minimal if the size t of the square matrix A is as small as possible. The minimal such t is called the degree of $W_1, \ldots, W_r \in \mathbb{C}^{p \times q}$, and is denoted by $\delta(W_1, \ldots, W_r)$. A realization always exists for $t = pr$ and

$$A = \begin{pmatrix} 0 & I_p & 0 & \cdot \\ \cdot & \cdot & \cdot & \cdot \\ \cdot & \cdot & 0 & I_p \\ 0 & \cdot & \cdot & 0 \end{pmatrix}, \qquad B = \begin{pmatrix} W_1 \\ \vdots \\ W_r \end{pmatrix}, \qquad C = (I \; 0 \; \cdots \; 0),$$

but it need not be minimal.

The following criteria for minimality is proved in [7, Theorem 0.1]. In the statement, we use the notation

$$N_j(\Sigma) \quad = \quad \cap_{\nu=0}^{j-1} \ker CA^\nu \tag{2.1}$$

$$R_j(\Sigma) = \bigvee_{\nu=0}^{j-1} \operatorname{Im} A^\nu B. \tag{2.2}$$

Theorem 2.1 *Let $\Sigma = (A, B, C)$ be a realization of the matrices $W_1, \ldots, W_r$, with state space X. It is minimal if and only if the following three conditions hold:*

(1) $N_r(\Sigma) = \{0\}$.

(2) $R_r(\sigma) = X$.

(3) *For $1 \leq j \leq r-1$, it holds that $N_j(\Sigma) \subset R_{r-j}(\Sigma)$.*

Two realizations of the sequence of matrices $W_1, \ldots, W_r$ are called similar if there is an invertible matrix S such that (1.7) holds. The next result from [7] gives a necessary and sufficient condition for two minimal realizations to be similar.

Proposition 2.2 *Let $W_1, \ldots, W_r \in \mathbb{C}^{p\times q}$, and for $i, j \leq r-1$, let $H_{i,j}$ denotes the block Hankel matrix with u, v entry W_{u+v-1}, with $u \in \{1, \ldots, i\}$ and $v \in \{1, \ldots, j\}$. Let α and β be defined by:*

$$\alpha = \operatorname{Min} \{i \,|\operatorname{Rank} H_{i,2n-i} = \operatorname{Rank} H_{i+1,2n-i}\} \tag{2.3}$$

$$\beta = \operatorname{Min} \{j \,|\operatorname{Rank} H_{2n-j,j} = \operatorname{Rank} H_{2n-j,j+1}\} \tag{2.4}$$

for $1 \leq i, j \leq r$. Then a necessary and sufficient condition for the minimal realization of the sequence $W_1, \ldots, W_r$ to be unique (up to a similarity matrix) is that

$$\alpha + \beta \leq r. \tag{2.5}$$

As a corollary of the preceding proposition, we have:

Lemma 2.3 *Let $W_1, \ldots, W_r$ be the first r coefficients of the Laurent expansion at infinity of the function $W(\lambda) = I_m + C(\lambda I_n - A)^{-1}B$. Then, $W_i = CA^{i-1}B$ and*

$$H_{ij} = \begin{pmatrix} C \\ CA \\ \vdots \\ CA^{i-1} \end{pmatrix} \begin{pmatrix} B & AB & \cdots & A^{j-1}B \end{pmatrix}. \tag{2.6}$$

Assume that the realization $W(\lambda) = I_m + C(\lambda I_n - A)^{-1}B$ of the function W is minimal. Then,

$$\operatorname{Rank} H_{ij} < n \quad \text{when} \quad i < n \text{ or } j < n \tag{2.7}$$

$$\operatorname{Rank} H_{ij} = n \quad \text{when} \quad i \geq n \text{ and } j \geq n. \tag{2.8}$$

In particular,

$$\alpha(W_1, \ldots, W_{2n}) = \beta(W_1, \ldots, W_{2n}) = n. \tag{2.9}$$

In [7] a compression algorithm is developped to obtain a minimal realization from any given realization. This algorithm is made out of three basic operations, which we now recall. We set $\Sigma = (A, B, C)$ and recall that the notations $N_j(\Sigma)$ and $R_j(\Sigma)$ were defined in (2.1) and (2.2). In the following propositions, the symbol $\oplus$ denotes a direct sum.

Proposition 2.4 *Let $\Sigma = (A, B, C)$ be a realization of the matrices $W_1, \dots, W_r$, with state space X. Let $j \in \{1, \dots, r-1\}$ and let $X_1 \subset N_j(\Sigma)$ be such that*

$$X_1 \oplus R_{r-j}(\Sigma) = R_{r-j}(\Sigma) + N_j(\Sigma).$$

Let $X_0 \subset X$ be such that $X_1 \oplus X_0 = X$. Consider the partitions of A, B, C according to this decomposition of X

$$A = \begin{pmatrix} A_{00} & A_{10} \\ A_{01} & A_{11} \end{pmatrix}, \qquad B = \begin{pmatrix} B_0 \\ B_1 \end{pmatrix}, \qquad C = (C_0 \ C_1). \tag{2.10}$$

Then, the system $\Sigma_0 = (A_{00}, B_0, C_0)$ is a realization of the matrices $W_1, \dots, W_r$ and $N_j(\Sigma_0) \subset R_{r-j}(\Sigma_0)$.

Proposition 2.5 *Let $\Sigma = (A, B, C)$ be a realization of the matrices $W_1, \dots, W_r$, with state space X. Set $X_1 = N_r(\Sigma)$, let X_0 be such that $X_1 \oplus X_0 = X$, and consider the associated partition (2.10) of the system Σ. Then, the system $\Sigma_0 = (A_{00}, B_0, C_0)$ is a realization of the matrices $W_1, \dots, W_r$ and $N_r(\Sigma_0) = \{0\}$.*

Proposition 2.6 *Let $\Sigma = (A, B, C)$ be a realization of the matrices $W_1, \dots, W_r$, with state space X. Set $X_0 = R_r(\Sigma)$, let X_1 be such that $X_1 \oplus X_0 = X$, and consider the associated partition (2.10) of the system Σ. Then, the system $\Sigma_0 = (A_{00}, B_0, C_0)$ is a realization of the matrices $W_1, \dots, W_r$ and $R_r(\Sigma_0) = X_0$.*

The compression algorithm of [7] consists of a repeated application of these three propositions. Starting from an arbitrary realization of the matrices $W_1, \dots, W_r$, one obtains, after a finite number of steps, a realization for which the conditions (1), (2) and (3) of Theorem 2.1 hold, i.e. a minimal realization.

The following theorem is proved using the results of [7] and will be needed in the sequel.

Theorem 2.7 *Let W be a $\mathbb{C}^{m \times m}$-valued rational function analytic at infinity with $W(\infty) = I_m$, and let $n = \deg W$. Let $W_1, \cdots, W_{2n}$ be the $2n$ first coefficients of the Laurent expansion of the matrix-valued rational function W. Then, $\delta(W_1, \dots, W_{2n}) = n$, and all minimal realizations of the matrices $W_1, \dots, W_{2n}$ are similar. In particular, the function W may be reconstructed from $W_1, \dots, W_{2n}$ by the formula $W(\lambda) = I_m + C(\lambda I_n - A)^{-1}B$ where (A, B, C) is any minimal realization of $W_1, \dots, W_{2n}$.*

Proof: Let $W(\lambda) = I_m + C(\lambda I_n - A)^{-1}B$ be a minimal realization of W. Therefore, as already noted, $W_i = CA^{i-1}B$ for $i \geq 1$. It follows that for every j,

$$\delta(W_1, \dots, W_j) \leq n$$

and in particular, $\delta(W_1, \ldots, W_{2n}) \leq n$. From the preceding lemma, we have $\alpha = n$ and $\beta = n$. In particular, $\alpha + \beta$ is equal to the number of matrices W_i considered. By Proposition 2.2, the minimal realization is then unique up to similarity. By Theorem 3.2 of [7], the degree of the sequence $W_1, \ldots, W_{2n}$ is equal to the rank of the matrix $H_{\alpha,\beta}$. Since, as shown in Lemma 2.3, Rank $H_{n,n} = n$, it follows that $\delta(W_1, \ldots, W_{2n}) = n$. The last claim follows easily from these facts. ■

3 The main theorems

Let $X_1, \ldots, X_r \in \mathbb{C}^{m\times m}$. A noncommutative polynomial in $X_1, \ldots, X_r$ is a finite linear combination with complex coefficients of finite products $Z_1 \cdots Z_s$, where $s \geq 0$ and where the Z_i are chosen among the X_i; if $s = 0$, the product is taken to be equal to the identity matrix I_m.

Theorem 3.1 *Let $(A, B, C) \in \mathbb{C}^{n\times n} \times \mathbb{C}^{n\times m} \times \mathbb{C}^{m\times n}$ be a triple of matrices such that both A and $A^\times = A - BC$ have no real spectrum, and let P denote the Riesz projection corresponding to the eigenvalues of A in the open upper half-plane $\mathbb{C}_+$. Assume that (1.11) holds for all $\ell \geq 0$ and define*

$$k(t) = 2C\left(Pe^{-2itA^\times}|_{\text{Ran } P}\right)^{-1} PB. \tag{3.1}$$

Then, $k(0) = 2CPB$ and for every $\ell \geq 0$, there exists a noncommutative polynomial $\mathbf{p}_\ell$ in the 2ℓ variables $k(0), k(0)^*, \ldots, k^{(\ell-1)}(0), k^{(\ell-1)}(0)^*$ (with coefficients independent of A, B, C) such that

$$CA^\ell PB = (-1)^\ell c_\ell k^{(\ell)}(0) + \mathbf{p}_\ell(k(0), k(0)^*, \ldots, k^{(\ell-1)}(0), k^{(\ell-1)}(0)^*) \tag{3.2}$$

where $c_\ell = -(2(-2i)^\ell)^{-1}$.
The equations (3.2) can be inverted, and for every $\ell \geq 1$, there exists a noncommutative polynomial $\mathbf{q}_\ell$ in the 2ℓ variables CA^jPB, CA^jB, $j = 0, \ldots, \ell - 1$ such that

$$k^{(\ell)}(0) = (-1)^\ell c_\ell^{-1} CA^\ell PB + \mathbf{q}_\ell(CB, (CPB), \cdots, CA^{\ell-1}B, CPA^{\ell-1}B). \tag{3.3}$$

Combining (1.11) and (3.2), it follows that $CA^\ell B$ is a noncommutative polynomial function of the 2ℓ variables $k^{(i)}(0), k^{(i)}(0)^*$, $i = 0, \ldots, \ell - 1$:

$$CA^\ell B = (c_\ell k^{(\ell)}(0) + \mathbf{p}_\ell(k(0), k(0)^*, \ldots)) + (c_\ell k^{(\ell)}(0) + \mathbf{p}_\ell(k(0), k(0)^*, \ldots))^*. \tag{3.4}$$

This last formula gives in fact the $(\ell + 1)$–th coefficient of the Laurent expansion of W at infinity, and thus gives a formula for this coefficient independent of the given realization of W.

Condition (1.11) holds in particular when the function $W(\lambda) = I_m + C(\lambda I_n - A)^{-1}B$ takes strictly positive values on the real line, as is shown in Proposition 4.1. The noncommutative polynomials $\mathbf{p}_\ell$ and $\mathbf{q}_\ell$ play a central role in the characterization of potentials associated to rational weights functions. The proof of Theorem 3.1 contains an algorithmic way to compute these polynomials, which is summarized in the next theorem; in the statement, A_N denotes the matrix

$$A_N = \begin{pmatrix} -CB & I & 0 & 0 & \cdot \\ -CAB & 0 & I & 0 & \cdot \\ \cdot & \cdot & \cdot & \cdot & \cdot \\ \cdot & \cdot & \cdot & 0 & I \\ -CA^{N-1}B & \cdot & \cdot & 0 & I \end{pmatrix}. \tag{3.5}$$

Theorem 3.2 *The noncommutative polynomials $\mathbf{q}_\ell$ are defined by*

$$\mathbf{q}_\ell(CB, CPB, \ldots) = c_\ell^{-1}(\ell!) \sum_{j=0}^{\ell-1} \alpha_{\ell j} CPA^j B, \tag{3.6}$$

where the $\alpha_{\ell j}$ are matrices which are computed in the following steps: $\alpha_{\ell\ell} = \frac{(-1)^\ell}{\ell!} I$ and, for $0 \leq j \leq \ell - 1$, $\alpha_{\ell j}$ is the noncommutative polynomials in the $2j$ variables F_v, G_v $v = 0, \ldots j-1$, defined by

$$\alpha_{\ell+1,j} = -t_{\ell+1,j} - - \sum_{u,v \in T_{\ell j}} \frac{\alpha_{vu}}{(\ell+1-v)!} - \sum_{v,u \in J_{\ell j}} \frac{\alpha_{vu}}{(\ell+1-v)!} r_{u,\ell-1-v,j}. \tag{3.7}$$

In this expression, $T_{\ell j}$ and $J_{\ell j}$ are defined by:

$$T_{\ell j} = \{u, v |\ u + \ell + 1 - v = j;\ 0 \leq u \leq j;\ 1 \leq v \leq \ell\}, \tag{3.8}$$

$$J_{\ell j} = \{u, v |\ 1 \leq v \leq \ell;\ 0 \leq u \leq v;\ j \leq u + \ell - v\}, \tag{3.9}$$

and $t_{N,j}$ and $s_{N,j}$, $N = 1, \ldots,$ $j = 0, \ldots, N-1$ are the noncommutative polynomials respectively in the $2N$ variables $F_j = CA^jB$ and $G_j = CPA^jB$, $j = 0, \ldots, N-1$, and in the N variables $F_j = CA^jB$, $j = 0, \ldots, N-1$ computed from the recurrence relations $t_{1,0} = -G_0$ and $s_{1,0} = -F_0$ and for $N \geq 1$

$$(t_{N+1,0} \cdots t_{N+1,N}) = (t_{N,0} \cdots t_{N,N-1}) A_N - (G_N\ 0 \cdots 0) \tag{3.10}$$

$$(s_{N+1,0} \cdots s_{N+1,N}) = (s_{N,0} \cdots s_{N,N-1}) A_N - (F_N\ 0 \cdots 0). \tag{3.11}$$

The $r_{N,M,j}$, $j = 0, \ldots, N+M-1$ are the noncommutative polynomials in the $2(N+M)$ variables F_j, G_j, $j = 0, \ldots, N+M-1$ defined by

$$r_{N,M,t} = t_{M+N-1,t} + \sum_{\ell \in I_t} G_{M-\ell-1} s_{N+\ell,t} \quad for \quad 0 \leq j < N, \tag{3.12}$$

$$r_{N,M,t} = t_{M+N,t} + \sum_{\ell \in I_t} G_{M-\ell-1} s_{N+\ell,t} + + G_{M+N-t-1} \quad for \quad N \leq t \leq N+M-2, \tag{3.13}$$

$$r_{N,M,N+M-1} = t_{N+M,N+M-1} + G_0, \tag{3.14}$$

where $I_t = \{\ell \in \mathbb{N};\ 0 \leq \ell \leq M-1 \text{ and } t < N + \ell\}$.

The noncommutative polynomials $\mathbf{p}_\ell$ are computed as follows from the $\mathbf{q}_\ell$:

Theorem 3.3 *The noncommutative polynomials* $\mathbf{p}_\ell$ *are given by*

$$\mathbf{p}_\ell = \sum_0^\ell \beta_{\ell j} k^{(j)}(0).$$

In this expression, the matrices $\beta_{\ell j}$ *are defined by*

$$\begin{pmatrix} \epsilon_{00} & \epsilon_{01} & \cdots & \epsilon_{0\ell} \\ 0 & \epsilon_{11} & \cdots & \epsilon_{1\ell} \\ 0 & 0 & \cdot & \cdot \\ 0 & \cdot & 0 & \epsilon_{\ell\ell} \end{pmatrix}^{-1} = \begin{pmatrix} \beta_{00} & \beta_{01} & \cdots & \beta_{0\ell} \\ 0 & \beta_{11} & \cdots & \beta_{1\ell} \\ 0 & 0 & \cdot & \cdot \\ 0 & \cdot & 0 & \beta_{\ell\ell} \end{pmatrix}$$

where

$$\epsilon_{\ell j} = \begin{cases} c_\ell^{-1}(-1)^\ell & \text{if } \ell = j, \\ c_\ell^{-1}\ell!\alpha_{\ell j} & \text{if } j = 0, 1, \ldots, \ell-1, \\ 0 & \text{if } \ell < j. \end{cases}$$

The $\beta_{\ell j}$ *are well defined since* $\epsilon_{\ell j} = 0$ *for* $\ell < j$.

In the next theorem we show how the potential can be constructed from the first coefficients of the Laurent expansion of the spectral function at infinity.

Theorem 3.4 *Let* W *be the spectral function of a differential operator of the form* (1.1), *with* $k(t) \in L_1^{m\times m}(0, \infty)$. *Assume that* W *is rational and analytic at infinity, with* $W(\infty) = I_m$, *and that the McMillan degree of* W *is* n. *Then, the potential can be expressed from the* $2n$ *matrices* $k(0), \ldots, k^{(2n-1)}(0)$ *and their adjoints as follows:*

1. *Set* $M_0 = \frac{k(0)}{2}$ *and compute the* $2n-1$ *matrices*

$$M_j = c_j k^{(j)}(0) + \mathbf{p}_j(k(0), k(0)^*, \ldots) \quad j = 1, \ldots, 2n-1, \tag{3.15}$$

   where the constants c_j *and the polynomials* $\mathbf{p}_j$ *are defined in Theorem* 3.1, *and set* $N_j = M_j + M_j^*$. *The matrix* N_j *is the* $(j+1)$*–th coefficient of the Laurent expansion of the spectral function* W *at infinity.*

2. *Using the reduction procedure of* [7] *(which is reviewed in the first section) compute a minimal triple* $(A, B, C) \in \mathbb{C}^{n\times n} \times \mathbb{C}^{n\times m} \times \mathbb{C}^{m\times n}$ *such that:*

$$N_j = CA^jB, \qquad j = 0, \ldots, 2n-1. \tag{3.16}$$

3. *Compute the Riesz projection corresponding to the spectrum of* A *in* $\mathbb{C}_+$, *and use the obtained formulas for* A, B, C, P *in formula* (3.1).

In the next theorem, we characterize functions arising as potential of differential operators of the form (1.1).

Theorem 3.5 *Let* $k(t)$ *be a* $\mathbb{C}^{m\times m}$*–valued function analytic in a neighborhood of the origin and define matrices* M_j, $j = 0, 1, \ldots$ *by* $M_0 = \frac{1}{2}k(0)$ *and by* (3.15) *for* $j \geq 1$. *Then, a necessary condition for* k *to be the potential associated to a differential equation of the form* (1.1) *with a rational weight is that there exists an integer* ℓ_0 *such that, for* $\ell \geq \ell_0$,

$$\text{Rank } (M_{i+j})_{i,j=0}^{\ell} = p \tag{3.17}$$

When condition (3.17) *is in force, the function*

$$Z(\lambda) = \frac{I_m}{2} + \sum_0^\infty \frac{M_j}{\lambda^{j+1}} \tag{3.18}$$

is rational and analytic at infinity, of McMillan degree p. Suppose that Z is analytic in $\mathbb{C}_+ \cup \mathbb{R}$ *and that*

$$W(\lambda) = 2\text{Re } Z(\lambda) > 0 \tag{3.19}$$

for all real λ. *Then, k is the potential of the differential equation with spectral function W. The function W can be computed from the procedure described in Theorem* 3.4.

4 Preliminary computations

We first show that condition (1.11) holds for minimal realizations of spectral functions.

Proposition 4.1 *Let W be a* $\mathbb{C}^{m\times m}$*-valued rational function analytic on the real line and at infinity, with* $W(\infty) = I_m$, *and assume that* $W(\lambda) > 0$ *for all real* λ. *Let* $W(\lambda) = I_m + C(\lambda I_n - A)^{-1}B$ *be a minimal realization of W, and let P denote the Riesz projection corresponding to the eigenvalues of W in* $\mathbb{C}_+$. *Then, for every* $j \in \mathbb{N}$, *condition* (1.11) *holds.*

Proof: Since W takes selfadjoint values on the real line,

$$W(\lambda) = I_m + C(\lambda I_n - A)^{-1}B = I_m + B^*(\lambda I_n - A^*)^{-1}C^*$$

are two minimal realizations of W. Hence there exists a uniquely defined invertible and hermitian matrix such that

$$A^* = HAH^{-1}, \quad B^* = CH^{-1}, \quad C^* = HB.$$

(See [1]). It follows that the Riesz projection P satisfies $P^* = H(I_n - P)H$. Therefore,

$$\begin{aligned} (CA^\ell PB)^* &= B^*P^*A^{\ell *}C^* \\ &= CH^{-1}H(I_n - P)H^{-1}HA^\ell H^{-1}HB \\ &= C(I_n - P)A^\ell B. \end{aligned}$$

Thus,

$$CA^\ell PB + (CA^\ell PB)^* = CA^\ell PB + CA^\ell(I_n - P)B = CA^\ell B.$$

■

The following propositions are technical; they will be needed in the proof of the existence of the polynomials $\mathbf{p}_\ell$ defined in the preceding section.

Proposition 4.2 *Let $(A,B,C) \in \mathbb{C}^{n\times n} \times \mathbb{C}^{n\times m} \times \mathbb{C}^{m\times n}$ be a triple of matrices and assume that A has no real spectrum. Let P denote the Riesz projection corresponding to the eigenvalues of A in $\mathbb{C}_+$. Then for every $N \geq 1$,*

$$CPA^{\times N}P = CA^NP + \sum_{j=0}^{N-1} t_{N,j}CA^jP \tag{4.1}$$

where the $t_{N,j}$, $N = 1,\ldots, j = 0,\ldots,N-1$ are the polynomials in the $2N$ variables CA^jB, CPA^jB, $j = 0,\ldots,N-1$, defined recursively by $t_{1,0} = -CPB$ and the recursion (3.10) for $N \geq 1$, and

$$CA^{\times N}P = CA^NP + \sum_{j=0}^{N-1} s_{N,j}CA^jP \tag{4.2}$$

where the $s_{N,j}$, $N = 1,\ldots, j = 0,\ldots,N-1$ are the polynomials in the N variables CA^jB, $j = 0,\ldots,N-1$ defined recursively by $s_{1,0} = -CB$ and the recursion (3.11) for $N \geq 1$.

Proof: Notice that

$$(\lambda I_n - A^\times)^{-1} = (\lambda I_n - A)^{-1} - (\lambda I_n - A^\times)^{-1}BC(\lambda I_n - A)^{-1}.$$

Hence

$$A^{\times N} = A^N - \sum_{k=0}^{N-1} A^{\times(N-1-k)}BCA^k. \tag{4.3}$$

Set

$$d_{N,k} = A^{\times(N-1-k)}B, \quad k = 0,\ldots,N-1, \quad N = 1,2\ldots$$

It follows from equation (4.3) that the matrices $d_{N,k}$ are defined recursively by $d_{1,0} = B$ and

$$(d_{N+1,0}\cdots d_{N+1,N}) = (d_{N,0}\cdots d_{N,N-1})A_N + (A^NB\ 0\cdots 0) \tag{4.4}$$

where A_N is defined by (3.5). To prove the first part of the proposition, we multiply both sides of (4.3) by CP on the left and by P on the right and observe that $t_{N,j} = -CPd_{N,j}$. Multiplying both sides of (4.4) by CP on the left we obtain the recursion (3.10). To prove the second part of the proposition, we multiply both sides of (4.3) by C on the left and by P on the right and take into account that $s_{N,j} = -Cd_{N,j}$. ■

We note that an explicit expression for the $t_{N,j}, j = 0,\cdots,N-1$ is available as follows: set $A_1 = (-CB\ I)$, $B_1 = -(0\ CPB)$, $X_1 = CPB$, and for $N \geq 2$, A_N be defined by (3.5) and

$$B_N = -(CPA^NB\ 0\cdots 0) \tag{4.5}$$
$$X_N = (t_{N,0}\cdots t_{N,N-1}). \tag{4.6}$$

Then,

$$\begin{aligned} X_{N+1} &= X_NA_N + B_N \\ X_N &= X_{N-1}A_{N-1} + B_{N-1} \\ &\vdots \\ X_2 &= X_1A_1 + B_1 \end{aligned}$$

and so,

$$X_{N+1} = B_N + B_{N-1}A_N + B_{N-2}A_{N-1}A_N + \cdots + B_1A_2A_3\cdots A_N + X_1A_1\cdots A_N. \quad (4.7)$$

A formula similar to (4.7) holds for the polynomials $s_{N,j}$. More precisely, set

$$\begin{aligned} C_N &= -(CA^NB\ 0\ \cdots\ 0) \\ Y_N &= (s_{N,0}\cdots s_{N,N-1}). \end{aligned}$$

Then,

$$Y_{N+1} = C_N + C_{N-1}A_N + C_{N-2}A_{N-1}A_N + \cdots + C_1A_2A_3\cdots A_N + Y_1A_1\cdots A_N. \quad (4.8)$$

Proposition 4.3 *Let $(A,B,C) \in \mathbb{C}^{n\times n} \times \mathbb{C}^{n\times m} \times \mathbb{C}^{m\times n}$ be a triple of matrices and assume that A has no real spectrum. Let P denote the Riesz projection corresponding to the eigenvalues of A in $\mathbb{C}_+$. Then, for every $N \geq 1$ and every $M \geq 0$,*

$$CA^MPA^{\times N}P = CA^{N+M}P + \sum_{j=0}^{N+M-1} r_{N,M,j}CA^jP \quad (4.9)$$

where the $r_{N,M,j}$, $j = 0,\ldots,N+M-1$ are the polynomials in the $2(N+M)$ variables CA^jB, CA^jPB, $j = 0,\ldots,N+M-1$ defined by (3.14)-(3.14).

Proof: Interchanging the roles of A and $A^\times$ we can rewrite (4.3) as

$$A^M = A^{\times M} + \sum_{k=0}^{M-1} A^{(M-1-k)}BCA^{\times k}, \quad (4.10)$$

and therefore

$$A^MA^{\times N} = A^{\times(M+N)} + \sum_{k=0}^{M-1} A^{(M-1-k)}BCA^{\times(k+N)}. \quad (4.11)$$

Taking into account (4.3) and the definition of the $d_{N,j}$ it follows that

$$\begin{aligned} A^MA^{\times N} &= A^{N+M} - \sum_{t=0}^{N+M-1} d_{N+M,t}CA^t + \\ &\quad + \sum_{k=0}^{M-1} A^{M-1-k}B\left(CA^{N+k} - \sum_{j=0}^{N+k-1} Cd_{N+k,j}CA^j\right) \\ &= A^{N+M} - \sum_{t=0}^{N+M-1} d_{N+M,t}CA^t + \\ &\quad + \sum_{\ell=N}^{N+M-1} A^{N+M-1-\ell}BCA^\ell + \\ &\quad + \sum_{j=0}^{N+M-2}\left(\sum_{k\in I_j} A^{M-k-1}Bs_{N+k,j}\right)CA^j \end{aligned}$$

where

$$I_t = \{\ell \in \mathbb{N};\ 0 \le \ell \le M-1 \text{ and } t < N+\ell\}. \tag{4.12}$$

To end the proof of the proposition it remains to multiply both sides of this equation by CP on the left and by P on the right. ■

5 Proofs of the main theorems

5.1 Proof of Theorem 3.1

It will be more convenient to work with the function $\kappa(t) = C\left(Pe^{tA^\times}|_{\mathrm{Im}\,P}\right)^{-1}PB$ defined in (1.9). We set $X_\ell = \frac{PA^{\times\ell}P}{\ell!}$ and define recursively $\mathbb{C}^{n\times n}$ matrices $Y_0, Y_1, \ldots$ by

$$Y_0 = P \tag{5.1}$$

$$Y_1 = -X_1 \tag{5.2}$$

$$Y_\ell = -X_\ell - \sum_{j=1}^{\ell-1} Y_j X_{\ell-j}, \qquad \ell \ge 2. \tag{5.3}$$

Note that

$$CY_\ell B = \frac{\kappa^{(\ell)}(0)}{\ell!}.$$

The proof of the theorem is divided into two steps:

STEP 1: *For every* $\ell \ge 1$,

$$CY_\ell = -\sum_{j=0}^{\ell} \alpha_{\ell j} CPA^j \tag{5.4}$$

where $\alpha_{\ell\ell} = \frac{(-1)^\ell}{\ell!}$ *and, for* $0 \le j \le \ell-1$, *the* $\alpha_{\ell j}$ *are the polynomials in the* $2j$ *variables* $CA^iB, CPA^iB\ i = 0, \ldots j-1$, *defined recursively by:*

$$\alpha_{\ell+1,j} = -t_{\ell+1,j} - -\sum_{u,i\in T_{\ell j}} \frac{\alpha_{iu}}{(\ell+1-i)!} - \sum_{i,u\in J_{\ell j}} \frac{\alpha_{iu}}{(\ell+1-i)!} r_{u,\ell-1-i,j}.$$

In this expression, the polynomials $t_{N,j}$ *and* $r_{N,M,j}$ *are defined in Theorem 3.2 and* $T_{\ell j}$ *and* $J_{\ell j}$ *are defined by* (3.8) *and* (3.9).

PROOF OF STEP 1: We proceed by induction. For $\ell = 1$,

$$\begin{aligned} Y_1 &= -X_1 - P \\ &= -PAP + PBCP - P \end{aligned}$$

and so

$$\begin{aligned} CY_1 &= -CPA + CPBCP - CP \\ &= -CPA + (CPB - I_n)CP. \end{aligned}$$

Assume the result true at rank ℓ. Then, using the induction for the ranks up to ℓ, we have

$$
\begin{aligned}
CY_{\ell+1} &= -CX_{\ell+1} - \sum_{i=1}^{\ell} CY_i X_{\ell+1-i} \\
&= -CX_{\ell+1} - \sum_{i=1}^{\ell}\left(\sum_{u=0}^{i}\alpha_{iu}CA^uP\right)X_{\ell+1-i} \\
&= \frac{-1}{(\ell+1)!}CPA^{\times(\ell+1)}P - \sum_{i=1}^{\ell}\left(\sum_{u=0}^{i}\alpha_{iu}CA^uP\right)\frac{1}{(\ell+1-i)!}PA^{\times(\ell+1-i)}P. \quad (5.5)
\end{aligned}
$$

From Proposition 4.2 we have

$$
\begin{aligned}
CPA^{\times(\ell+1)}P &= CPA^{\ell+1} + \sum_{j=0}^{\ell} t_{\ell+1,j}(CPA^j) \\
CA^uPA^{\times(\ell+1-i)}P &= CA^{u+\ell+1-i}P + \sum_{j=0}^{\ell+u-i} r_{u,\ell+1-i,j}(CA^jP).
\end{aligned}
$$

Thus, (5.5) becomes

$$
\begin{aligned}
CY_{\ell+1} &= -\frac{1}{(\ell+1)!}CPA^{\ell+1} - \sum_{j=0}^{\ell} t_{\ell+1,j}CPA^j - \\
&\quad -\sum_{ii=1}^{\ell}\left(\sum_{u=0}^{i}\frac{1}{(\ell+1-i)!}\alpha_{iu}\left(CA^{u+\ell+1-i}P + \sum_{j=0}^{u+\ell-i} r_{u,\ell+1-i,j}CPA^j\right)\right) \\
&= -\frac{1}{(\ell+1)!}CPA^{\ell+1} - \sum_{j=0}^{\ell} t_{\ell+1,j}CPA^jP - \\
&\quad -\sum_{i=1}^{\ell}\left(\sum_{u=0}^{i}\frac{1}{(\ell+1-i)!}\alpha_{iu}CA^{u+\ell+1-i}P\right) - \\
&\quad -\sum_{i=0}^{\ell}\sum_{u=0}^{i}\sum_{j=0}^{u+k-i}\frac{1}{(\ell+1-i)!}\alpha_{iu}r_{u,\ell-1-i,j}CPA^j.
\end{aligned}
$$

But,

$$
\sum_{i=1}^{\ell}\sum_{u=0}^{i}\frac{1}{(\ell+1-i)!}\alpha_{iu}CA^{u+\ell+1-i}P = \sum_{j=0}^{\ell+1}\left(\sum_{u,i\in T_{\ell j}}\frac{\alpha_{iu}}{(k+1-i)!}\right)CA^jP
$$

where $T_{\ell j}$ is defined by (3.8), and

$$
\sum_{i=1}^{\ell}\sum_{u=0}^{i}\sum_{j=0}^{u+k-i}\frac{1}{(\ell+1-i)!}\alpha_{iu}r_{u,\ell-1-i,j}CPA^j = \sum_{j=0}^{\ell}\left(\sum_{i,u\in J_{\ell j}}\frac{1}{(\ell+1-i)!}\alpha_{iu}r_{u,\ell-1-i,j}\right)CPA^j
$$

where $J_{\ell j}$ is defined by (3.9). Thus (3.7) holds, and in particular, we have

$$\alpha_{\ell+1,\ell+1} = -\frac{1}{(\ell+1)!} - \sum_{i=1}^{\ell} \frac{\alpha_{i,i}}{(\ell+1-i)!}$$

and so $\alpha_{\ell,\ell} = \frac{(-1)^\ell}{\ell!}$.

STEP 2: *Formulas* (3.2) *and* (3.3) *hold.*

PROOF OF STEP 2: Multiplying both sides of (5.4) by B on the right, we obtain

$$\kappa^{(\ell)}(0) = -CA^\ell P - \sum_{i=0}^{\ell-1} \ell! \alpha_{\ell,i} CPA^i B.$$

Using (1.10), we have

$$\frac{1}{2(-2i)^\ell} k^{(\ell)}(0) = -CA^\ell P - \sum_{i=0}^{\ell-1} \ell! \alpha_{\ell,i} CPA^i B.$$

Formula (3.3) follows, with

$$\mathbf{q}_\ell = -2(-2i)^\ell \sum_{i=0}^{\ell-1} \ell! \alpha_{\ell i} CPA^i B. \tag{5.6}$$

Thus, $\mathbf{q}_\ell$ is a polynomial in the variables CA^iB, CPA^iB, $i = 0 \ldots \ell-1$ and so in the variables CA^iB and their adjoints, since (1.11) holds.

Theorem 3.2 is a gathering of the main points of the proof of Theorem 3.1. Theorem 3.3 consists in inverting the equations (5.6).

5.2 Proof of Theorem 3.4

Let $W(\lambda) = I_m + C(\lambda I - A)^{-1}B$ be a minimal realization of W. The Laurent expansion of W at infinity is

$$W(\lambda) = I_m + \sum_{\ell=1}^{\infty} \frac{CA^{\ell-1}B}{\lambda^\ell}.$$

Furthermore, by the preceding proposition, the coefficients M_ℓ computed in (3.4) are equal to $M_\ell = CA^\ell PB$, where P denotes the Riesz projection corresponding to the eigenvalues of A in $\mathbb{C}_+$. By Proposition 4.1, condition (1.11) holds and so

$$N_\ell = M_\ell + M_\ell^* = CA^\ell B.$$

By Theorem 2.7, one can obtain uniquely, up to a similarity matrix, the matrices A, B and C from $N_0 \ldots, N_{2\ell-1}$ and thus obtain the function W in terms of the first 2ℓ terms of its Laurent expansion at infinity.

5.3 Proof of Theorem 3.5

Assume that k is the potential associated to a rational spectral function W, analytic at infinity with $W(\infty) = I_m$. By Theorem 3.4, $M_j = CA^{j-1}PB$, where $W(\lambda) = I_m + C(\lambda I_n - A)^{-1}B$ is a minimal realization of W, and so (3.17) holds, since the McMillan degree of W is equal to the rank of the block Hankel matrix $(W_{i+j-1})_{i,j=1\dots,m}$ for m large enough (see [12]). Furthermore, $Z(\lambda) = \sum_{j=1}^{\infty} \frac{M_j}{\lambda^j}$ is rational and is analytic in $\mathbb{C}_+$ and at infinity, and for $\lambda \in \mathbb{R}$ we have $W(\lambda) = \text{Re } Z(\lambda)$. It is thus also necessary that $\text{Re } Z(\lambda) > 0$ for real λ. Let $Z(\lambda) = \frac{I_m}{2} + c(\lambda I_p - a)^{-1}b$ be a minimal realization of Z. Then, $W(\lambda) = I_m + C_o(\lambda I_n - A_o)^{-1}B_o$ is a minimal realization of W, with

$$C_o = (c \ \ b^*) \qquad B_o = \begin{pmatrix} b \\ c^* \end{pmatrix} \qquad A_o = \begin{pmatrix} a & 0 \\ 0 & a^* \end{pmatrix}. \tag{5.7}$$

For this realization, the Riesz projection is equal to

$$T_o = \begin{pmatrix} I_s & 0 \\ 0 & 0 \end{pmatrix}. \tag{5.8}$$

Assume that $\text{Re } Z(\lambda) > 0$ for all $\lambda \in \mathbb{R}$. By the result of [2] mentioned in the introduction, W is the spectral function of the operator (1.1) with potential $\tilde{k}$ defined by (1.8) with A_o, B_o, C_o, P_o defined by (5.7) and (5.8). In order to show that $k = \tilde{k}$, we show that all the derivatives of these two functions at the origine coincide. We note that, for all $j \geq 0$,

$$C_o A_o^j P_o B_o = c a^j b.$$

Thus, $C_o A_o^j P_o B_o = M_j$, and so

$$k^{(\ell)}(0) = c_\ell^{-1} M_\ell + \mathbf{q}_\ell(M_0, M_0^*, \dots)$$

But, by Theorem 3.1,

$$\tilde{k}^{(\ell)}(0) = c_\ell^{-1} M_\ell + \mathbf{q}_\ell(CB, (CB)^*, \dots)$$

and thus $\tilde{k}^{(\ell)}(0) = k^{(\ell)}(0)$ for all $\ell \geq 0$, and k and $\tilde{k}$ coincide.

References

[1] D. Alpay and I. Gohberg. *Unitary rational matrix functions*, volume 33 of *Operator Theory: Advances and Applications*, pages 175–222. Birkhäuser Verlag, Basel, 1988.

[2] D. Alpay and I. Gohberg. Inverse spectral problem for differential operators with rational scattering matrix functions. *Journal of Differential Equations*, 118:1–19, 1995.

[3] H. Bart, I. Gohberg, and M. Kaashoek. *Minimal factorization of matrix and operator functions*, volume 1 of *Operator Theory: Advances and Applications*. Birkhäuser Verlag, Basel, 1979.

[4] H. Bart, I. Gohberg, and M. Kaashoek. Convolution equations and linear systems. *Integral Equations and Operator Theory*, 5:283–340, 1982.

[5] A. Bruckstein, B. Levy, and T. Kailath. Differential methods in inverse scattering. *SIAM journal of applied mathematics*, 45:312–335, 1985.

[6] H. Dym and A. Iacob. *Positive definite extensions, canonical equations and inverse problems*, volume 12 of *Operator Theory: Advances and Applications*, pages 141–240. Birkhäuser Verlag, Basel, 1984.

[7] I. Gohberg, M. Kaashoek, and L. Lerer. On minimality in the partial realization problem. *Systems and Control Letters*, 9:97–104, 1987.

[8] M.G. Kreĭn. Continuous analogues of propositions for polynomials orthogonal on the unit circle. *Dokl. Akad. Nauk. SSSR*, 105:637–640, 1955.

[9] M.G. Kreĭn and F.E. Melik-Adamyan. On the theory of S–matrices of canonical equations with summable potentials. *Dokl. Akad. Nauk. SSSR*, 16:150–159, 1968.

[10] F.E. Melik-Adamyan. Canonical differential operators in Hilbert space. *Izvestya Akademii Nauk. Armyanskoi SSR Matematica*, 12:10–31, 1977.

[11] F.E. Melik-Adamyan. On a class of canonical differential operators. *Izvestya Akademii Nauk. Armyanskoi SSR Matematica*, 24:570–592, 1989. English translation in: Soviet Journal of Contemporary Mathematics, vol. 24, pages 48–69 (1989).

[12] M.W. Wonham. *Linear Multivariable Control: Geometric Approach.* Springer–Verlag, New–York, 1979.

Daniel Alpay
Department of Mathematics
Ben–Gurion University of the Negev
POB 653. 84105 Beer-Sheva
Israel

Israel Gohberg
School of Mathematical Sciences
Raymond and Beverly Sackler Faculty
of Exact Sciences
Tel–Aviv University
Tel–Aviv, Ramat–Aviv 69989, Israel

MSC: 34L25, 81U40, 47A56

Operator Theory
Advances and Applications, Vol. 98

MULTIDIMENSIONAL FUNCTIONAL EQUATIONS GENERATED BY AFFINE TRANSFORMATIONS.

G.BELITSKII*

Solvability conditions for the equation

$$(T\varphi)(x) \equiv \sum_{j=1}^{t} A_j(x)\varphi(\lambda_j x + e_j) = \gamma(x), \quad \lambda_j \in \mathbb{R}^1, \quad e_j \in \mathbb{R}^n$$

in classes of continuous or smooth functions $\varphi(x)$ in $\mathbb{R}^n$ are investigated. We establish that this equation is normally solvable in some class of smooth functions. Moreover, we prove that the operator T with constant coefficients is semi-Fredholm with dim Ker$T^* < \infty$, and it is Fredholm with ind$T = 0$ in the case $\lambda_i \neq \lambda_j$.

1 Introduction

We consider a multidimensional functional equation

$$(T\varphi)(x) \equiv \sum_{j=1}^{t} A_j(x)\varphi(F_j x) = \gamma(x) \tag{1.1}$$

where

$$F_j x = \lambda_j x + e_j, \quad \lambda_j \in \mathbb{R}_+, \quad e_j \in \mathbb{R}^n, \quad j \in \overline{1,t},$$

are given transformations in $\mathbb{R}^n$. We assume that $F_i \neq F_j, \quad i \neq j$. The mappings

$$A_j : \mathbb{R}^n \to \mathrm{Hom}(\mathbb{C}^p, \mathbb{C}^r), \quad \gamma : \mathbb{R}^n \to \mathbb{C}^r,$$

are given and are supposed to be of class C^k for some $k \in [0, \infty]$. The vector function $\varphi : \mathbb{R}^n \to \mathbb{C}^p$ is unknown.

The operator

$$T : C^k(\mathbb{R}^n, \mathbb{C}^p) \to C^k(\mathbb{R}^n, \mathbb{C}^r)$$

*Supported by Guastello Foundation

acts in the spaces of corresponding vector functions. These spaces are endowed with the topology of uniform convergence on each compact subset of all derivatives up to k.

We are interested in solvability conditions of Equation (1.1), in particular, the operator properties of T : normal solvability, Fredholm property, etc.

Let us remind that the operator T is called *semi-Fredholm,* if the subspace $\operatorname{Im} T$ is closed and

$$\min(\dim \operatorname{Ker} T,\ \dim \operatorname{Ker} T^*) < \infty.$$

If, moreover, both spaces $\operatorname{Ker} T$, $\operatorname{Ker} T^*$ are finite dimensional, then T is called *Fredholm.* The number

$$\operatorname{ind} T \equiv \dim \operatorname{Ker} T - \dim \operatorname{Ker} T^*$$

is called *index* of T.

We also remind the reader that $\operatorname{Im} T$ is closed if and only if it is normally solvable (see [3]).

At first, consider the following simplest

Example 1.1. Let

$$(T\varphi)(x) \equiv \varphi(x) - a\varphi(\lambda x + \beta), \quad c \neq 0, \quad x \in \mathbb{R}^1. \tag{1.2}$$

Here $p = r = 1$, $t = 2$, $A_1(x) = 1$, $A_2(x) = -a$.

Let us consider three cases.

1. $\lambda = 1$. Without loss of generality one can assume that $\beta > 0$. Then *the operator* (1.2) *is surjective in the space* C^k *for each* $k \in [0, \infty]$. Indeed, fix some $\varepsilon > 0$ and decompose a given C^k-function γ in a sum $\gamma = \gamma_+ + \gamma_-$ where $\gamma_\pm$ are C^k-functions such that

$$\gamma_+(x) = 0, \quad x > \varepsilon, \quad \gamma_-(x) = 0, \quad x < -\varepsilon.$$

Set

$$\varphi_+(x) = \sum_{n=0}^{\infty} a^n \gamma_+(x + n\beta)$$

and

$$\varphi_-(x) = -\sum_{n=-\infty}^{-1} a^n \gamma_-(x + n\beta).$$

Then the function $\varphi = \varphi_+ + \varphi_-$ is a C^k-solution of (1.1).

Also note that the function

$$\varphi(x) = e^{\alpha x} h\left(\frac{x}{\beta}\right),$$

with $\alpha = i\frac{\pi}{2\beta} + \frac{1}{\beta}\ln a$ and an arbitrary 1-periodic function h, belongs to $\operatorname{Ker} T$. Thus,

$$\dim \operatorname{Ker} T = \infty, \quad \operatorname{Ker} T^* = \{0\}.$$

Let now $\lambda \neq 1$. Then the mapping $Fx = \lambda x + \beta$ has the unique fixed point $z = \frac{\beta}{1-x}$. Without loss of generality one can assume that $\beta = 0$, $\lambda \in (0,1)$. Let us prove

that *if* (1.1) *has a local* C^k*-solution in a neighborhood of the fixed point* $z = 0$, *then it has a* C^k*-solution on the whole* $\mathbb{R}^1$.

Indeed, let $\varphi_0 \in C^k$ be such that

$$\varphi_0(x) - a\varphi_0(\lambda x) = \gamma(x), \quad x \in (-\varepsilon, \varepsilon)$$

for some $\varepsilon > 0$. Then the function

$$\varphi(x) = \varphi_0(x) + \sum_{n=0}^{\infty} a^n[\gamma(\lambda^n x) - \varphi_0(\lambda^n x) + a\varphi_0(\lambda^{n+1}x)]$$

is a C^k-solution on $\mathbb{R}^1$.

2. $| a\lambda^k | \neq 1$. We will prove that the *operator* T *is normally solvable.* Namely, if

$$1 - a\lambda^s \neq 0, \quad s = 0, 1, \dots, k-1 \tag{1.3}$$

then T is surjective, and if

$$1 - a\lambda^s = 0 \tag{1.4}$$

for some $s \in \mathbb{Z}_+$, $s \leq k$, then (1.1) has a C^k-solution, if and only if

$$\gamma^{(s)}(0) = 0. \tag{1.5}$$

Indeed, let either (1.3) or (1.4) and (1.5) be fulfilled. Then there is a polynomial $P_\gamma(x)$ of degree k such that

$$\tilde{\gamma}(x) \equiv \gamma(x) - P_\gamma(x) + aP_\gamma(\lambda x) = o(x^k), \quad x \to 0.$$

The function

$$\varphi(x) = P_\gamma(x) + \sum_{n=0}^{\infty} c^n \tilde{\gamma}(\lambda^n x)$$

is a C^k-solution of (1.1). Hence the space $\operatorname{Im} T$ is closed.

Note that

$$\operatorname{Ker} T = \operatorname{Ker} T^* = \{0\}$$

in case (1.3). If (1.4) holds, then

$$\operatorname{Ker} T = \{\varphi \mid \varphi(x) = cx^s, \quad c \in \mathbb{R}\}$$

$$\operatorname{Ker} T^* = \{ch \mid (h, \gamma) = \gamma^{(s)}(0), \quad c \in \mathbb{R}\}.$$

Hence, $\dim \operatorname{Ker} T = \dim \operatorname{Ker} T^* = 1$.

3. Assume that

$$1 - a\lambda^k = e^{i\delta}, \quad \delta \in \mathbb{R}$$

One can show that T *is not normally solvable in the class* C^k. Indeed, it is sufficient to construct a C^k-function γ, $\gamma(x) = o(x^k)$, $x \to 0$ such that (1.1) has no C^k-solutions. To this end, let us set

$$\gamma(x) = \int_0^x \frac{1}{\ln |t|} \cdot \exp(i\theta \ln |t|) dt^k, \quad \theta = -\frac{\delta}{\ln\lambda}.$$

Thus, we obtain that if $\lambda = 1$, then the operator T of Example 1.1 is semi-Fredholm with $\operatorname{Ker} T^* = \{0\}$, $\dim \operatorname{Ker} T = \infty$. In case 2 it is Fredholm. More exactly, if (1.3) holds, then $\operatorname{Ker} T = \operatorname{Ker} T^* = \{0\}$. If (1.4) holds, then $\dim \operatorname{Ker} T = \dim \operatorname{Ker} T^* = 1$. Finally, in case 3 this operator is not normally solvable.

We show that a similar situation occurs for a general operator of type (1.1). Assume, for simplicity, that

$$\operatorname{rank} A_j(x) = r, \quad j \in \overline{1,t}, \quad x \in \mathbb{R}^n \tag{1.6}$$

Then there is a compact subset $I \subset \mathbb{R}^n$ which depends only on the system $\{F_j\}$ and such that the following result is true.

Theorem 1.2. *Let Equation* (1.1) *have a C^k-solution in a neighborhood of I. Then it has a C^k-solution on the whole space $\mathbb{R}^n$.*

We give below a construction of the compact subset I and some more exact conditions related to the coefficients $A_j(x)$. Theorem 1.2 follows from Theorem 3.1 on continuation of solutions.

It may happen that $I = \emptyset$, and we obtain an existence theorem similar to the case $\lambda = 1$ of Example 1.1. The compact I is reduced to the fixed point $z = \frac{\beta}{1-\lambda}$ in the case $\lambda < 1$ of this Example.

It follows from Theorem 1.2 that the supports of all functionals $h \in \operatorname{Ker} T^*$ are located in I (see Corollary 3.3 below). However, the implication

$$\frac{\partial^s \gamma}{\partial x_1^{s_1} \dots \partial x_n^{s_n}} |_I = 0, \quad s = \sum s_i \le k \Rightarrow \gamma \in \operatorname{Im} T$$

fails, because $\operatorname{Im} T$ is not closed as in case 3 in Example 1.1. We will prove that if the order of smoothness k is "big enough", then Equation (1.1) is normally solvable and, moreover, the operator T is semi-Fredholm at least in the case of constant coefficients. Namely, *assume that $A_j(x) = A_j$ are constant.* Set

$$\theta = \max_j(\|A_j\|, \|(A_j^* A_j)^{-1}\|).$$

Theorem 1.3. *There is $k_0 = k_0(\theta)$ such that if $k \ge k_0$, then the operator T is semi-Fredholm with finite dimensional space $\operatorname{Ker} T^*$.*

This theorem follows from Theorem 3.4. The special case of this theorem on the line $\mathbb{R}^1$ with constant coefficients was investigated in [1].

It may happen that $\dim \operatorname{Ker} T = \infty$ (see case 1 in Example 1.1), and this is a case of non-Fredholm property of the operator T.

However, in a generic case *the operator T is Fredholm with* $\operatorname{ind} T = 0$ (see Theorem 3.6 below), as in case 2 of Example 1.1.

2 Some preliminary constructions.

Assume that $\lambda_1 = \dots = \lambda_q \equiv \lambda > \lambda_j$, $j \notin \overline{1,q}$ with some $q \in \overline{1,t}$. It may happen that $q = t$ or $q = 1$.

Select all vectors from the system $\{ e_1 \ldots e_q \}$ which do not belong to the convex hull of other vectors of this system. For the sake of definitness, let

$$e_\nu \notin \operatorname{conv}\{ e_1 \ldots \hat{e}_\nu, \ldots e_q \}, \quad \nu = 1, \ldots l, \tag{2.1}$$

with some $l \in \overline{1,q}$. If $q = 1$, then we set $l = 1$. Note that $q \geq 2$ implies $l \geq 2$. It may happen that $l = q$.

Denote by K_ν, $\nu \in \overline{1,l}$ a *wedge*, generated by vectors $\{ e_1 - e_\nu, \ldots e_q - e_\nu \}$:

$$K_\nu = \{ x : x = \sum_{j=1}^{q} c_j(e_j - e_\nu), \ c_j \geq 0 \}.$$

It follows from (2.1) that K_ν is a *cone* (see [2]). We set $K_1 = \{0\}$ in the case $q = l = 1$.

Proposition 2.1. $\bigcap\limits_{\nu=1}^{l} K_\nu = \{ 0 \}$.

Proof. Let $l \geq 2$. Since $e_\mu \neq e_\nu$, $\mu \neq \nu$, $\mu, \nu \in \overline{1,q}$, there is some linear functional $h \in (\mathbb{R}^n)'$ such that

$$h(e_\mu) \neq h(e_\nu), \quad \mu_\nu \neq \nu, \quad \mu, \nu \in \overline{1,q}.$$

Assume that

$$h(e_{\nu_1}) = \min_{j \leq q} h(e_j), \quad h(e_{\nu_2}) = \max_{j \leq q} h(e_j).$$

It follows from (2.1) that $\nu_1, \nu_2 \in \overline{1,l}$. Let, for the sake of definitness, $\nu_1 = 1$, $\nu_2 = l$. Then

$$h(e_1) < h(e_j) < h(e_l), \quad j \in \overline{1,q}, \quad j \neq 1, l.$$

Now let us suppose that $x \in \bigcap\limits_{\nu=1}^{l} K_\nu$. In particular,

$$x = \sum_{j=1}^{q} \alpha_j(e_j - e_1) = \sum_{j=1}^{q} \beta_j(e_j - e_l)$$

with some $\alpha, \beta \geq 0$. Then

$$h(x) = \sum \alpha_j h(e_j - e_1) = \sum \beta_j h(e_j - e_l).$$

Since

$$h(e_j - e_l) < 0 < h(e_j - e_1),$$

we obtain that $\alpha_j = \beta_j = 0$, hence $x = 0$. □

Now consider the *dual cones*

$$K_\nu' = \{ h \in (\mathbb{R}^n)' : h(x) \geq 0, \ x \in K_\nu \}.$$

Since

$$(K_1' + \ldots + K_\nu')' = \bigcap K_\nu'' = \bigcap K_\nu = \{ 0 \}$$

(see [2]), we obtain

Corollary 2.2. $\sum_{\nu=1}^{l} K_\nu' = (\mathbb{R}^n)'$.

Choose some $\delta \in \mathbb{R}$ and consider the closed subsets

$$M_\nu(\delta) = \{ x : h(x) \geq \delta, \quad h \in K_\nu', \quad \|h\| = 1\}.$$

Proposition 2.3. *The intersection* $M(\delta) \equiv \bigcap_{\nu=1}^{l} M_\nu(\delta)$ *is a compact subset.*

Proof. It follows from Corollary 2.2 that

$$\operatorname{conv}(\bigcap_{\nu=1}^{l} S_\nu) \supset \{ h \in (\mathbb{R}^n)' : \|h\| \leq \rho \}$$

for some $\rho > 0$. Here

$$S_\nu = \{ h \in (\mathbb{R}^n)' : h \in K_\nu', \|h\| = 1 \}.$$

Therefore,

$$x \in M(\delta) \Rightarrow \{ h(x) \geq \delta, \quad \|h\| \leq \rho \}, \tag{2.2}$$

and in the case $\delta \geq 0$ it follows that either $M(\delta) = \emptyset$ or $M(\delta) = \{0\}$. Now assume that $\delta < 0$. Then we have from (2.2)

$$x \in M(\delta) \Rightarrow \|x\| \leq |\delta| \cdot \rho.$$

□

Finally, let us denote

$$I_\nu = \{ x : h(x) \geq \min_\mu h(z_{\mu\nu}), \quad h \in K_\nu' \}$$

where

$$z_{\mu\nu} = \frac{1}{\lambda - \lambda_\mu}(e_\mu - e_\nu), \quad \mu \notin 1, q, \quad \nu \in \overline{1, l}$$

is a fixed point of the mapping

$$F_\nu^{-1} F_\mu(x) = \frac{\lambda_\mu}{\lambda} x + \frac{1}{\lambda}(e_\mu - e_\nu).$$

Note that I_ν is a closed subset invariant with respect to all mappings $F_\nu^{-1} F_j$, $j \in \overline{1, t}$, and if

$$\delta \leq \min_{\mu,\nu} \min_h \{ h(z_{\mu,\nu}), \quad h \in K_\nu', \quad \|h\| = 1 \},$$

then $I_\nu \subset M_\nu(\delta)$. Hence Proposition 2.3 implies

Corollary 2.4. *The interesection*

$$I = \bigcap_{\nu=1}^{l} I_\nu$$

is a convex compact set.

Remark 2.5. If $q = 1$, thus $\lambda_1 > \lambda_j$, $j \in \overline{2, t}$, then $I = I_1$. It is easy to see that in this case I is a convex hull of points z_{j1}, $j \in \overline{2, t}$.

3 Statement of results.

Our assumption related to coefficients A_j is *that the matrix function* $A_\nu(x)$, $\nu \in \overline{1,l}$ *is surjective on the complement* $\mathbb{R}^n \setminus I_\nu$. It means that

$$\operatorname{rank} A_\nu(x) = r, \quad x \notin I_\nu, \quad \nu \in \overline{1,l} \tag{3.1}$$

We will say that a C^k-vector function $\varphi : \mathbb{R}^n \to \mathbb{C}^p$ is a *solution on an open subset* $V \subset \mathbb{R}^n$, if (1.1) is fulfilled for all $x \in V$. In particular, *each* C^k*-function* φ *is a solution on empty subset* $V = \emptyset$.

Theorem 3.1. *For each* C^k*-solution* φ_0 *of Equation (1.1) in a neighborhood of* I *there is a* C^k*-solution on the whole* $\mathbb{R}^n$ *which coincides with* φ_0 *in a neighborhood of the intersection*

$$I' = \bigcap_{\nu=1}^{l} F_\nu(I_\nu).$$

Corollary 3.2. *Let* $I = \emptyset$. *Then for every* $\gamma \in C^k$ *there is a* C^k*-solution* φ *of Equation (1.1).*

It follows from Theorem 3.1 in a general case that all obstacles to the solvability are located on I. More exactly, denote by

$$T^* : (C^k(\mathbb{R}^n, \mathbb{C}^r))^* \to (C^k(\mathbb{R}^n, \mathbb{C}^p))^*$$

a *conjugate operator* which acts in dual spaces. It is well known that

$$\gamma \in \operatorname{Im} T \Rightarrow (h, \gamma) = 0, \quad h \in \operatorname{Ker} T^*.$$

Moreover, $\operatorname{cl}(\operatorname{Im} T) = (\operatorname{Ker} T^*)_\perp$.

Corollary 3.3. *Let* $h \in \operatorname{Ker} T^*$. *Then*

$$\operatorname{supp} h \subset I.$$

Indeed, assume that $x_0 \notin I$. Let $U \ni x_0$ be a neighborhood of x_0 such that

$$\operatorname{cl}(U) \bigcap I = \emptyset.$$

It follows from Theorem 3.1 that

$$\gamma(x) = 0, \quad x \notin U \Rightarrow \gamma \in \operatorname{Im} T.$$

Hence

$$\gamma(x) = 0, \quad x \notin U \Rightarrow (h, \gamma) = 0, \quad h \in \operatorname{Ker} T^*. \qquad \square$$

Generally speaking, it is not true that

$$\frac{\partial^{|s|}\gamma}{\partial x_1^{s_1} \ldots \partial x_n^{s_n}} \Big|_I = 0, \quad |s| \le k \Rightarrow \gamma \in \operatorname{Im} T$$

because the space Im T may be not closed. However, if a smoothness k is "big enough" then the equation (1.1) is normally solvable. Hence $\gamma \in \mathrm{Im}T$, if and only if $(h,\gamma) = 0,\ \ h \in \mathrm{Ker}T^*$.

Let us formulate an exact theorem.

Assume, that $A_j(x) = A_j$ *are constant,* and

$$\mathrm{rank}A_\nu = r, \quad \nu \in \overline{1,l} \tag{3.2}$$

Then A_l is a surjective matrix. The matrix

$$B = A_l^*(A_l A_l^*)^{-1}$$

is the right-hand inverse to A_l. Set

$$c = \max_j \|B \cdot A_j\|, \quad j \in \overline{1,t}, \quad j \neq l.$$

Theorem 3.4. *There exists* $k_0 = k_0(c)$ *such that if* $k \geq k_0$ *then the operator*

$$T : C^k(\mathbb{R}^n, \mathbb{C}^p) \to C^k(\mathbb{R}^n, \mathbb{C}^r)$$

is semi-Fredholm, with closed Im T *and finite dimensional* Ker T^*.

We say that C^k-function φ_0 *is a solution of (1.1) on a closed subset* $V \subset \mathbb{R}^n$, if

$$\frac{\partial^{|s|}}{\partial x_1^{s_1} \ldots \partial x_n^{s_n}}(T\varphi_0 - \gamma)\,|_V = 0, \quad |s| \leq k.$$

Corollary 3.5. *Let* $k \geq k_0(c)$ *and Equation* (1.1) *has a* C^k*-solution on* I. *Then it has a* C^k*-solution on the whole space* $\mathbb{R}^n$.

As we saw, the operator T, is generally speaking not Fredholm even if $k = \infty$ because Ker T is infinite dimensional. However, in the case $q = 1$ it is Fredholm:

Theorem 3.6. *Assume that* $p = r$ *and* $q = 1$, *i.e.,* $\lambda_1 > \lambda_j,\ \ j \in \overline{2,t}$. *If* $k \geq k_0(c)$ *then the operator* T *is Fredholm with* ind $T = 0$.

4 Proofs.

4.1 Proof of Theorem 3.1.

Choose cones $\tilde{K}_\nu \subset \mathbb{R}^n, \ \nu \in \overline{1,l}$ such that

$$K_\nu \subset \mathrm{int}\tilde{K}_\nu, \quad \bigcap_{\nu=1}^{l} \tilde{K}_\nu = \{0\} \tag{4.1}$$

Here int V means interior of a set V. It follows from (4.1) that

$$\tilde{K}_\nu' \subset K_\nu'$$

and, moreover,

$$x \in K_\nu \Rightarrow h(x) > 0, \quad h \in \tilde{K}_\nu'.$$

Besides,

$$\sum \tilde{K}'_\nu = (\mathbb{R}^n)'.$$

Further, let $\varepsilon > 0$, $\nu \in \overline{1,l}$. Set

$$\tilde{I}_\nu(\varepsilon) = \{ x : h(x) \geq \min_\mu h(z_{\mu,\nu}) - \varepsilon, \quad h \in \tilde{K}'_\nu \quad \|h\| = 1 \}$$

Note the following properties of these closed subsets:

1. $\tilde{I}_\nu(\varepsilon) \supset I_\nu$,

2. The intersection $\bigcap_{\nu=1}^{l} \tilde{I}_\nu(\varepsilon)$ is compact. Moreover for each neighborhood $U \supset I$ it is possible to choose $\tilde{K}_\nu$ and $\varepsilon > 0$ such that

$$\bigcap_{\nu=1}^{l} \tilde{I}_\nu(\varepsilon) \subset U.$$

3. Subset $\tilde{I}_\nu(\varepsilon)$ is a *common absorber* of the family of mappings $\{G_j = F_\nu^{-1}F_j\}_{j=1}^{t}$ in the following sense. For any point $x \in \mathbb{R}^n$ there is a neighborhood $W \ni x$ and a number $N = N(x)$ such that

$$G_{j_1} \circ G_{j_2} \circ \ldots \circ G_{j_s} \subset \tilde{I}_\nu(\varepsilon), \quad s \geq N.$$

This property follows from the equality

$$h(G_j x) - h(x) = \frac{\lambda_j - \lambda}{\lambda} h(x) + \frac{1}{\lambda} h(e_j - e_\nu), \quad h \in (\mathbb{R}^n)'$$

Therefore, if

$$h(x) \leq \min_\mu h(z_{\mu\nu}) - \varepsilon$$

for some $h \in \tilde{K}'_\nu$, $\|h\| = 1$, then

$$h(G_j x) \geq h(x) + \delta$$

where

$$\delta = \min[\min_{\mu \notin \overline{1,q}} \frac{\lambda - \lambda_\mu}{\lambda} \cdot \varepsilon; \ \min_{j \in \overline{1,q}} \frac{1}{\lambda} h(e_j - e_\nu)].$$

Note that one can choose $N(x) = 1$ for $x \in \tilde{I}_\nu(\varepsilon)$, i.e., $\tilde{I}(\varepsilon)$ is invariant with respect to all mappings $F_\nu^{-1}F_j$, $j \in \overline{1,t}$.

Now let φ_0 be a C^k-solution of (1.1) in a neighborhood $U \supset I$. Choose $\tilde{K}_\nu$ and $\varepsilon > 0$ such that

$$\bigcap_{\nu=1}^{l} \tilde{I}_\nu(\varepsilon) \subset U.$$

Substitution $\varphi \mapsto \varphi + \varphi_0$ reduces Equation (1.1) to the form

$$T\varphi = \tilde{\gamma}, \quad \tilde{\gamma} = \gamma - T\varphi_0.$$

Obviously, $\tilde{\gamma}|U = 0$.

Choose neighborhoods $U_\nu \supset \tilde{I}_\nu(\varepsilon)$ such that

$$\bigcap_\nu U_\nu \subset U,$$

and consider some "resolution of identity". Namely, let C^∞-functions $\tau_1, \dots, \tau_l$ be such that

$$\tau_\nu \mid U_\nu = 0, \quad \tau_1(x) + \dots + \tau_l(x) = 1, \quad x \notin U.$$

Setting $\gamma_\nu = \tau_\nu \cdot \tilde{\gamma}$, we obtain

$$\tilde{\gamma} = \sum \gamma_\nu, \quad \gamma_\nu | U_\nu = 0.$$

Now it is sufficient to construct a C^k-solutions φ_ν of the equations

$$T\varphi_\nu = \gamma_\nu, \quad \nu = \overline{1,l} \tag{4.2}$$

such that φ_ν vanishes in a neighborhood of $I_\nu(\varepsilon)$. Then the sum $\varphi = \varphi_1 + \dots + \varphi_l + \varphi_0$ will be a solution that we need.

To find φ_ν, let us consider the operator

$$L_\nu : C^k(\mathbb{R}^n, \mathbb{C}^r) \to C^k(\mathbb{R}^n, \mathbb{C}^r)$$

defined by the formula

$$(L_\nu g)(x) = \begin{cases} -\sum\limits_j B_{\nu,j}(x) g(F_\nu^{-1} F j x) = \tilde{\gamma}_\nu(x), & x \notin \tilde{I}_\nu(\varepsilon) \\ 0 \quad , & x \in \tilde{I}_\nu(\varepsilon) \end{cases}$$

Here

$$B_{\nu,j}(x) = (A_\nu(x) A_\nu^*(x))^{-1} A_j(x)(F_\nu^{-1} F_j x), \quad x \notin \tilde{I}_\nu(\varepsilon)$$

and

$$\tilde{\gamma}_\nu(x) = \begin{cases} (A_\nu(x) A_\nu^*(x))^{-1} \gamma_\nu(x), & x \notin \tilde{I}_\nu(\varepsilon) \\ 0 \quad , & x \in \tilde{I}_\nu(\varepsilon). \end{cases}$$

By virtue of (4.1), the C^k-functions $B_{j\nu}$ and $\tilde{\gamma}_\nu$ are well defined.

It follows from the properties 1-3 of the subsets $\tilde{I}_\nu(\varepsilon)$ that the series

$$g(x) = \sum_{n=0}^{\infty} (L_\nu^n \tilde{\gamma}_\nu)(x)$$

converges to some C^k-solution of the equation $g = L_\nu g$. It is easy to check that the function

$$\varphi_\nu(x) = A_\nu^*(F_\nu^{-1} x) g(F_\nu^{-1} x)$$

is a solution of (4.2). Obviously, φ_ν vanishes in a neighborhood of $\tilde{I}_\nu(\varepsilon)$. □

Remark 4.1. In fact, we constructed a continuous operator R_U which depends on a neighborhood $U \supset I$ and relates every solution φ_0 on U to the solution

$$R_U(\varphi_0, \gamma) = \varphi_1 + \ldots + \varphi_l + \varphi_0$$

on the whole $\mathbb{R}^n$. This operator is a right inverse to T in the sense that

$$TR_U(\varphi_0, \gamma) = \gamma$$

for each solution φ_0 on U.

Remark 4.2. In the case $q = l = 1$ conditions (3.1) and (3.2) mean that

$$\operatorname{rank} A_1(x) = r, \quad x \in \mathbb{R}^n.$$

If, besides, $p = r$ then extension of φ_0 to the global solution is unique.

4.2 Proofs of Theorems 3.4, 3.6.

We start with a special case of the operator T. Namely, consider the operator

$$(T_\circ \varphi)(x) = \varphi(x) - \sum_{j=1}^{t} B_j \varphi(\alpha_j x + e_j)$$

where $\alpha_j \in (0,1)$, $e_j \in \mathbb{R}^n$ and $B_j \in \operatorname{Hom}(\mathbb{C}^r, \mathbb{C}^r)$.

In this case

$$I = \operatorname{conv}\{ z_j \}, \quad z_j = \frac{1}{1-\alpha_j} e_j.$$

Set

$$c = \max_j \|B_j\|, \quad k_\circ = 1 + \max_{j \in \overline{1,t}} \frac{\ln 2 - \ln(ct)}{\ln \alpha_j}.$$

Then we have an inequality

$$\sum_{j=1}^{t} \|B_j\| \cdot \alpha_j^k < \frac{1}{2}, \quad k \geq k_0 \tag{4.3}$$

Lemma 4.3. *Let $k \geq k_0$. Then for every $\gamma \in C^k$ there is a polynomial mapping $P_\gamma : \mathbb{R}^n \to \mathbb{C}^r$ of a degree $\leq k_\circ$ such that*

$$\gamma + P_\gamma = T_\circ \varphi, \quad \varphi \in C^k.$$

Moreover, the correspondence $\gamma \mapsto P_\gamma$ may be chosen to be continuous. Besides, the space Ker $T_\circ$ is of finite dimension.

Proof. We will find a derivative $\varphi^{(k_0)}$ of the solution of the equation $T_0 \varphi = \gamma$. Namely, consider the equation

$$\psi(x) = \sum_{j=1}^{t} B_j \cdot \alpha_j^{k_0} \psi(\alpha_j x + e_j) + \gamma^{(k_0)}(x), \tag{4.4}$$

where

$$\psi : \mathbb{R}^n \to \mathbb{C}^r \otimes (\mathbb{R}^n)^{\otimes k_0} \tag{4.5}$$

is a tensor of the corresponding rank.

Denote by $D^{(k_0)}$ the space of all C^{k-k_0} mappings (4.5) such that $\psi = \varphi^{(k_0)}$ for some C^k-function $\varphi : \mathbb{R}^n \to \mathbb{C}^r$. The subspace $D^{(k_0)}$ is closed in the space of all C^{k-k_0} mappings (4.5). Obviously, the operator $L = \sum B_j \alpha_j^{k_0} \psi(\alpha_j x + e_j)$ acts in the space D^{k_0}.

Since $\alpha_j < 1$, the compact I reduces to the convex hull of the points

$$z_{j1} = \frac{1}{1-\alpha_j} \cdot e_j, \quad j \in \overline{1,t}.$$

under lemma's conditions. Fix some closed neighborhood $U \supset I$ invariant with respect to $F_j(x) = \alpha_j x + e_j$.

Because of (4.3) the operator L is a contraction in U. Hence equation (4.4) has the unique solution

$$\psi(x) = \sum_{n=0}^{\infty} L^n(\gamma^{(k_0)}(x)) \in D^{(k_0)}.$$

Now assume that $\psi = \varphi^{(k_0)}$. Then the difference

$$P_\gamma = \varphi - T_0\varphi - \gamma$$

is a polynomial mapping of a degree $\leq k_0$. Obviously, P_γ is continous with respect to γ.

Further, let $\varphi \in \mathrm{Ker} T_0$. Then $\psi = \varphi^{(k_0)}$ satisfies the equation $\psi = L\psi$. Therefore $\psi = 0$ and φ is a polynomial mapping of a degree $\leq k_0$. □

To complete a proof of Theorem 3.4, it is sufficient to show that for any $\gamma \in C^k$, $k \geq k_0$ there is a polynomial mapping P_γ of degree $\leq k_0$ such that

$$\gamma(x) + \sum_{j=1}^{q} A_j P_\gamma(\alpha_j x + e_j) = (T\varphi)(x), \quad \varphi \in C^k,$$

and that the correspondence $\gamma \mapsto P_\gamma$ is continous. To this end, let us fix some neighborhood $U \supset I$ and choose a C^∞-function τ satisfying conditions

$$\tau(x) = \begin{cases} 1, & x \in V, \\ 0, & x \notin U, \end{cases}$$

where $V \supset I$ is some smaller neighborhood, $\mathrm{cl}(V) \subset U$.

Now consider the operators

$$(Nv)(x) = \begin{cases} \tau(x) \sum_{j=2}^{q} B_j v(F_1^{-1} F_j x), & x \in U \\ 0, & x \notin U \end{cases}$$

$$(Lv)(x) = \begin{cases} \tau(x) \sum_{j=q+1}^{t} B_j v(F_1^{-1} F_j x), & x \in U \\ 0, & x \notin U \end{cases}$$

Here

$$B_j = (A_1 A_1^*)^{-1} A_j A_1^*.$$

Note that the operator N is nilpotent. Indeed, let $p \in \mathbb{Z}_+$ be such that

$$\frac{1}{\lambda}\sum_{i=1}^{s}(e_{j_i} - e_1) \notin U, \quad s \geq p, \quad j_i \in \overline{2,q}.$$

Then $N^s = 0, \ s \geq p$. Therefore, one can apply Lemma 4.3 to the operator

$$T_0 \equiv (\mathrm{id} + N)^{-1}(\mathrm{id} + N + L) = \mathrm{id} + \sum_{i=0}^{p} N^i L.$$

Namely, consider the vector function

$$\tilde{\gamma} = (\mathrm{id} + N)^{-1}(A_1 A_1^*)^{-1}\gamma.$$

It follows from Lemma 4.3 that there exists of a polynomial mapping P_γ of degree $\leq k_0$ such that

$$\tilde{\gamma} + P_\gamma = (\mathrm{id} + N)^{-1}(\mathrm{id} + N + L)v, \quad v \in C^k.$$

Hence

$$(A_1 A_1^*)^{-1}\gamma + (\mathrm{id} + N)P_\gamma = (\mathrm{id} + N + L)v,$$

and the C^k-function

$$\varphi_0(x) = A_1^* v(F_1^{-1}x)$$

is a solution of the equation

$$(T\varphi)(x) = \gamma(x) + \sum_{j=1}^{q} A_j P_\gamma(\alpha_j x + e_j)$$

in a neighborhood of I. By Theorem 3.1 there exists a C^k-solution defined on the whole space $\mathbb{R}^n$, which completes the proof of Theorem 3.4.

Now let us prove Theorem 3.6.

It follows from Lemma 4.3 that Equation (1.1) is normally solvable, if $k \geq k_0(c)$. and, besides,

$$\dim \mathrm{Ker} T < \infty, \dim \mathrm{Ker} T^* < \infty.$$

It remains to prove that these dimensions are equal. Let us use a homotopical invariance of an index.

Consider the family of operators

$$(T_\delta\varphi)(x) = A_1\varphi(F_1 x) + \delta\sum_{j=2}^{t} A_j\varphi(F_j x)$$

with the same A_j and F_j. Obviously,

$$T_1 = T, \quad (T_o\varphi)(x) = A_1(x)\varphi(F_1 x).$$

Since Ker $T_\circ$ = Ker$T_\circ^*$ = $\{0\}$, we obtain

$$\mathrm{ind}T = 0. \qquad \square$$

References.

1. G.Belitskii, V.Nicolaevsky, Linear functional equations on the line, Integr. Equat. and Oper. Th., V.21 (1995), 212-223.

2. I.Glazman, Yu.Lyubich, Finite dimensional linear analysis. MIT press 1974.

3. Yu.Lyubich, Linear functional analysis, Enc. of Math. Sci., v.19, Springer Verlag, 1992.

Department of Mathematics
Ben-Gurion Univ. of the Negev
Beer Sheva 84105
Israel

AMS classification 39B Functional equations

Operator Theory
Advances and Applications, Vol. 98

REALIZATION THEOREMS FOR OPERATOR-VALUED R-FUNCTIONS

S.V. BELYI AND E.R. TSEKANOVSKII

Dedicated to the memory of Professor Israel Glazman

In this paper we consider realization problems for operator-valued R-functions acting on a Hilbert space E ($\dim E < \infty$) as linear-fractional transformations of the transfer operator-valued functions (characteristic functions) of linear stationary conservative dynamic systems (Brodskii-Livšic rigged operator colligations). We give complete proofs of both the direct and inverse realization theorems announced in [6], [7].

1. INTRODUCTION

Realization theory of different classes of operator-valued (matrix-valued) functions as transfer operator-functions of linear systems plays an important role in modern operator and systems theory. Almost all realizations in the modern theory of non-selfadjoint operators and its applications deal with systems (operator colligations) in which the main operators are *bounded* linear operators [8], [10-14], [17], [21]. The realization with an *unbounded* operator as a main operator in a corresponding system has not been investigated thoroughly because of a number of essential difficulties usually related to unbounded non-selfadjoint operators.

We consider realization problems for operator-valued R-functions acting on a finite dimensional Hilbert space E as linear-fractional transformations of the transfer operator-functions of linear stationary conservative dynamic systems (l.s.c.d.s.) θ of the form

$$\begin{cases} (\mathbb{A} - zI) = KJ\varphi_- \\ \varphi_+ = \varphi_- - 2iK^*x \end{cases} \qquad (\text{Im } \mathbb{A} = KJK^*),$$

or

$$\theta = \begin{pmatrix} \mathbb{A} & K & J \\ \mathfrak{H}_+ \subset \mathfrak{H} \subset \mathfrak{H}_- & & E \end{pmatrix}.$$

In the system θ above $\mathbb{A}$ is a bounded linear operator, acting from $\mathfrak{H}_+$ into $\mathfrak{H}_-$, where $\mathfrak{H}_+ \subset \mathfrak{H} \subset \mathfrak{H}_-$ is a rigged Hilbert space, $\mathbb{A} \supset T \supset A$, $\mathbb{A}^* \supset T^* \supset A$, A is a Hermitian operator in $\mathfrak{H}$, T is a non-Hermitian operator in $\mathfrak{H}$, K is a linear bounded operator from E into $\mathfrak{H}_-$, $J = J^* = J^{-1}$, $\varphi_\pm \in E$, φ_- is an input vector, φ_+ is an output vector, and $x \in \mathfrak{H}_+$ is a vector of the inner state of the system θ. The operator-valued function

$$W_\theta(z) = I - 2iK^*(\mathbb{A} - zI)^{-1}KJ \qquad (\varphi_+ = W_\theta(z)\varphi_-),$$

is the transfer operator-function of the system θ.

We establish criteria for a given operator-valued R-function $V(z)$ to be realized in the form

$$V(z) = i[W_\theta(z) + I]^{-1}[W_\theta(z) - I]J.$$

It is shown that an operator-valued R-function

$$V(z) = Q + F \cdot z + \int_{-\infty}^{+\infty} \left(\frac{1}{t - z} - \frac{t}{1 + t^2} \right) dG(t),$$

acting on a Hilbert space E ($\dim E < \infty$) with some invertibility condition can be realized if and only if

$$F = 0 \quad \text{and} \quad Qe = \int_{-\infty}^{+\infty} \frac{t}{1 + t^2}\, dG(t)e,$$

for all $e \in E$ such that

$$\int_{-\infty}^{+\infty} (dG(t)e, e)_E < \infty.$$

Moreover, if two realizable operator-valued R-functions are different only by a constant term then they can be realized by two systems θ_1 and θ_2 with corresponding non-selfadjoint operators that have the same Hermitian part A.

The rigged operator colligation θ mentioned above is exactly an unbounded version of the well known Brodskiĭ-Livšic bounded operator colligation α of the form [11]

$$\alpha = \begin{pmatrix} T & K & J \\ \mathfrak{H} & & E \end{pmatrix} \qquad (\operatorname{Im} T = KJK^*),$$

with a bounded linear operator T in $\mathfrak{H}$ (and without rigged Hilbert spaces).

To prove the direct and inverse realization theorems for operator-valued R-functions we build a functional model which generally speaking is an unbounded version of the Brodskiĭ-Livšic model with diagonal real part. This model for bounded linear operators was constructed in [11].

When this paper was submitted for publication, an article by D. Arov and M. Nudelman [5] appeared considering realization problem for another class of operator-valued functions (contractive) but not in terms of rigged operator colligations. At the end of this paper there is an example showing how a given R-function can be realized by a rigged operator colligation.

Acknowledgement. The authors express their gratitude to the referees and to G. Androlakis, P. Casazza, M. Lammers, and V. Peller for their valuable suggestions that helped to improve the presentation of this paper.

2. Preliminaries

In this section we recall some basic definitions and results that will be used in the proof of the realization theorem.

The Rigged Hilbert Spaces. Let $\mathfrak{H}$ denote a Hilbert space with inner product (x, y) and let A be a closed linear Hermitian operator, i.e. $(Ax, y) = (x, Ay)$ $(\forall x, y \in \mathfrak{D}(A))$, acting in the Hilbert space $\mathfrak{H}$ with generally speaking, non-dense domain $\mathfrak{D}(A)$. Let $\mathfrak{H}_0 = \overline{\mathfrak{D}(A)}$ and A^* be the adjoint to the operator A (we consider A acting from $\mathfrak{H}_0$ into $\mathfrak{H}$).

Now we are going to equip $\mathfrak{H}$ with spaces $\mathfrak{H}_+$ and $\mathfrak{H}_-$ called, respectively, spaces with positive and negative norms [9]. We denote $\mathfrak{H}_+ = \mathfrak{D}(A^*)$ $((\overline{\mathfrak{D}(A^*)} = \mathfrak{H})$ with inner product

$$(f, g)_+ = (f, g) + (A^* f, A^* g) \qquad (f, g \in \mathfrak{H}_+), \tag{1}$$

and then construct the *rigged* Hilbert space $\mathfrak{H}_+ \subset \mathfrak{H} \subset \mathfrak{H}_-$. Here $\mathfrak{H}_-$ is the space of all linear functionals over $\mathfrak{H}_+$ that are continuous with respect to $\|\cdot\|_+$. The norms of these spaces are connected by the relations $\|x\| \leq \|x\|_+$ $(x \in \mathfrak{H}_+)$, and $\|x\|_- \leq \|x\|$ $(x \in \mathfrak{H})$. It is well known that there exists an isometric operator $\mathcal{R}$ which maps $\mathfrak{H}_-$ onto $\mathfrak{H}_+$ such that

$$\begin{aligned} (x, y)_- &= (x, \mathcal{R}y) = (\mathcal{R}x, y) = (\mathcal{R}x, \mathcal{R}y)_+ \qquad (x, y \in \mathfrak{H}_-), \\ (u, v)_+ &= (u, \mathcal{R}^{-1}v) = (\mathcal{R}^{-1}u, v) = (\mathcal{R}^{-1}u, \mathcal{R}^{-1}v)_- \qquad (u, v \in \mathfrak{H}_+). \end{aligned} \tag{2}$$

The operator $\mathcal{R}$ will be called the Riesz-Berezanskii operator. In what follows we use symbols $(+)$, $(\cdot)$, and $(-)$ to indicate the norms $\|\cdot\|_+$, $\|\cdot\|$, and $\|\cdot\|_-$ by which geometrical and topological concepts are defined in $\mathfrak{H}_+$, $\mathfrak{H}$, and $\mathfrak{H}_-$.

Analogues of von Neumann's formulae. It is easy to see that for a Hermitian operator A in the above settings $\mathfrak{D}(A) \subset \mathfrak{D}(A^*)(= \mathfrak{H}_+)$ and $A^* y = PAy$ $(\forall y \in \mathfrak{D}(A))$, where P is an orthogonal projection of $\mathfrak{H}$ onto $\mathfrak{H}_0$. We put

$$\mathfrak{L} := \mathfrak{H} \ominus \mathfrak{H}_0 \quad \mathfrak{M}_\lambda := (A - \lambda I)\mathfrak{D}(A) \quad \mathfrak{N}_\lambda := (\mathfrak{M}_{\bar\lambda})^\perp \tag{3}$$

The subspace $\mathfrak{N}_\lambda$ is called a *defect subspace* of A for the point $\bar\lambda$. The cardinal number $\dim\mathfrak{N}_\lambda$ remains constant when λ is in the upper half-plane. Similarly, the number $\dim\mathfrak{N}_\lambda$ remains constant when λ is in the lower half-plane. The numbers $\dim\mathfrak{N}_\lambda$ and $\dim\mathfrak{N}_{\bar\lambda}$

($\mathrm{Im}\lambda < 0$) are called the *defect numbers* or *deficiency indices* of operator A [1]. The subspace $\mathfrak{N}_\lambda$ which lies in $\mathfrak{H}_+$ is the set of solutions of the equation $A^*g = \lambda Pg$.

Let now P_λ be the orthogonal projection onto $\mathfrak{N}_\lambda$, set

$$\mathfrak{B}_\lambda = P_\lambda \mathfrak{L}, \qquad \mathfrak{N}'_\lambda = \mathfrak{N}_\lambda \ominus \overline{\mathfrak{B}_\lambda} \tag{4}$$

It is easy to see that $\mathfrak{N}'_\lambda = \mathfrak{N}_\lambda \cap \mathfrak{H}_0$ and $\mathfrak{N}'_\lambda$ is the set of solutions of the equation $A^*g = \lambda g$ (see [25]), when $A^* : \mathfrak{H} \to \mathfrak{H}_0$ is the adjoint operator to A.

The subspace $\mathfrak{N}'_\lambda$ is the defect subspace of the densely defined Hermitian operator PA on $\mathfrak{H}_0$ ([22]). The numbers $\dim \mathfrak{N}'_\lambda$ and $\dim \mathfrak{N}'_{\bar\lambda}$ ($\mathrm{Im}\lambda < 0$) are called *semi-defect numbers* or the *semi-deficiency indices* of the operator A [16]. The von Neumann formula

$$\mathfrak{H}_+ = \mathfrak{D}(A^*) = \mathfrak{D}(A) + \mathfrak{N}_\lambda + \mathfrak{N}_{\bar\lambda}, \qquad (\mathrm{Im}\lambda \neq 0), \tag{5}$$

holds, but this decomposition is not direct for a non-densely defined operator A. There exists a generalization of von Neumann's formula [3], [24] to the case of a non-densely defined Hermitian operator (direct decomposition).

We call an operator A *regular*, if PA is a closed operator in $\mathfrak{H}_0$. For a regular operator A we have

$$\mathfrak{H}_+ = \mathfrak{D}(A) + \mathfrak{N}'_\lambda + \mathfrak{N}'_{\bar\lambda} + \mathfrak{N}, \qquad (\mathrm{Im}\lambda \neq 0) \tag{6}$$

where $\mathfrak{N} := \mathcal{R}\mathfrak{L}$. This is a generalization of von Neumann's formula. For $\lambda = \pm i$ we obtain the (+)-orthogonal decomposition

$$\mathfrak{H}_+ = \mathfrak{D}(A) \oplus \mathfrak{N}'_i \oplus \mathfrak{N}'_{-i} \oplus \mathfrak{N}. \tag{7}$$

Let $\tilde{A}$ be a closed Hermitian extension of the operator A. Then $\mathfrak{D}(\tilde{A}) \subset \mathfrak{H}_+$ and $P\tilde{A}x = A^*x$ ($\forall x \in \mathfrak{D}(\tilde{A})$). According to [25] a closed Hermitian extension $\tilde{A}$ is said to be *regular* if $\mathfrak{D}(\tilde{A})$ is (+)-closed. According to the theory of extensions of closed Hermitian operators A with non-dense domain [16], an operator U ($\mathfrak{D}(U) \subseteq \mathfrak{N}_i$, $\mathfrak{R}(U) \subseteq \mathfrak{N}_{-i}$) is called an *admissible operator* if $(U - I)f_i \in \mathfrak{D}(A)$ ($f_i \in \mathfrak{D}(U)$) only for $f_i = 0$. Then (see [4]) any symmetric extension $\tilde{A}$ of the non-densely defined closed Hermitian operator A, is defined by an isometric admissible operator U, $\mathfrak{D}(U) \subseteq \mathfrak{N}_i$, $\mathfrak{R}(U) \subseteq \mathfrak{N}_{-i}$ by the formula

$$\tilde{A}f_{\tilde{A}} = Af_A + (-if_i - iUf_i), \quad f_A \in \mathfrak{D}(A) \tag{8}$$

where $\mathfrak{D}(\tilde{A}) = \mathfrak{D}(A) \dotplus (U - I)\mathfrak{D}(U)$. The operator $\tilde{A}$ is self-adjoint if and only if $\mathfrak{D}(U) = \mathfrak{N}_i$ and $\mathfrak{R}(U) = \mathfrak{N}_{-i}$.

Let us denote now by $P^+_{\mathfrak{N}}$ the orthogonal projection operator in $\mathfrak{H}_+$ onto $\mathfrak{N}$. We introduce a new inner product $(\cdot,\cdot)_1$ defined by

$$(f, g)_1 = (f, g)_+ + (P^+_{\mathfrak{N}} f, P^+_{\mathfrak{N}} g)_+ \tag{9}$$

for all $f, g \in \mathfrak{H}_+$. The obvious inequality

$$\|f\|_+^2 \le \|f\|_1^2 \le 2\|f\|_+^2$$

shows that the norms $\|\cdot\|_+$ and $\|\cdot\|_1$ are topologically equivalent. It is easy to see that the spaces $\mathfrak{D}(A)$, $\mathfrak{N}'_i$, $\mathfrak{N}'_{-i}$, $\mathfrak{N}$ are (1)-orthogonal. We write $\mathfrak{M}_1$ for the Hilbert space $\mathfrak{M} = \mathfrak{N}'_i \dotplus \mathfrak{N}'_{-i} \oplus \mathfrak{N}$ with inner product $(f,g)_1$. We denote by $\mathfrak{H}_{+1}$ the space $\mathfrak{H}_+$ with norm $\|\cdot\|_1$, and by $\mathcal{R}_1$ the corresponding Riesz-Berezanskii operator related to the rigged Hilbert space $\mathfrak{H}_{+1} \subset \mathfrak{H} \subset \mathfrak{H}_{-1}$. The following theorem gives a characterization of the regular extensions for a regular closed Hermitian operator A (see [4]).

Theorem 1. *I. For each closed Hermitian extension $\tilde{A}$ of a regular operator A there exists a (1)-isometric operator $V = V(\tilde{A})$ on $\mathfrak{M}_1$ with the properties: a) $\mathfrak{D}(V)$ is (+)-closed and belongs to $\mathfrak{N} \oplus \mathfrak{N}'_i$, $\mathfrak{R}(V) \subset \mathfrak{N} \oplus \mathfrak{N}'_{-i}$; b) $Vh = h$ only for $h = 0$, and $\mathfrak{D}(\tilde{A}) = \mathfrak{D}(A) \oplus (I+V)\mathfrak{D}(V)$.*

Conversely, for each (1)-isometric operator V with the properties a) and b) there exists a closed Hermitian extension $\tilde{A}$ in the sense indicated.

II. The extension $\tilde{A}$ is regular if and only if the manifold $\mathfrak{R}(I+V)$ is (1)-closed.

III. The operator $\tilde{A}$ is self-adjoint if and only if $\mathfrak{D}(V) = \mathfrak{N} \oplus \mathfrak{N}'_i$, $\mathfrak{R}(V) = \mathfrak{N} \oplus \mathfrak{N}'_{-i}$.

The following theorem can be found in [16].

Theorem 2. *Let $\tilde{A}$ be a regular self-adjoint extension of a regular Hermitian operator A, that is determined by an admissible operator U and let*

$$\hat{\mathfrak{N}}_i = \{f_i \in \mathfrak{N}_i, (U - I)f_i \in \mathfrak{H}_0\}. \tag{10}$$

Then

$$\mathfrak{H}_+ = \mathfrak{D}(\tilde{A}) \dotplus (U+I)\hat{\mathfrak{N}}_i. \tag{11}$$

Bi-extensions. Denote by $[\mathfrak{H}_1, \mathfrak{H}_2]$ the set of all linear bounded operators acting from the Hilbert space $\mathfrak{H}_1$ into the Hilbert space $\mathfrak{H}_2$.

Definition. An operator $\mathbb{A} \in [\mathfrak{H}_+, \mathfrak{H}_-]$ is a *bi-extension* of A if both $\mathbb{A} \supset A$ and $\mathbb{A}^* \supset A$.

If $\mathbb{A} = \mathbb{A}^*$, then $\mathbb{A}$ is called a self-adjoint bi-extension of the operator A. We write $\mathfrak{S}(A)$ for the class of bi-extensions of A. This class is closed in the weak topology and is invariant under taking adjoints. The following theorem from [4], [25] gives a description of $\mathfrak{S}(A)$.

Theorem 3. *Every bi-extension $\mathbb{A}$ of a regular Hermitian operator A has the form:*

$$\mathbb{A} = AP^+_{\mathfrak{D}(A)} + [A^* + \mathcal{R}_1^{-1}(Q - \frac{i}{2}P^+_{\mathfrak{N}'_i} + \frac{i}{2}P^+_{\mathfrak{N}'_{-i}})]P^+_{\mathfrak{M}} \tag{12}$$

$$\mathbb{A}^* = AP^+_{\mathfrak{D}(A)} + [A^* + \mathcal{R}_1^{-1}(Q^* - \frac{i}{2}P^+_{\mathfrak{N}'_i} + \frac{i}{2}P^+_{\mathfrak{N}'_{-i}})]P^+_{\mathfrak{M}} \tag{13}$$

where Q is an arbitrary operator in $[\mathfrak{M}, \mathfrak{M}]$ and Q^* is its adjoint with respect to the (1)-metric.

Corollary 1. *Every self-adjoint bi-extension $\mathbb{A}$ of the regular Hermitian operator A is of the form:*

$$\mathbb{A} = AP^+_{\mathfrak{D}(A)} + [A^* + \mathcal{R}_1^{-1}(S - \frac{i}{2}P^+_{\mathfrak{N}'_i} + \frac{i}{2}P^+_{\mathfrak{N}'_{-i}})]P^+_{\mathfrak{M}}, \tag{14}$$

where S is an arbitrary (1)-self-adjoint operator in $[\mathfrak{M}, \mathfrak{M}]$.

Let $\mathbb{A}$ be a bi-extension of a Hermitian operator A. The operator $\hat{A}f = \mathbb{A}f$, $\mathfrak{D}(\hat{A}) = \{f \in \mathfrak{H}, \mathbb{A}f \in \mathfrak{H}\}$ is called the *quasi-kernel* of $\mathbb{A}$. If $\mathbb{A} = \mathbb{A}^*$ and $\hat{A}$ is a quasi-kernel of $\mathbb{A}$ such that $A \neq \hat{A}$, $\hat{A}^* = \hat{A}$ then $\mathbb{A}$ is said to be a *strong* self-adjoint bi-extension of A.

Classes Ω_A and Λ_A. (∗)-extensions. Let A be a closed Hermitian operator.

Definition. We say that a closed densely defined linear operator T acting on the Hilbert space $\mathfrak{H}$ belongs to the class Ω_A if:

(1) $T \supset A$ and $T^* \supset A$;
(2) $(-i)$ is a regular point of T.[1]

It was mentioned in [4] that sets $\mathfrak{D}(T)$ and $\mathfrak{D}(T^*)$ are (+)-closed, the operators T and T^* are $(+,\cdot)$-bounded. The following theorem [25] is an analogue of von Neumann's formulae for the class Ω_A.

Theorem 4. *I. To each operator of the class Ω_A there corresponds an operator M on the space $\mathfrak{M}_1$ with the following properties:*

(1) $\mathfrak{D}(M) = \mathfrak{N}'_i \oplus \mathfrak{N}$, and $\mathfrak{R}(M) = \mathfrak{N}'_{-i} \oplus \mathfrak{N}$;
(2) $Mx + x = 0$ only for $x = 0$, and $M^*x + x = 0$ only for $x = 0$. Moreover, the following hold:

$$\mathfrak{D}(T) = \mathfrak{D}(A) \oplus (M + I)(\mathfrak{N}'_i \oplus \mathfrak{N}), \tag{15}$$

$$\mathfrak{D}(T^*) = \mathfrak{D}(A) \oplus (M^* + I)(\mathfrak{N}'_{-i} \oplus \mathfrak{N}). \tag{16}$$

[1] The condition, that $(-i)$ is a regular point in the definition of the class Ω_A is not essential. It is sufficient to require the existence of some regular point for T.

II. Conversely, for each pair of (1)-adjoint operators M and M^* in $[\mathfrak{M}_1, \mathfrak{M}_1]$ satisfying (1) and (2) above, formulas (15) and (16) give a corresponding operator T in the class Ω_A. Moreover, if $f = g + (M+I)\varphi$, $g \in \mathfrak{D}(A)$, and $\varphi \in \mathfrak{N}'_i \oplus \mathfrak{N}$ then

$$Tf = Ag + A^*(I+M)\varphi + i\mathcal{R}_1^{-1}P_{\mathfrak{N}}^+(I-M)\varphi \quad (f \in \mathfrak{D}(T)). \tag{17}$$

Similarly, if $f = g + (M^*+I)\psi$, $g \in \mathfrak{D}(A)$, and $\psi \in \mathfrak{N}'_{-i} \oplus \mathfrak{N}$, then

$$T^*f = Ag + A^*(I+M^*)\psi + i\mathcal{R}_1^{-1}P_{\mathfrak{N}}^+(M^*-I)\psi \quad (f \in \mathfrak{D}(T)), \tag{18}$$

Definition. An operator $\mathbb{A}$ in $[\mathfrak{H}_+, \mathfrak{H}_-]$ is called a $(*)$-extension of an operator T from the class Ω_A if both $\mathbb{A} \supset T$ and $\mathbb{A}^* \supset T^*$.

This $(*)$-extension is called *correct*, if an operator $\mathbb{A}_R = \frac{1}{2}(\mathbb{A} + \mathbb{A}^*)$ is a strong self-adjoint bi-extension of an operator A. It is easy to show that if $\mathbb{A}$ is a $(*)$-extension of T, then T and T^* are quasi-kernels of $\mathbb{A}$ and $\mathbb{A}^*$, respectively.

Definition. We say that the operator T of the class Ω_A belongs to the class Λ_A if

(1) T admits a correct $(*)$-extension;
(2) A is the maximal common Hermitian part of T and T^*.

Theorem 5. *Let an operator T belong to Ω_A and let M be an operator in $[\mathfrak{M}, \mathfrak{M}]$ that is related to T by Theorem 4. Then T belongs to Λ_A if and only if there exists either (1)-isometric operator or a $(\cdot)$-isometric operator U in $[\mathfrak{N}'_i, \mathfrak{N}'_{-i}]$ such that*

$$\begin{cases} (U+I)\mathfrak{N}'_i + (M+I)(\mathfrak{N}'_i \oplus \mathfrak{N}) = \mathfrak{M}, \\ (U+I)\mathfrak{N}'_i + (M+I)(\mathfrak{N}'_i \oplus \mathfrak{N}) = \mathfrak{M}. \end{cases} \tag{19}$$

Corollary 2. *If a closed Hermitian operator A has finite and equal defect indices, then the class Ω_A coincides with the Λ_A.*

Extended Resolvents and Extended Spectral Functions of a Hermitian Operator. Let A be a closed Hermitian operator on $\mathfrak{H}$ and $\mathfrak{h}$ be a Hilbert space such that $\mathfrak{H}$ is a subspace of $\mathfrak{h}$. Let $\tilde{A}$ be a self-adjoint extension of A on $\mathfrak{h}$, and $\tilde{E}(t)$ be the spectral function of $\tilde{A}$. An operator function $R_\lambda = P_{\mathfrak{H}}(\tilde{A} - \lambda I)^{-1}|_{\mathfrak{H}}$ is called a *generalized resolvent* of A, and $E(t) = P_{\mathfrak{H}}\tilde{E}(t)|_{\mathfrak{H}}$ is the corresponding *generalized spectral function.* Here

$$R_\lambda = \int_{-\infty}^{\infty} \frac{dE(t)}{t-\lambda} \quad (\mathrm{Im}\lambda \neq 0). \tag{20}$$

If $\mathfrak{h} = \mathfrak{H}$ then R_λ and $E(t)$ are called *canonical resolvent* and *canonical spectral function,* respectively. According to [19] we denote by $\hat{R}_\lambda$ the $(-,\cdot)$-continuous operator from $\mathfrak{H}_-$ into $\mathfrak{H}$ which is adjoint to $R_{\bar{\lambda}}$:

$$(\hat{R}_\lambda f, g) = (f, R_{\bar{\lambda}g}) \quad (f \in \mathfrak{h}_-, g \in \mathfrak{H}). \tag{21}$$

It follows that $\hat{R}_\lambda f = R_\lambda f$ for $f \in h$, so that $\hat{R}_\lambda$ is an extension of R_λ from $\mathfrak{H}$ to $\mathfrak{H}_-$ with respect to $(-,\cdot)$-continuity. The function $\hat{R}_\lambda$ of the parameter λ, (Im$\lambda \neq 0$) is called the *extended generalized (canonical) resolvent* of the operator A. We write $\aleph$ for the family of all finite intervals on the real axis. It is known [19] that if $\Delta \in \aleph$ then $E(\Delta)\mathfrak{H} \subset \mathfrak{H}_+$ and the operator $E(\Delta)$ is $(\cdot,+)$-continuous. We denote by $\hat{E}(\Delta)$ the $(-,\cdot)$-continuous operator from $\mathfrak{H}_-$ to $\mathfrak{H}$ that is adjoint to $E(\Delta) \in [\mathfrak{H}, \mathfrak{H}_+]$. Similarly,

$$(\hat{E}(\Delta)f, g) = (f, E(\Delta)g) \quad (f \in \mathfrak{H}_-,\, g \in \mathfrak{H}), \tag{22}$$

One can easily see that $\hat{E}(\Delta)f = E(\Delta)f$, $\forall f \in \mathfrak{H}$, so that $\hat{E}(\Delta)$ is the extension of $E(\Delta)$ by continuity. We say that $\hat{E}(\Delta)$, as a function of $\Delta \in \aleph$, is the *extended generalized (canonical) spectral function* of A corresponding to the self-adjoint extension $\tilde{A}$ (or to the original spectral function $E(\Delta)$). It is known [19] that $\hat{E}(\Delta) \in [\mathfrak{H}_-, \mathfrak{H}_+]$, $\forall \Delta \in \aleph$, and $(\hat{E}(\Delta)f, f) \geq 0$ for all $f \in \mathfrak{H}_-$. It is also known [19] that the complex scalar measure $(E(\Delta)f, g)$ is a complex function of bounded variation on the real axis. However, $(\hat{E}(\Delta)f, g)$ may be unbounded for $f, g \in \mathfrak{H}_-$.

Now let $\hat{R}_\lambda$ be an extended generalized (canonical) resolvent of a closed Hermitian operator A and let $\hat{E}(\Delta)$ be the corresponding extended generalized (canonical) spectral function. It was shown in [19] that for any $f, g \in \mathfrak{H}_-$,

$$\int_{-\infty}^{+\infty} \frac{|d(\hat{E}(\Delta)f, g)|}{1+t^2} < \infty, \tag{23}$$

and the following integral representation holds

$$\hat{R}_\lambda - \frac{\hat{R}_i + \hat{R}_{-i}}{2} = \int_{-\infty}^{+\infty} \left(\frac{1}{t-\lambda} - \frac{t}{1+t^2}\right) d\hat{E}(t). \tag{24}$$

Lemma 6. *Let $\mathbb{A} = A^* + \mathcal{R}^{-1}(S - \frac{i}{2}P^+_{\mathfrak{N}_i} + \frac{i}{2}P^+_{\mathfrak{N}_{-i}})P^+_{\mathfrak{M}}$ be a strong self-adjoint bi-extension of a regular Hermitian operator A with the quasi-kernel $\hat{A}$ and let $\hat{E}(\Delta)$ be the extended canonical spectral function of $\hat{A}$. Then for every $f \in \mathfrak{H} \oplus L$, $f \neq 0$, and for every $g \in \mathfrak{H}_-$ there is an integral representation*

$$(\bar{R}_\lambda f, g) = \int_{-\infty}^{+\infty} \left(\frac{1}{t-\lambda} - \frac{t}{1+t^2}\right) d(\hat{E}(t)f, g) + \frac{1}{2}((\hat{R}_i + \hat{R}_{-i})f, g). \tag{25}$$

Here $F = \mathfrak{H}_+ \ominus \mathfrak{D}(A)$, $L = \mathcal{R}^{-1}(S - \frac{i}{2}P^+_{\mathfrak{N}_i} + \frac{i}{2}P^+_{\mathfrak{N}_{-i}})F$, $\bar{R}_\lambda = \overline{(\mathbb{A} - \lambda I)^{-1}}$.

Theorem 7. *Let* $\mathbb{A} = A^* + \mathcal{R}^{-1}(S - \frac{i}{2}P^+_{\mathfrak{N}_i} + \frac{i}{2}P^+_{\mathfrak{N}_{-i}})P^+_{\mathfrak{M}}$ *be a strong self-adjoint bi-extension of a regular Hermitian operator* A *with the quasi-kernel* $\hat{A}$ *and let* $\hat{E}(\Delta)$ *be the extended canonical spectral function of* $\hat{A}$. *Also, let* $F = \mathfrak{H}_+ \ominus \mathfrak{D}(A)$ *and* $L = \mathcal{R}^{-1}(S - \frac{i}{2}P^+_{\mathfrak{N}_i} + \frac{i}{2}P^+_{\mathfrak{N}_{-i}})F$. *Then for every* $f \in L \dotplus \mathfrak{L}$ *with* $f \neq 0$ *and* $f \in \mathfrak{R}(\mathbb{A} - \lambda I)$, *we have*

$$\int_{-\infty}^{+\infty} d(\hat{E}(t)f, f) = \infty, \qquad \text{if } f \notin \mathfrak{L}, \tag{26}$$

and

$$\int_{-\infty}^{+\infty} d(\hat{E}(t)f, f) < \infty, \qquad \text{if } f \in \mathfrak{L}. \tag{26'}$$

Moreover, there exist real constants b *and* c *such that*

$$c\|f\|_-^2 \leq \int_{-\infty}^{+\infty} \frac{d(\hat{E}(t)f, f)}{1+t^2} \leq b\|f\|_-^2, \tag{27}$$

for all $f \in L \dotplus \mathfrak{L}$.

Corollary 3. *In the settings of Theorem 7 for all* $f, g \in L \dotplus \mathfrak{L}$

$$\left|\left(\frac{\hat{R}_i + \hat{R}_{-i}}{2} f, g\right)\right| \leq a\sqrt{\int_{-\infty}^{+\infty} \frac{d(\hat{E}(t)f, f)}{1+t^2}} \cdot \sqrt{\int_{-\infty}^{+\infty} \frac{d(\hat{E}(t)g, g)}{1+t^2}}, \tag{28}$$

where $a > 0$ *is a constant (see* [2]*).*

3. Linear Stationary Conservative Dynamic Systems

In this section we consider linear stationary conservative dynamic systems (l. s. c. d. s.) θ of the form

$$\begin{cases} (\mathbb{A} - zI) = KJ\varphi_- \\ \varphi_+ = \varphi_- - 2iK^*x \end{cases} \qquad (\mathrm{Im}\ \mathbb{A} = KJK^*). \tag{29}$$

In a system θ of the form (29) $\mathbb{A}$, K and J are bounded linear operators in Hilbert spaces, φ_- is an input vector, φ_+ is an output vector, and x is an inner state vector of the system θ. For our purposes we need the following more precise definition:

Definition. *The array*

$$\theta = \begin{pmatrix} \mathbb{A} & K & J \\ \mathfrak{H}_+ \subset \mathfrak{H} \subset \mathfrak{H}_- & & E \end{pmatrix} \tag{30}$$

is called a linear stationary conservative dynamic system or Brodskiĭ-Livšic rigged operator colligation if

(1) $\mathbb{A}$ *is a correct* $(*)$*-extension of an operator* T *of the class* Λ_A.
(2) $J = J^* = J^{-1} \in [E, E]$, $\dim E < \infty$
(3) $\mathbb{A} - \mathbb{A}^* = 2iKJK^*$, *where* $K \in [E, \mathfrak{H}_-]$ $\quad (K^* \in [\mathfrak{H}_+, E])$

In this case, the operator K is called a *channel operator* and J is called a *direction operator*. A system θ of the form (30) will be called a *scattering* system (*dissipative* operator colligation) if $J = I$. We will associate with the system θ the operator-valued function

$$W_\theta(z) = I - 2iK^*(\mathbb{A} - zI)^{-1}KJ \tag{31}$$

which is called the *transfer operator-valued function* of the system θ or the characteristic operator-valued function of Brodskiĭ-Livšic rigged operator colligation. According to Theorem 7, $\mathfrak{R}(K) \subset \mathfrak{R}(\mathbb{A} - \lambda I)$ and therefore $W_\theta(z)$ is well-defined. It may be shown [10], [25] that the transfer operator-function of the system θ of the form (30) has the following properties:

$$\begin{aligned} &W_\theta^*(z)JW_\theta(z) - J \geq 0 \quad (\operatorname{Im} z > 0, z \in \rho(T)), \\ &W_\theta^*(z)JW_\theta(z) - J = 0 \quad (\operatorname{Im} z = 0, z \in \rho(T)), \\ &W_\theta^*(z)JW_\theta(z) - J \leq 0 \quad (\operatorname{Im} z < 0, z \in \rho(T)), \end{aligned} \tag{32}$$

where $\rho(T)$ is the set of regular points of an operator T. Similar relations take place if we change $W_\theta(z)$ to $W_\theta^*(z)$ in (32). Thus, the transfer operator-valued function of the system θ of the form (30) is J-contractive in the lower half-plane on the set of regular points of an operator T and J-unitary on real regular points of an operator T.

Let θ be a l.s.c.d.s. of the form (30). We consider the operator-valued function

$$V_\theta(z) = K^*(\mathbb{A}_R - zI)^{-1}K. \tag{33}$$

The transfer operator-function $W_\theta(z)$ of the system θ and an operator-function $V_\theta(z)$ of the form (33) are connected with the relation

$$V_\theta(z) = i[W_\theta(z) + I]^{-1}[W_\theta(z) - I]J. \tag{34}$$

As it is known [11] an operator-function $V(z) \in [E, E]$ is called an *operator-valued R-function* if it is holomorphic in the upper half-plane and $\operatorname{Im} V(z) \geq 0$ whenever $\operatorname{Im} z > 0$.

It is known [11,17] that an operator-valued R-function acting on a Hilbert space E ($\dim E < \infty$) has an integral representation

$$V(z) = Q + F \cdot z + \int_{-\infty}^{+\infty} \left(\frac{1}{t-z} - \frac{t}{1+t^2} \right) dG(t), \tag{35}$$

where $Q = Q^*$, $F \geq 0$ in the Hilbert space E, and $G(t)$ is a non-decreasing operator-function on $(-\infty, +\infty)$ for which

$$\int_{-\infty}^{+\infty} \frac{dG(t)}{1+t^2} \in [E, E].$$

Definition. *We call an operator-valued R-function $V(z)$ acting on a Hilbert space E, ($\dim E < \infty$) realizable if in some neighborhood of the point $(-i)$, the function $V(z)$ can be represented in the form*

$$V(z) = i[W_\theta(z) + I]^{-1}[W_\theta(z) - I]J, \tag{36}$$

where $W_\theta(z)$ is the transfer operator-function of some l.s.c.d.s. θ with the direction operator J ($J = J^* = J^{-1} \in [E, E]$).

Definition. *An operator-valued R-function $V(z) \in [E, E]$, ($\dim E < \infty$) is said to be a member of the class $N(R)$ if in the representation (35) we have*

$$i) \quad F = 0,$$

$$ii) \quad Qe = \int_{-\infty}^{+\infty} \frac{t}{1+t^2} dG(t)e,$$

for all $e \in E$ with

$$\int_{-\infty}^{+\infty} (dG(t)e, e)_E < \infty.$$

We now establish the next result.

Theorem 8. *Let θ be a l.s.c.d.s. of the form (30) with $\dim E < \infty$. Then the operator-function $V_\theta(z)$ of the form (33), (34) belongs to the class $N(R)$.*

Proof. Let G_{-i} be a neighborhood of $(-i)$ and $\lambda, \mu \in G_{-i}$. Then,

$$\begin{aligned} V_\theta(\lambda) - V_\theta(\mu) &= K^*(\mathbb{A}_R - \lambda I)^{-1}K - K^*(\mathbb{A}_R - \mu I)^{-1}K \\ &= (\mu - \lambda)K^*(\mathbb{A}_R - \lambda I)^{-1}(\mathbb{A}_R - \mu I)^{-1}K, \end{aligned} \tag{37}$$

and

$$\frac{V_\theta(\lambda) - V_\theta(\mu)}{\mu - \lambda} = K^*(\mathbb{A}_R - \lambda I)^{-1}(\mathbb{A}_R - \mu I)^{-1}K, \tag{38}$$

for all $\lambda, \mu \in G_{-i}$. Therefore, letting $\lambda \to \mu$ we can say that $V_\theta(z)$ is holomorphic in G_{-i}. Without loss of generality (see [25]) we can conclude that $V_\theta(z)$ is holomorphic in any one of the half-planes.

It is obvious that $V_\theta^*(z) = \overline{V_\theta(z)} = V_\theta(\bar{z})$. Furthermore,

$$\text{Im}V_\theta(z) = \frac{1}{2i} K^*(\mathbb{A}_R - \bar{z}I)^{-1}(\mathbb{A}_R - zI)^{-1}K. \tag{39}$$

Since $(-i)$ is a regular point of the operator T in the system (30) then (see [10]) $I + iV(\lambda)J$ is invertible in G_{-i}.

Let now $D_z = (\mathbb{A}_R - zI)^{-1}K$, then it is easy to see that the adjoint operator D_z^* is given by $D_z^* = K^*(\mathbb{A}_R - \bar{z}I)^{-1}$. Therefore, we have $\text{Im}V_\theta(z) = \text{Im}z D_z^* D_z$ which implies that $\text{Im}V_\theta(z) \geq 0$ when $\text{Im}z > 0$. Hence we can conclude that $V_\theta(z)$ is an operator R-function and admits representation (35).

Let now $B = K^*(\mathbb{A}_R + iI)^{-1}(\mathbb{A}_R - iI)^{-1}K$. It follows from (39) that $B = \frac{1}{2i}(V_\theta(i) - V_\theta^*(i))$. Using Theorem 7 and representation (35) one can show that

$$Bf = \int_{-\infty}^{\infty} \frac{dG(t)}{1+t^2} f, \qquad f \in E \tag{40}$$

and $B \in [E, E]$.

Let $\hat{E}(\Delta)$ be the canonical extended spectral function of the quasi-kernel $\hat{A}$ of the operator $\mathbb{A}_R = \frac{1}{2}(\mathbb{A} + \mathbb{A}^*)$. Then relying on Lemma 6 for all $f, g \in E$ we have

$$(V_\theta(\lambda)f, g)_E = \int_{-\infty}^{+\infty} \left(\frac{1}{t-\lambda} - \frac{t}{1+t^2} \right) d(\hat{G}(t)f, g)_E + (\hat{Q}f, g)_E, \tag{41}$$

where $\hat{G}(\Delta) = K^* \hat{E}(\Delta) K$, $\Delta \in \aleph$ and

$$\hat{Q} = \frac{1}{2} K^*[(\mathbb{A}_R - iI)^{-1} + (\mathbb{A}_R + iI)^{-1}]K = \frac{1}{2}[V_\theta(-i) + V_\theta^*(-i)]. \tag{42}$$

From Theorem 7 (see also [19]), we have for all $f \in E$ with $Kf \in \mathfrak{L}$,

$$\int_{-\infty}^{\infty} d(\hat{G}(t)f, f)_E < \infty, \tag{43}$$

and

$$c\|Kf\|_-^2 \leq \int_{-\infty}^{+\infty} \frac{d(\hat{G}(t)f, f)_E}{1+t^2} \leq b\|Kf\|_-^2. \tag{44}$$

Moreover, (28) implies that

$$\left|\left(\hat{Q}f,g\right)_E\right| \le C\sqrt{\int_{-\infty}^{+\infty}\frac{d(\hat{G}(t)f,f)_E}{1+t^2}}\cdot\sqrt{\int_{-\infty}^{+\infty}\frac{d(\hat{G}(t)g,g)_E}{1+t^2}}. \tag{45}$$

By (41) we have for any $f,g\in E$

$$(V_\theta(\lambda)f,g)_E = (\hat{Q}f,g)_E + \int_{-\infty}^{+\infty}\left(\frac{1}{t-\lambda}-\frac{t}{1+t^2}\right)d(\hat{G}(t)f,g)_E. \tag{46}$$

On the other hand (35) implies

$$(V_\theta(\lambda)f,g)_E = (Qf,g)_E + \lambda(Ff,g)_E + \int_{-\infty}^{+\infty}\left(\frac{1}{t-\lambda}-\frac{t}{1+t^2}\right)d(G(t)f,g)_E. \tag{47}$$

Comparing (46) and (47) we get $(Qf,g)_E = (\hat{Q}f,g)_E$, $(Ff,g)_E = 0$, and $(G(\Delta)f,g) = (\hat{G}(\Delta)f,g)$ $(\Delta\in\aleph)$, for all $f,g\in E$. Taking into account the continuity and positivity of F, $G(\Delta)$, and $\hat{G}(\Delta)$, we find that $F=0$ and $G(\Delta)=\hat{G}(\Delta)$ $(\Delta\in\aleph)$.

Thus,

$$V(\lambda) = Q + \int_{-\infty}^{+\infty}\left(\frac{1}{t-\lambda}-\frac{t}{1+t^2}\right)dG(t), \tag{48}$$

holds.

Let $E_\infty = K^{-1}\mathfrak{L}$, $E_\infty\subset E$. Since $\hat{E}(\Delta)$ coincides with $E(\Delta)$ on $\mathfrak{L}$, then for any $e\in E_\infty$, we have

$$\int_{-\infty}^{+\infty} d(\hat{G}(t)e,e)_E < \infty. \tag{49}$$

If $e\notin E_\infty$, then $Ke\notin\mathfrak{L}$ (see Theorem 7) and

$$\int_{-\infty}^{+\infty} d(\hat{G}(t)e,e)_E = \infty. \tag{50}$$

Further, since

$$Q = \frac{1}{2}\left[V_\theta(i)+V_\theta(-i)\right] = \frac{1}{2}\left[K^*((\mathbb{A}_R+iI)^{-1}+(\mathbb{A}_R-iI)^{-1})K\right], \tag{51}$$

we have $\mathfrak{R}(Q) \subseteq \mathfrak{R}(K^*) \subseteq E$. Now formula (45) yields

$$|(Qf,g)_E| \le C\|f\|_E \cdot \|g\|_E, \qquad f,g \in E. \tag{52}$$

On the other hand, if $e \in E_\infty$ then

$$\begin{aligned} Qe &= \frac{1}{2}\left[K^*(\hat{A}_R + iI)^{-1} + (\hat{A}_R - iI)^{-1})Ke\right] \\ &= K^* \int_{-\infty}^{+\infty} \frac{t}{1+t^2}\, dE(t)Ke = \int_{-\infty}^{+\infty} \frac{t}{t^2+1}\, d\hat{G}(t)e. \end{aligned}$$

This completes the proof.

Next, we establish the converse.[2]

Theorem 9. *Let an operator-valued function $V(z)$ act on a finite-dimensional Hilbert space E and belong to the class $N(R)$. Then $V(z)$ admits a realization by the system θ of the form (30) with a preassigned direction operator J for which $I + iV(-i)J$ is invertible.*

Proof. We will use several steps to prove this theorem.

STEP 1. Let $C_{00}(E,(-\infty,+\infty))$ be the set of continuous compactly supported vector-valued functions $f(t)$ $(-\infty < t < +\infty)$ with values in a finite dimensional Hilbert space E. We introduce an inner product $(\cdot,\cdot)$ defined by

$$(f,g) = \int_{-\infty}^{+\infty} (G(dt)f(t), g(t))_E \tag{53}$$

for all $f,g \in C_{00}(E,(-\infty,+\infty))$. In order to construct a Hilbert space, we identify with zero all functions $f(t)$ such that $(f,f) = 0$. Then we make the completion and obtain the new Hilbert space $L^2_G(E)$. Let us note that the set $C_{00}(E,(-\infty,+\infty))$ is dense in $L^2_G(E)$. Moreover, if $f(t)$ is continuous and

$$\int_{-\infty}^{+\infty} (G(dt)f(t), f(t))_E < \infty, \tag{54}$$

then $f(t)$ belongs to $L^2_G(E)$.

Let $\mathfrak{D}_0$ be the set of the continuous vector-valued (with values in E) functions $f(t)$ such that in addition to (54), we have

$$\int_{-\infty}^{+\infty} t^2 (G(dt)f(t), f(t))_E < \infty. \tag{55}$$

[2] The method of rigged Hilbert spaces for solving inverse problems in the theory of characteristic operator-valued functions was introduced in [23] and was developed further in [2].

Since $C_{00} \subset \mathfrak{D}_0$, it follows that $\mathfrak{D}_0$ is dense in $L^2_G(E)$. We introduce an operator $\hat{A}$ on $\mathfrak{D}_0$ in the following way:

$$\hat{A}f(t) = tf(t). \tag{56}$$

Below we denote again by $\hat{A}$ the closure of the Hermitian operator $\hat{A}$ (56). It is easy to see that this operator is Hermitian. Now $\hat{A}$ is a self-adjoint operator in $L^2_G(E)$ (see [9]).

Let $\tilde{\mathfrak{H}}_+ = \mathfrak{D}(\hat{A})$ and define the inner product

$$(f,g)_{\tilde{\mathfrak{H}}_+} = (f,g) + (\hat{A}f, \hat{A}g) \tag{57}$$

for all $f, g \in \tilde{\mathfrak{H}}_+$. It is clear that $\tilde{\mathfrak{H}}_+$ is a Hilbert space with norm $\|\cdot\|_{\tilde{\mathfrak{H}}_+}$ generated by the inner product (57). We equip the space $L^2_G(E)$ with spaces $\tilde{\mathfrak{H}}_+$ and $\tilde{\mathfrak{H}}_-$:

$$\tilde{\mathfrak{H}}_+ \subset L^2_G(E) \subset \tilde{\mathfrak{H}}_-. \tag{58}$$

Let us denote by $\tilde{\mathcal{R}}$ the corresponding Riesz-Berezanskii operator, $\tilde{\mathcal{R}} \in [\tilde{\mathfrak{H}}_-, \tilde{\mathfrak{H}}_+]$.

Consider the following subspaces of the space E:

$$\begin{aligned} E_\infty &= \{e \in E : \int_{-\infty}^{+\infty} d(G(t)e,e)_E < \infty\} \\ F_\infty &= E_\infty^\perp. \end{aligned} \tag{59}$$

If $e \in E_\infty$, then (54) implies that the function $e(t) = e$ is an element of the space $L^2_G(E)$. On the other hand, if $e \in E$ and $e \notin E_\infty$ then $e(t)$ does not belong to $L^2_G(E)$. It can be shown that any function $e(t) = e \in E$ can be identified with an element of $\tilde{\mathfrak{H}}_-$. Indeed, since for all $e \in E$

$$\int_{-\infty}^{+\infty} \frac{d(G(t)e,e)_E}{1+t^2} < \infty, \tag{60}$$

the function

$$\tilde{e}(t) = \frac{e}{\sqrt{1+t^2}} \tag{61}$$

belongs to the space $L^2_G(E)$. Letting $f(t) \in \mathfrak{D}_0$, we have

$$\int_{-\infty}^{+\infty} (1+t^2)(G(dt)f(t), f(t))_E < \infty. \tag{62}$$

Therefore, the function $\tilde{f}(t) = \sqrt{1+t^2}f(t)$ belongs to the space $L^2_G(E)$ and hence

$$(\tilde{f}(t), \tilde{e}(t)) = \int_{-\infty}^{+\infty} (G(dt)\tilde{f}(t), \tilde{e}(t))_E.$$

Furthermore,

$$
\begin{aligned}
|(\tilde{f}(t),\tilde{e}(t))| &\le \|\tilde{f}(t)\|\cdot\|\tilde{e}(t)\| \\
&= \sqrt{\int_{-\infty}^{+\infty}(1+t^2)(G(dt)f(t),f(t))_E}\cdot\sqrt{\int_{-\infty}^{+\infty}\frac{d(G(t)\tilde{e}(t),\tilde{e}(t))}{1+t^2}e} \qquad (63)\\
&= \|f\|_{\tilde{\mathfrak{H}}_+}\cdot\|e\|_E.
\end{aligned}
$$

Also,

$$
\begin{aligned}
\int_{-\infty}^{+\infty}(G(dt)f(t),e(t))_E &= \int_{-\infty}^{+\infty}\left(\sqrt{1+t^2}G(dt)f(t),\frac{e}{\sqrt{1+t^2}}\right)_E \\
&= \int_{-\infty}^{+\infty}(G(dt)\tilde{f}(t),\tilde{e}(t))_E \\
&= (\tilde{f}(t),\tilde{e}(t)).
\end{aligned}
$$

Therefore,

$$
e(f)=\int_{-\infty}^{+\infty}(G(dt)f(t),e(t))_E \qquad (64)
$$

is a continuous linear functional over $\tilde{\mathfrak{H}}_+$, for $f\in\mathfrak{D}_0$. Since $\mathfrak{D}_0$ is dense in $\tilde{\mathfrak{H}}_+$, $e(t)=e$ belongs to $\tilde{\mathfrak{H}}_-$.

We calculate the Riesz-Berezanskii mapping on the vectors $e(t)=e$, $e\in E$. By the definition of $\tilde{\mathcal{R}}$, for all $f\in\tilde{\mathfrak{H}}_+$ we have $(f,e)=(f,\tilde{\mathcal{R}}e)_{\tilde{\mathfrak{H}}_+}$. Hence, for all $f\in\mathfrak{D}_0$ (see also [2])

$$
\begin{aligned}
(f,e) &= \int_{-\infty}^{+\infty}(G(dt)f(t),e(t))_E = \int_{-\infty}^{+\infty}(1+t^2)\left(G(dt)f(t),\frac{e(t)}{1+t^2}\right)_E \\
&= \left(f,\frac{e(t)}{1+t^2}\right)_{\tilde{\mathfrak{H}}_+} = (f,\tilde{\mathcal{R}}e)_{\tilde{\mathfrak{H}}_+}.
\end{aligned}
$$

Thus

$$
\tilde{\mathcal{R}}e=\frac{e(t)}{1+t^2},\quad e\in E. \qquad (65)
$$

Let us note some properties of the operator $\hat{A}$. It is easy to see that for all $g\in\tilde{\mathfrak{H}}_+$, we have that $\|\hat{A}g\|\le\|g\|_{\tilde{\mathfrak{H}}_+}$. Taking this into account we obtain

$$
\|\hat{A}f\|_{\tilde{\mathfrak{H}}_-}=\sup_{g\in\tilde{\mathfrak{H}}_+}\frac{|(\hat{A}f,g)|}{\|g\|_{\tilde{\mathfrak{H}}_+}}=\sup_{g\in\tilde{\mathfrak{H}}_+}\frac{|(f,\hat{A}g)|}{\|g\|_{\tilde{\mathfrak{H}}_+}}\le\sup_{g\in\tilde{\mathfrak{H}}_+}\frac{\|f\|\cdot\|\hat{A}g\|}{\|g\|_{\tilde{\mathfrak{H}}_+}}\le\|f\|. \qquad (66)
$$

Hence, the operator $\hat{A}$ is $(\cdot,-)$-continuous. Let $\overline{\hat{A}}$ be the extension of the operator $\hat{A}$ to $\mathfrak{H}$ with respect to $(\cdot,-)$-continuity. Now,

$$
(\overline{\hat{A}}-\lambda I)^{-1}g-(\overline{\hat{A}}-\mu I)^{-1}g=(\lambda-\mu)(\overline{\hat{A}}-\lambda I)^{-1}(\overline{\hat{A}}-\mu I)^{-1}g \qquad (67)
$$

holds for all $g \in \tilde{\mathfrak{H}}_-$. Note in particular that

$$(\overline{\hat{A}} - iI)^{-1}g - (\overline{\hat{A}} + iI)^{-1}g = 2i(\overline{\hat{A}} - iI)^{-1}(\overline{\hat{A}} + iI)^{-1}g \tag{68}$$

and

$$\|(\overline{\hat{A}} - iI)^{-1}g\|^2 = \|(\overline{\hat{A}} + iI)^{-1}g\|^2 \tag{69}$$

for all g in $\tilde{\mathfrak{H}}_-$. It follows from (60) that the element

$$f(t) = \frac{f}{t-\lambda}, \quad f \in E \tag{70}$$

belongs to the space $L^2_G(E)$. It is easy to show that, for all $e \in E$,

$$(\overline{\hat{A}} - \lambda I)^{-1}e = \frac{e}{t-\lambda}, \qquad (\text{Im}\lambda \neq 0). \tag{71}$$

STEP 2. Now let $\tilde{\mathfrak{H}}_+$ be the Hilbert space constructed in Step 1 and let

$$\mathfrak{D}(A) = \tilde{\mathfrak{H}}_+ \ominus \tilde{\mathcal{R}}E, \tag{72}$$

where by $\ominus$ we mean orthogonality in $\tilde{\mathfrak{H}}_+$. We define an operator A on $\mathfrak{D}(A)$ by the following expression:

$$A = \hat{A}\Big|_{\mathfrak{D}(A)}. \tag{73}$$

Obviously, A is a closed Hermitian operator.

Let us note that if $E_\infty = 0$ then $\mathfrak{D}(A)$ is dense in $L^2_G(E)$. Define $\mathfrak{H}_0 = \overline{\mathfrak{D}(A)}$ and let P be the orthogonal projection of $\mathfrak{H} = L^2_G(E)$ onto $\mathfrak{H}$. We shall show that PA and $P\hat{A}$ are closed operators in $\mathfrak{H}$. Let

$$A_1 = \hat{A}\Big|_{\mathfrak{D}(A_1)}, \qquad \mathfrak{D}(A_1) = \tilde{\mathfrak{H}}_+ \ominus \tilde{\mathcal{R}}E_\infty. \tag{74}$$

The following obvious inclusions hold: $A \subset A_1 \subset \hat{A}$. It is easy to see that $\mathfrak{D}(A_1) = \mathfrak{D}(A) \oplus \tilde{\mathcal{R}}F_\infty$, $\overline{\mathfrak{D}(A_1)} = \mathfrak{H}_0$ and A_1 is a closed Hermitian operator. Indeed, if we identify the space E with the space of functions $e(t) = e$, $e \in E$ we would obtain $L^2_G(E) \ominus \mathfrak{H}_0 = E_\infty$. Since

$$\int_{-\infty}^{+\infty} \frac{d(G(t)e,h)_E}{1+t^2} = 0$$

and

$$\tilde{\mathcal{R}}\tilde{e} = \frac{\tilde{e}}{1+t^2}, \qquad \tilde{e} \in F_\infty$$

for all $e \in E_\infty$, $h \in F_\infty$, we find that E_∞ is $(\cdot)$-orthogonal to $\mathcal{R}F_\infty$ and hence $\overline{\mathfrak{D}(A_1)} = \mathfrak{H}_0$.

We denote by A_1^* the adjoint of the operator A_1. Now we are going to find the defect subspaces $\mathfrak{N}_i$ and $\mathfrak{N}_{-i}$ of the operator A. Since the subspace $E \in \tilde{\mathfrak{H}}_-$ is $(\cdot)$-orthogonal to $\mathfrak{D}(A)$, we have that $(\overline{\hat{A}} \pm iI)^{-1}E = \mathfrak{N}_{\pm i}$. Moreover, by (71) we have

$$(\overline{\hat{A}} \pm iI)^{-1}e = \frac{e}{t \pm i}, \qquad e \in E. \tag{75}$$

Therefore

$$\mathfrak{N}_{\pm i} = \left\{ f(t) \in L^2_G(E),\ f(t) = \frac{e}{t \pm i}, \quad e \in E \right\}. \tag{76}$$

Similarly, the defect subspaces of the operator A_1 are

$$\mathfrak{N}^0_{\pm i} = \left\{ f(t) \in L^2_G(E),\ f(t) = \frac{e}{t \pm i},\ e \in E_\infty \right\}. \tag{77}$$

Obviously, $\mathfrak{N}^0_\lambda \subset \mathfrak{D}_0$ because

$$\int\limits_{-\infty}^{+\infty} \frac{t}{|t-\lambda|^2}(G(dt)e, e)_E \le K(\lambda) \int\limits_{-\infty}^{+\infty} (G(dt)e, e)_E < \infty, \quad e \in E_\infty.$$

Taking into account that

$$\mathfrak{D}(A_1^*) = \mathfrak{D}(A) \dotplus \mathfrak{N}^0_i \dotplus \mathfrak{N}^0_{-i}, \tag{78}$$

we can conclude that $\mathfrak{D}(A_1^*) \subseteq \mathfrak{D}(\hat{A})$. At the same time, the inclusion $A_1 \subset \hat{A}$ implies that $\mathfrak{D}(A_1^*) \supset \mathfrak{D}(\hat{A})$. Combining these two we obtain $\mathfrak{D}(A_1^*) = \mathfrak{D}(\hat{A})$ and $P\hat{A} = A_1^*$. Since A_1^* is a closed operator, $P\hat{A}$ is also closed. Consequently, $\hat{A}$ is the regular self-adjoint extension of the operator A which implies A is a regular Hermitian operator.

Since $\hat{A}$ is the self-adjoint extension of operator A we find by (10) that

$$\mathfrak{D}(\hat{A}) = \mathfrak{D}(A) \dotplus (I - U)\mathfrak{N}_i \tag{79}$$

for some admissible isometric operator U acting from $\mathfrak{N}_i$ into $\mathfrak{N}_{-i}$. It is easy to check that $U(\overline{\hat{A}} - iI)^{-1}e = (\overline{\hat{A}} + iI)^{-1}e$, for all e in E. Consequently, the operator U has the form:

$$U\left(\frac{e}{t-i}\right) = \frac{e}{t+i}, \quad e \in E. \tag{80}$$

Straightforward calculations show that

$$\hat{A}(I-U)\left(\frac{e}{t-i}\right) = t\frac{e}{t-i} - t\frac{e}{t+i} = \frac{2ite}{t^2+1}.$$

Let A^* be the adjoint of the operator A. In the space $\mathfrak{D}(A^*) = \mathfrak{H}_+$ we introduce an inner product

$$(f,g)_+ = (f,g) + (A^*f, A^*g), \tag{81}$$

and construct the rigged space $\mathfrak{H}_+ \subset \mathfrak{H} \subset \mathfrak{H}_-$ with corresponding Riesz-Berezanskii operator $\mathcal{R}$. Since $P\hat{A}$ is a closed Hermitian operator, $\tilde{\mathfrak{H}}_+$ is a subspace of $\mathfrak{H}_+$.

By Theorem 2, $\mathfrak{H}_+ = \mathfrak{D}(\hat{A}) \dot{+} (U-I)\hat{\mathfrak{N}}_i$, where

$$\hat{\mathfrak{N}}_i = \{f_i \in \mathfrak{N}_i, \ (U-I)f_i \in \mathfrak{H}_0\}.$$

Taking into account that

$$(U-I)\left(\frac{e}{t-i}\right) = \frac{-2ie}{t^2+1}, \qquad e \in E,$$

we can conclude that

$$\hat{\mathfrak{N}}_i = \left\{\frac{\tilde{e}}{t-i}, \ e \in F_\infty = E \ominus E_\infty\right\}.$$

Therefore,

$$\mathfrak{D}(A^*) = \mathfrak{D}(\hat{A}) \dot{+} \left\{\frac{t\tilde{e}}{t^2+1}\right\}, \qquad e \in F_\infty. \tag{82}$$

STEP 3. In this Step we will construct a special self-adjoint bi-extension whose quasi-kernel coincides with the operator $\hat{A}$. Then applying (7), we will have

$$\mathfrak{H}_+ = \mathfrak{D}(A) \oplus \mathfrak{N}'_i \oplus \mathfrak{N}'_{-i} \oplus \mathfrak{N},$$

where $\mathfrak{N}'_{\pm i}$ are semidefect spaces of the operator A, $\mathfrak{N} = \mathcal{R}E_\infty$, and

$$\mathfrak{D}(A) \oplus E_\infty = \mathfrak{H} = L^2_G(E).$$

We begin by setting

$$(f,g)_1 = (f,g)_+ + (P^+_{\mathfrak{N}} f, P^+_{\mathfrak{N}} g)_+, \quad \text{for all } \ f,g \in \mathfrak{H}_+. \tag{83}$$

Here $P^+_{\mathfrak{N}}$ is an orthoprojection of $\mathfrak{H}_+$ onto $\mathfrak{N}$. Obviously, the norm $\|\cdot\|_1$ is equivalent to $\|\cdot\|_+$. We denote by $\mathfrak{H}_{+1}$ the space $\mathfrak{H}_+$ with the norm $\|\cdot\|_1$, so that $\mathfrak{H}_{+1} \subset \mathfrak{H} \subset \mathfrak{H}_{-1}$ is the corresponding rigged space with Riesz-Berezanskii operator $\mathcal{R}_1$.

By Theorem 1 there exists a (1)-isometric operator V such that

$$\mathfrak{D}(\hat{A}) = \mathfrak{D}(A) \oplus (V+I)(\mathfrak{N}'_i \oplus \mathfrak{N}), \tag{84}$$

where $\mathfrak{D}(V) = \mathfrak{N}'_i \oplus \mathfrak{N}$, $\mathfrak{R}(V) = \mathfrak{N}'_{-i} \oplus \mathfrak{N}$ and (-1) is a regular point for the operator V. Moreover,

$$\begin{cases} \varphi = i(I + P^{+}_{\mathfrak{N}'_i})(A^* + iI)^{-1} f_i, \\ V\varphi = i(I + P^{+}_{\mathfrak{N}'_{-i}})(A^* - iI)^{-1} U f_i, \\ \text{where } \varphi \in \mathfrak{D}(V),\ f_i \in \mathfrak{N}_i. \end{cases} \tag{85}$$

Here U is the isometric operator described in Step 2. Consequently we obtain

$$\begin{cases} f_i = \frac{i}{2}(A^* + iI)(I + P^{+}_{\mathfrak{N}})\varphi, \\ U f_i = -\frac{i}{2}(A^* - iI)(I + P^{+}_{\mathfrak{N}})V\varphi, \\ \text{where } \varphi \in \mathfrak{D}(V),\ f_i \in \mathfrak{N}_i. \end{cases} \tag{86}$$

It follows that

$$\begin{aligned} f_i - U f_i &= \varphi + V\varphi + iA^* P^{+}_{\mathfrak{N}}(V - I)\varphi \\ \hat{A}(f_i - U f_i) &= i(I + U) f_i = A^*(\varphi + V\varphi) + iP^{+}_{\mathfrak{N}}(I - V)\varphi \\ f_i + U f_i &= \varphi - V\varphi - iA^* P^{+}_{\mathfrak{N}}(I - V)\varphi \end{aligned}$$

Applying formula (11) we get

$$\mathfrak{H}_+ = \mathfrak{D}(\hat{A}) \dotplus (U + I)\tilde{\mathfrak{N}}_i, \text{ and } \tilde{\mathfrak{N}}_i = \{f_i \in \tilde{\mathfrak{N}}_i\ (U - I) f_i \in \mathfrak{H}\}.$$

Since $f_i - U f_i = \varphi + V\varphi + iA^* P^{+}_{\mathfrak{N}}(V - I)\varphi$, we find that $f_i - U f_i \in \mathfrak{H}$ if and only if $P^{+}_{\mathfrak{N}}(V + I)\varphi = 0$. (This follows from the fact that $A^* P^{+}_{\mathfrak{N}}(V - I)\varphi \in \mathfrak{D}(A) \subset \mathfrak{H}$ and from the formula $\mathfrak{H} = \mathfrak{H}_0 \dotplus \mathfrak{N}$ (see [4])). Let us note that if $P^{+}_{\mathfrak{N}}(V + I)\varphi = 0$ then $f_i + U f_i = \varphi - V\varphi$. Thus,

$$\tilde{\mathfrak{N}}_i = \{f = (A^* + iI)(I + P^{+}_{\mathfrak{N}})\varphi,\ P^{+}_{\mathfrak{N}}(V + I)\varphi = 0\}. \tag{87}$$

Let $N = \operatorname{Ker} P^{+}_{\mathfrak{N}}(I + V)$. Then we have

$$\mathfrak{H}_+ = \mathfrak{D}(\hat{A}) \dotplus (I - V)N. \tag{88}$$

We denote by P_0 the projection operator of $\mathfrak{H}_+$ onto $\mathfrak{D}(\hat{A})$ along $(I - V)N$, $P_1 = I - P_0$. Since $\mathfrak{D}(\hat{A}) = \tilde{\mathfrak{H}}_+$, we have $P_0 \in [\mathfrak{H}_+, \tilde{\mathfrak{H}}_+]$. We will denote by $P_0^* \in [\tilde{\mathfrak{H}}_-, \mathfrak{H}_-]$ the adjoint operator to P_0, i.e. $(P_0 f, g) = (f, P_0^* g)$, for all $f \in \mathfrak{H}_+$, $g \in \tilde{\mathfrak{H}}_-$. If $\tilde{f}_i \in \tilde{\mathfrak{N}}_i$, then $\tilde{f}_i + U\tilde{f}_i = (I - V)\varphi$, for $\varphi \in N$, and

$$\begin{aligned} A^*(I - V)\varphi &= iP^{+}_{\mathfrak{N}'_i}\varphi + iP^{+}_{\mathfrak{N}'_{-i}}V\varphi + AP^{+}_{\mathfrak{N}}(I - V)\varphi = i(V + I)\varphi + A^* P^{+}_{\mathfrak{N}}(I - V)\varphi \\ &= i[(I + V)\varphi - iA^* P^{+}_{\mathfrak{N}}(I - V)\varphi]. \end{aligned}$$

This implies

$$A^*(I+U)\tilde{f}_i = i(\tilde{f}_i - U\tilde{f}_i).$$

Hence

$$A^*\left(\frac{t\tilde{e}}{t^2+1}\right) = -\frac{\tilde{e}}{t^2+1}, \quad \tilde{e} \in F_\infty. \tag{89}$$

Let $Q \in [E, E]$ be the operator in the definition of the class $N(R)$. We introduce a new operator R_0 acting in the following way:

$$R_0 f = iQ\tilde{\mathcal{R}}^{-1}A^*P_1 f, \quad f \in \mathfrak{H}_+. \tag{90}$$

In order to show that $R_0 \in [\mathfrak{H}_+, E]$, we consider the following calculation for $f \in \mathfrak{H}_+$:

$$\begin{aligned}
\|R_0 f\|_E &= \sup_{g\in E}\frac{|(R_0 f, g)_E|}{\|g\|_E} = \sup_{g\in E}\frac{|(Q\tilde{\mathcal{R}}^{-1}A^*P_1 f, g)_E|}{\|g\|_E} \\
&= \sup_{g\in E}\frac{|(\tilde{\mathcal{R}}^{-1}A^*P_1 f, Qg)_E|}{\|g\|_E} \le \sup_{g\in E}\frac{\|\tilde{\mathcal{R}}^{-1}A^*P_1 f\|_E\cdot\|Qg\|_E}{\|g\|_E} \\
&\le c\|A^*P_1 f\|_{\tilde{\mathfrak{H}}_+} \le b\|A^*P_1 f\|_{\mathfrak{H}_+}, \quad b, c \text{ - constants.}
\end{aligned}$$

Here we used that $P_1 f \subset \mathfrak{D}(\hat{A})$, for all $f \in \tilde{\mathfrak{H}}_+$, formulas (65) and (89), and the equivalence of the norms $\|\cdot\|_{\tilde{\mathfrak{H}}_+}$ and $\|\cdot\|_+$.

For $f \in \mathfrak{H}_+$, we have $P_1 f = (I-V)\varphi$, $\varphi \in N$ and

$$A^*P_1 f = i(V+I)\varphi + iA^*P^+_{\mathfrak{N}}(V-I)\varphi.$$

We now have

$$\begin{aligned}
\|A^*P^+_{\mathfrak{N}}(V-I)\varphi\|^2_+ &= \|A^*P^+_{\mathfrak{N}}(V-I)\varphi\|^2 + \|A^*A^*P^+_{\mathfrak{N}}(V-I)\varphi\|^2 \\
&= \|A^*P^+_{\mathfrak{N}}(V-I)\varphi\|^2 + \|PP^+_{\mathfrak{N}}(V-I)\varphi\|^2 \\
&\le \|A^*P^+_{\mathfrak{N}}(V-I)\varphi\|^2 + \|P^+_{\mathfrak{N}}(V-I)\varphi\|^2 \\
&= \|P^+_{\mathfrak{N}}(V-I)\varphi\|^2_+,
\end{aligned}$$

and

$$\begin{aligned}
\|i(V+I)\varphi + iA^*P^+_{\mathfrak{N}}(V-I)\varphi\|^2_+ &= \|A^*P^+_{\mathfrak{N}}(V-I)\varphi\|^2_+ + \|\varphi+V\varphi\|^2_+ \\
&\le \|P^+_{\mathfrak{N}}(V-I)\varphi\|^2_+ + \|\varphi+V\varphi\|^2_+ \\
&= \|\varphi - V\varphi\|^2_+.
\end{aligned}$$

This implies that there exists a constant k such that

$$\|A^*P_1 f\| \le \|P_1 f\|_+ \le k\|f\|_+, \quad \forall f \in \mathfrak{H}_+. \tag{91}$$

Therefore, for some constant $d > 0$ we have $\|R_0 f\|_{\le} d\|f\|_+$, $\forall f \in \mathfrak{H}_+$. Thus, $R_0 \in [\mathfrak{H}_+, E]$.

Let R_0^* be the adjoint operator to R_0, i.e. $R_0^* \in [E, \mathfrak{H}_-]$ and for all $f \in \mathfrak{H}_+$, $e \in E$, $(R_0 f, e)_E = (f, R_0^* e)$. Since $R_0(\mathfrak{D}(\hat{A})) = 0$, $\mathfrak{R}(R_0^*)$ is $(\cdot)$-orthogonal to $\mathfrak{D}(\hat{A})$. Letting $\mathfrak{M} = \mathfrak{N}'_{-i} \oplus \mathfrak{N}'_i \oplus \mathfrak{N}$, we obtain from (88)

$$\mathfrak{M} = (V + I)(\mathfrak{N}'_i \oplus \mathfrak{N}) \dotplus (I - V)N. \tag{92}$$

In the space $\mathfrak{M}$ we define an operator S in the following way

$$\begin{aligned} S(\varphi + V\varphi) &= \frac{i}{2}(I - V)\varphi, \quad \varphi \in \mathfrak{N}'_i \oplus \mathfrak{N}, \\ S(\varphi_N - V\varphi_N) &= \left[-\mathcal{R}_1(R_0^* + P_0^*)\tilde{\mathcal{R}}^{-1}A^* + \frac{i}{2}(P^+_{\mathfrak{N}'_i} - P^+_{\mathfrak{N}'_{-i}})\right](\varphi_N - V\varphi_N), \end{aligned} \tag{93}$$

where $\varphi_N \in N$. In order to show that S is a (1)-self-adjoint operator on $\mathfrak{M}$, we first check that

$$(S(\varphi + V\varphi), \varphi + V\varphi)_1 = (\varphi + V\varphi, S(\varphi + V\varphi))_1, \quad \varphi \in \mathfrak{N}'_i \oplus \mathfrak{N}. \tag{94}$$

It is easy to see that

$$(P^+_{\mathfrak{N}'_i} - P^+_{\mathfrak{N}'_{-i}})(\varphi_N - V\varphi_N) = \varphi_N + V\varphi_N, \quad \varphi_N \in N.$$

This follows from the definition of the space N and the fact that φ_N belongs to $\mathfrak{N}'_i$. Furthermore, since $\varphi_N \in \mathfrak{N}'_i$, and $V\varphi_N \in \mathfrak{N}'_{-i}$ we have that $P^+_{\mathfrak{N}'_{-i}}\varphi_N = \varphi_N$, $P^+_{\mathfrak{N}'_{-i}}V\varphi_N = \varphi_N$, and $P^+_{\mathfrak{N}'_i}V\varphi_N = P^+_{\mathfrak{N}'_{-i}}\varphi_N = 0$. Consequently,

$$\begin{aligned} ((\varphi_N + V\varphi_N), \varphi_N - V\varphi_N)_1 &= \|\varphi_N\|_1^2 - \|V\varphi_N\|_1^2 \\ &= \|P^+_{\mathfrak{N}'_i}\varphi_N\|_1^2 - \|P^+_{\mathfrak{N}'_{-i}}V\varphi_N\|_1^2 = 0. \end{aligned} \tag{95}$$

Since $P_0(I - V)N = 0$, we have

$$(\mathcal{R}_1 P_0^* \tilde{\mathcal{R}}^{-1}A^*(\varphi_N - V\varphi_N), \varphi_N - V\varphi_N) = (\tilde{\mathcal{R}}^{-1}A^*(\varphi_N - V\varphi_N), P_0(\varphi_N - V\varphi_N)) = 0. \tag{96}$$

This allows us to consider only the R_0^*-containing part of (93), i.e.

$$\begin{aligned} (S(\varphi_N - V\varphi_N), \varphi_N - V\varphi_N)_1 &= (-\mathcal{R}_1 R_0^* \tilde{\mathcal{R}}^{-1}A^*(\varphi_N - V\varphi_N), (\varphi_N - V\varphi_N)_1 \\ &= (\tilde{\mathcal{R}}^{-1}A^*(\varphi_N - V\varphi_N), -R_0(\varphi_N - V\varphi_N))_E \\ &= (\tilde{\mathcal{R}}^{-1}A^*(\varphi_N - V\varphi_N), iQ\tilde{\mathcal{R}}^{-1}A^*P_1(\varphi_N - V\varphi_N))_E \\ &= (-iQ\tilde{\mathcal{R}}^{-1}A^*(\varphi_N - V\varphi_N), \tilde{\mathcal{R}}^{-1}A^*(\varphi_N - V\varphi_N))_E \\ &= ((\varphi_N - V\varphi_N), R_0^*\mathcal{R}^{-1}A^*(\varphi_N - V\varphi_N))_E \\ &= ((\varphi_N - V\varphi_N), \mathcal{R}_1 R_0^*\mathcal{R}^{-1}A^*(\varphi_N - V\varphi_N))_1 \\ &= ((\varphi_N - V\varphi_N), S(\varphi_N - V\varphi_N))_1. \end{aligned}$$

Now we will show that

$$(97) \qquad (S(\varphi + V\varphi), \varphi_N - V\varphi_N)_1 = (\varphi + V\varphi, S(\varphi + V\varphi))_1, \quad \varphi_N \in N,\ \varphi \in \mathfrak{N}_i' \oplus \mathfrak{N}.$$

Let us note that $P_{\mathfrak{N}}^+(\varphi_N + V\varphi_N) = 0$ implies $P_{\mathfrak{N}}^+\varphi_N = -P_{\mathfrak{N}}^+V\varphi_N$. Also, $(\varphi, \varphi_N)_1 = (V\varphi, V\varphi_N)_1$, since V is a (1)-isometric mapping. We will now show that the orthogonality relations yield $(\varphi, V\varphi_N)_1 = (\varphi, P_{\mathfrak{N}}^+V\varphi_N)_1 = 0$. First we need a calculation

$$\begin{aligned}
(S(\varphi + V\varphi), \varphi_N - V\varphi_N)_1 &= \frac{i}{2}((I - V)\varphi, \varphi_N - V\varphi_N)_1 \\
&= i(\varphi, \varphi_N)_1 - \frac{i}{2}(\varphi, V\varphi_N)_1 - \frac{i}{2}(V\varphi, \varphi_N)_1 \\
&= i(\varphi, \varphi_N)_1 - \frac{i}{2}(\varphi, V\varphi_N)_1 - \frac{i}{2}(\varphi, P_{\mathfrak{N}}^+V\varphi_N)_1 \\
&= i(\varphi, \varphi_N)_1 - \frac{i}{2}(\varphi, V\varphi_N)_1 + \frac{i}{2}(\varphi, P_{\mathfrak{N}}^+V\varphi_N)_1 \\
&= i(\varphi, \varphi_N)_1 + \frac{i}{2}(P_{\mathfrak{N}}^+(I - V)\varphi, \varphi_N)_1.
\end{aligned}$$

Also, note that

$$\left(\varphi + V\varphi, \frac{i}{2}(P_{\mathfrak{N}_i'}^+ - P_{\mathfrak{N}_{-i}'}^+)(\varphi_N - V\varphi_N)\right)_1 = -\frac{i}{2}\left(\varphi + V\varphi, \varphi_N + V\varphi_N\right)_1,$$

and

$$\begin{aligned}
(\varphi + V\varphi, S(\varphi_N - V\varphi_N))_1 &= (\varphi + V\varphi, -\mathcal{R}_1(R_0^* + P_0^*)\tilde{\mathcal{R}}^{-1}A^*(\varphi_N - V\varphi_N))_1 \\
&\quad - \frac{i}{2}(\varphi + V\varphi, \varphi_N + V\varphi_N)_1.
\end{aligned}$$

Next, recall that $\mathfrak{R}(R_0^*)$ is $(\cdot)$-orthogonal to $\mathfrak{D}(\hat{A})$ and

$$\varphi + V\varphi \in \mathfrak{D}(\hat{A}) = \mathfrak{D}(A) \oplus (V + I)(\mathfrak{N}_i' \oplus \mathfrak{N}).$$

It follows that

$$(\varphi + V\varphi, \mathcal{R}_1 R_0^* \tilde{\mathcal{R}}^{-1}A^*(\varphi_N - V\varphi_N))_1 = (\varphi + V\varphi, R_0^* \tilde{\mathcal{R}}^{-1}A^*(\varphi_N - V\varphi_N)) = 0,$$

$$\begin{aligned}
(\varphi + V\varphi, -\mathcal{R}_1 P_0^* \tilde{\mathcal{R}}^{-1}A^*(\varphi_N - V\varphi_N))_1 &= -(\varphi + V\varphi, A^*(\varphi_N - V\varphi_N))_{\mathfrak{H}_+} \\
&= -(\varphi + V\varphi, A^*(\varphi_N - V\varphi_N)) \\
&\quad - (\hat{A}(\varphi + V\varphi), \tilde{A}_0 A^*(\varphi_N - V\varphi_N)).
\end{aligned}$$

Applying Theorem 1 we obtain:

$$\hat{A}(\varphi + V\varphi) = A^*(\varphi + V\varphi) + \frac{i}{2}\mathcal{R}^{-1}P_{\mathfrak{N}}^+(I - V)\varphi,$$

$$A^*(\varphi_N - V\varphi_N) = i(I+V)\varphi_N + A^*P^+_{\mathfrak{N}}(I-V)\varphi_N,$$

$$\begin{aligned}\hat{A}A^*(\varphi_N - V\varphi_N) &= AA^*P^+_{\mathfrak{N}}(I-V)\varphi_N + iA^*(V+I)\varphi_N - \frac{i}{2}\mathcal{R}_1^{-1}P^+_{\mathfrak{N}}(I-V)\varphi_N \\ &= iA^*(V+I)\varphi_N - P^+_{\mathfrak{N}}(I-V)\varphi_N.\end{aligned}$$

Here we used the following relations:

$$A^*(I-V) \in \mathfrak{D}(A),$$

$$\hat{A}(f_i - Uf_i) = A^*(\varphi + V\varphi) + iP^+_{\mathfrak{N}}(I-V)\varphi,$$

$$f_i - Uf_i = \varphi + V\varphi + iA^*P^+_{\mathfrak{N}}(V-I)\varphi,$$

$$\hat{A}(\varphi + V\varphi) = A^*(\varphi + V\varphi) + \frac{i}{2}\mathcal{R}^{-1}P^+_{\mathfrak{N}}(I-V)\varphi,$$

and

$$AA^*P^+_{\mathfrak{N}}(I - V\varphi_N - \frac{1}{2}\mathcal{R}^{-1}(I-V)\varphi = -P^+_{\mathfrak{N}}(I-V)\varphi_N.$$

The above identities yield that

$$(\varphi + V\varphi, A^*(\varphi_N - V\varphi_N))_{\tilde{\mathfrak{H}}_+} = (\varphi + V\varphi, i(\varphi_N + V\varphi_N))_1 - i(P^+_{\mathfrak{N}}(I-V)\varphi, \varphi_N)_1.$$

Thus,

$$\begin{aligned}(\varphi + V\varphi, -\mathcal{R}_1 P_0^+ \tilde{\mathcal{R}}^{-1}A^*(\varphi_N - V\varphi_N))_0 &= i(\varphi + V\varphi, \varphi_N + V\varphi_N) \\ &\quad + i(P^+_N(I-V)\varphi, \varphi_N),\end{aligned}$$

$$(\varphi + V\varphi, \frac{i}{2}(\varphi_N + V\varphi_N))_1 = -\frac{i}{2}(\varphi + V\varphi, \varphi_N + V\varphi_N)_1,$$

and

$$\begin{aligned}(\varphi + V\varphi, S(\varphi_N - V\varphi_N))_1 &= i(\varphi + V\varphi, \varphi_N + V\varphi_N) \\ &\quad + i(P^+_{\mathfrak{N}}(I-V)\varphi, \varphi_N)_1 - \frac{i}{2}(\varphi + \varphi, \varphi_N + V\varphi_N)_1 \\ &= i(\varphi, \varphi_N)_1 + \frac{i}{2}(V\varphi, \varphi_N)_1 \\ &\quad + \frac{i}{2}(\varphi, V\varphi_N)_1 + i(P^+_{\mathfrak{N}}(I-V)\varphi, \varphi_N)_1 \\ &= i(\varphi, \varphi_N)_1 + \frac{i}{2}(P^+_{\mathfrak{N}}(I-V)\varphi, \varphi_N)_1 \\ &= (S(\varphi + V\varphi), \varphi_N - V\varphi_N).\end{aligned}$$

This shows that S is a (1)-self-adjoint operator in $\mathfrak{M}$.

By Corollary 2, a self-adjoint bi-extension of the operator A is defined by the formula

$$\mathbb{B} = AP^+_{\mathfrak{D}(A)} + \left[A^* + \mathcal{R}^{-1}\left(S - \frac{i}{2}P^+_{\mathfrak{N}'_i} + \frac{i}{2}P^+_{\mathfrak{N}'_{-i}}\right)\right]P^+_{\mathfrak{M}}, \tag{98}$$

where S is defined by (97). Obviously, if $f = f_A + (V+I)\varphi$, $\varphi \in \mathfrak{N}_i' \oplus \mathfrak{N}$, and $f_A \in \mathfrak{D}(A)$ then $\mathbb{B}f = \hat{A}f$. This means that the quasi-kernel of the operator $\mathbb{B}$ coincides with $\hat{A}$.

STEP 4. In this Step we will construct a $(*)$-extension of some operator of the class Λ_A. First, we introduce the bounded linear operator K acting from the space E into the space $\mathfrak{H}_-$ as follows:

$$Ke = (P_0^* + R_0^*)P_{F_\infty} + \hat{I}P_{E_\infty}e, \quad e \in E, \tag{99}$$

where P_{F_∞} and P_{E_∞} are orthogonal projections of the space E onto F_∞ and E_∞ respectively, and $\hat{I}$ is an embedding of E_∞ in $\mathfrak{H}_-$.

Let $K^* \in [\mathfrak{H}_+, E]$ be an adjoint of the operator K, i.e.

$$(Kf, g) = (f, K^*g), \quad f \in E,\ g \in \mathfrak{H}_+.$$

Let

$$\mathbb{C} = K^*JK, \tag{100}$$

where $J \in [E, E]$ satisfies $J = J^* = J^{-1}$. Since $\mathfrak{R}(K)$ is orthogonal to $\mathfrak{D}(A)$, $\mathbb{C}(\mathfrak{D}(A)) = 0$. Moreover, $(\mathbb{C}f, g) = (f, \mathbb{C}g)$ for all $f \in \mathfrak{H}_+$, $g \in \mathfrak{H}_+$.

We define an operator $\mathbb{A}$ by

$$\mathbb{A} = \mathbb{B} + i\mathbb{C}. \tag{101}$$

We now show that $\mathbb{A}$ is a $(*)$-extension of some operator T of the class Λ_A.

Let λ be a regular point of the operator $\hat{A}$ and let $\hat{R}_\lambda = \overline{(\mathbb{B} - \lambda I)^{-1}}$. Also, note that

$$(\hat{R}_\lambda f, g) = (f, (\hat{A} - \bar{\lambda}I)^{-1}g), \quad \forall f \in \mathfrak{H}_-,\ g \in \mathfrak{H}.$$

As it was shown in Step 1 (see (71))

$$\overline{(\hat{A} - \lambda I)^{-1}} = \frac{e}{t - \lambda}, \quad \forall e \in E,$$

where E is considered as a subspace of $\tilde{\mathfrak{H}}_-$. Clearly,

$$\begin{aligned}(\hat{R}_\lambda P_0^* e, g) &= (P_0^* e, (\hat{A} - \bar{\lambda}I)^{-1}g) \\ &= (e, (\hat{A} - \bar{\lambda}I)^{-1}g) = (\overline{(\hat{A} - \lambda I)^{-1}}e, g),\ \forall e \in E,\ g \in \mathfrak{H} = L_G^2(E).\end{aligned}$$

It follows that

$$\hat{R}_\lambda P_0^* e = \frac{e}{t - \lambda}, \quad \forall e \in E. \tag{102}$$

It follows that

$$\hat{R}_\lambda P_0^* e = \frac{e}{t-\lambda}, \quad \forall e \in E. \tag{102}$$

Since $R_0(\mathfrak{D}(\hat{A})) = 0$, $R_0(\hat{A} - \bar{\lambda} I)^{-1} g = 0$, for all $g \in \mathfrak{H}$, and we have

$$(\hat{R}_\lambda R_0^* e, g) = (R_0^* e, (\hat{A} - \bar{\lambda} I)^{-1} g) = (e, R_0(\hat{A} - \bar{\lambda} I)^{-1} g) = 0,$$

$$\hat{R}_\lambda K e_1 = \hat{R}_\lambda P_0^* e_1 = \frac{e_1}{t-\lambda}, \quad e_1 \in F_\infty,$$
$$\hat{R}_\lambda K e_2 = \hat{R}_\lambda e_2 \frac{e_2}{t-\lambda} \in \tilde{\mathfrak{H}}_+, \quad e_2 \in E_\infty,$$

This implies that the operator K is invertible. Indeed, if $Ke = 0$, then $(P_0^* + R_0^*) e_1 = -\hat{I} e_2$ and $\hat{R}_\lambda K e = 0$. Hence, $\hat{R}_\lambda (P_0^* + R_0^*) \tilde{e} = -\hat{R}_\lambda e_2$. That is,

$$\frac{e_1}{t-\lambda} = \frac{e_2}{t-\lambda}, \quad e = \hat{e} + e_1,$$

which implies that $e = 0$.

We should also note that $\hat{R}_\lambda K \in [E, \mathfrak{H}_+]$, since $\hat{R}_\lambda$ maps $\mathfrak{R}(K)$ into $\mathfrak{H}_+$ continuously.

Let us consider now the operator-valued function V defined by

$$V(\lambda) = K^* \hat{R}_\lambda K, \quad \operatorname{Im}\lambda \neq 0. \tag{103}$$

Obviously, $(V(\lambda)e, h)_E = (\hat{R}_\lambda K e, K h)$ for $e \in E$, $h \in E$, $e = e_1 + e_2$, $h = h_1 + h_2$. Therefore,

$$\begin{aligned}
(\hat{R}_\lambda K e, K h) &= (\hat{R}_\lambda (P_0^* + R_0^*) e_1 + \hat{R}_\lambda e_2, (P_0^* + R_0^*) h_1 + \hat{I} h_2) \\
&= (\hat{R}_\lambda P_0^* e_1 + \hat{R}_\lambda e_2, (P_0^* + R_0) h_1 + \hat{I} h_2) \\
&= (\hat{R}_\lambda P_0^* e_1, P_0^* h_1) + (\hat{R}_\lambda P_0^* e_1, R_0^* h_1) + (\hat{R}_\lambda P_0^* e_1, h_2) + (\hat{R}_\lambda e_2, P_0^* h_1) \\
&+ (\hat{R}_\lambda e_2, R_0^* h_2) + (\hat{R}_\lambda e_2, h_2) \\
&= (P_0 \hat{R}_\lambda P_0^* e_1, h_1) + (P_0 \hat{R}_\lambda P_0^* e_1, h_2) + (\hat{R}_\lambda P_0^* e_1, h_2) + (\hat{R}_\lambda e_2, h_2) \\
&+ (R_0 \hat{R}_\lambda e_2, h_2)_E + (\hat{R}_\lambda e_2, h_2).
\end{aligned}$$

We also have

$$\hat{R}_\lambda P_0^* e_1 = \frac{e_1}{t-\lambda} \notin \tilde{\mathfrak{H}}_-.$$

Consider an element

$$\frac{e_1}{t-\lambda} - \frac{t e_1}{t^2+1} = -\frac{\lambda t e_1}{(t-\lambda)(t^2+1)}, \quad e_1 \in F_\infty.$$

Clearly

$$\int_{-\infty}^{+\infty} \frac{|\lambda|^2 t^4}{|t-\lambda|^2(t^2+1)} \cdot \frac{d(G(t)e_1, e_1)_E}{1+t^2} < \infty,$$

and hence

$$\frac{e_1}{t-\lambda} - \frac{te_1}{t^2+1} \in \mathfrak{D}(\hat{A}).$$

Moreover,

$$\frac{te_1}{t^2+1} \in (I-V)N, \quad e_1 \in F_\infty.$$

This implies

$$P_0\left\{\frac{e_1}{t-\lambda}\right\} = \frac{e_1}{t-\lambda} - \frac{te_1}{t^2+1},$$

$$P_1\left\{\frac{e_1}{t-\lambda}\right\} = \frac{te_1}{t^2+1}.$$

Consequently,

$$(P_0 \hat{R}_\lambda P_0^* e_1, h_2) = \int_{-\infty}^{+\infty} \left(\frac{1}{t-\lambda} - \frac{t}{t^2+1}\right) d(G(t)e_1, h_2)_E.$$

We also have that

$$(R_0 \hat{R}_\lambda P_0^*, h_1)_E = -(Q\tilde{\mathcal{R}}^{-1} A^* P_1 \hat{R}_\lambda P_0^* e_1, h_1)_E = -(\tilde{\mathcal{R}}^{-1} A^* P_1 \hat{R}_\lambda P_0 e_1, Qh_1)_E.$$

From (65) and (89) we obtain

$$\tilde{\mathcal{R}}^{-1} A^* P_1 \hat{R}_\lambda P_0^* e_1 = \tilde{\mathcal{R}}^* \left(\frac{e_1}{t^2+1}\right) = -e_1,$$

from which it follows that

$$(R_0 \hat{R}_\lambda P_0^*, h_2)_E = (e_1, Qh_2)_E = (Qe_1, h_2)_E.$$

Furthermore we obtain

$$\begin{aligned}
(\hat{R}_\lambda P_0^* e_1, h_2) &= \int_{-\infty}^{+\infty} \left(\frac{1}{t-\lambda}\right) d(G(t)e_2, h_2)_E \\
&= \int_{-\infty}^{+\infty} \left(\frac{1}{t-\lambda}\right) d(G(t)e_2, h_2)_E - (Qe_1, h_2)_E + (Qe_1, h_2)_E \\
&= \int_{-\infty}^{+\infty} \frac{t}{t^2+1} d(G(t)e_1, h_2)_E + (Qe_1, h_2)_E \\
&= \int_{-\infty}^{+\infty} \left(\frac{1}{t-\lambda} - \frac{t}{t^2+1}\right) d(G(t)e_1, h_2)_E + (Qe_1, h_2)_E.
\end{aligned}$$

Since $R_0\hat{R}_\lambda e_2 = 0$, we have

$$(\hat{R}_\lambda e_2, h_1) = \int_{-\infty}^{+\infty} \left(\frac{1}{t-\lambda}\right) d(G(t)e_2, h_1)_E - (Qe_2, h_1)_E + (Qe_2, h_1)_E$$
$$= \int_{-\infty}^{+\infty} \left(\frac{1}{t-\lambda} - \frac{t}{t^2+1}\right) d(G(t)e_2, h_1)_E + (Qe_2, h_1)_E$$

Thus,

$$(\hat{R}_\lambda e_2, h_2) = \int_{-\infty}^{+\infty} \left(\frac{1}{t-\lambda} - \frac{t}{t^2+1}\right) d(G(t)e_2, h_2)_E + (Qe_2, h_2)_E \tag{104}$$

These calculations imply

$$(\hat{R}_\lambda e, h) = \int_{-\infty}^{+\infty} \left(\frac{1}{t-\lambda} - \frac{t}{t^2+1}\right) d(G(t)e, h)_E + (Qe, h)_E,$$

hence,

$$(V(\lambda)e, h) = \int_{-\infty}^{+\infty} \left(\frac{1}{t-\lambda} - \frac{t}{t^2+1}\right) d(G(t)e, h)_E + (Qe, h)_E \tag{105}$$

Next, we show that $(\mathbb{B} + iI)\hat{R}_{\pm i}Ke = Ke$, for all $e \in E$, where $\mathbb{B}$ is the strong self-adjoint bi-extension defined by (98). By Theorem 7, the equation $(\mathbb{B} - \lambda I)x = f$ has a unique solution x for any

$$f \in \mathfrak{R}\left[\mathcal{R}_1^{-1}\left(S - \frac{i}{2}P^+_{\mathfrak{N}'_i} + \frac{i}{2}P^+_{\mathfrak{N}'_{-i}}\right)\right] + E_\infty.$$

We will now show that in fact

$$\mathfrak{R}(K) = \mathfrak{R}\left[\mathcal{R}_1^{-1}\left(S - \frac{i}{2}P^+_{\mathfrak{N}'_i} + \frac{i}{2}P^+_{\mathfrak{N}'_{-i}}\right)\right] + E_\infty.$$

If $\varphi_N \in N$, then

$$\left(S - \frac{i}{2}P^+_{\mathfrak{N}'_i} + \frac{i}{2}P^+_{\mathfrak{N}'_{-i}}\right)(\varphi_N - V\varphi_N) = \mathcal{R}_1(R_0^* + P_)^*)\tilde{\mathcal{R}}^{-1}A^*(\varphi_N - V\varphi_N).$$

Using (89) we can conclude that $\tilde{\mathcal{R}}^{-1}(I - V)N = F_\infty$, and hence

$$\mathfrak{R}\left[\mathcal{R}_1^{-1}\left(S - \frac{i}{2}P^+_{\mathfrak{N}'_i} + \frac{i}{2}P^+_{\mathfrak{N}'_{-i}}\right)\right](I - V)N = (P_0^* + R_0^*)F_\infty.$$

Letting $P^+ = P^+_{\mathfrak{N}'_i} + P^+_{\mathfrak{N}'_{-i}}$, we have

$$P^+\left(S - \frac{i}{2}P^+_{\mathfrak{N}'_i} + \frac{i}{2}P^+_{\mathfrak{N}'_{-i}}\right)(I+V)\varphi = 0,\ \varphi \in \mathfrak{M}.$$

Therefore,

$$E_\infty + \mathfrak{R}\left[\tilde{\mathcal{R}}^{-1}\left(S - \frac{i}{2}P^+_{\mathfrak{N}'_i} + \frac{i}{2}P^+_{\mathfrak{N}'_{-i}}\right)\right] = \mathfrak{R}(K).$$

Since $\hat{R}_\lambda = \overline{(\mathbb{B} - \lambda I)^{-1}}$, the above calculations imply

$$(\mathbb{B} - \lambda I)^{-1}Ke = \hat{R}_\lambda Ke, \tag{106}$$

for all $e \in E$. For $\mathrm{Im}\lambda \neq 0$ we have that $\hat{R}_\lambda KE = \mathfrak{N}_\lambda$ is the defect space of the operator A. Therefore $(\mathbb{B} + iI)\hat{R}_{\pm i}Ke = Ke$ and $\hat{R}_{\pm i}KE = \mathfrak{N}_{\pm i}$.

Taking into account (105) we get

$$\begin{aligned} V(-i) &= \int_{-\infty}^{+\infty}\left(\frac{1}{t+i} - \frac{t}{t^2+1}\right)dG(t) + Q \\ &= -i\int_{-\infty}^{+\infty}\frac{dG(t)}{1+t^2} + Q \\ &= -iB + Q. \end{aligned} \tag{107}$$

Therefore,

$$iV(-i)J + I = BJ + iQJ + I. \tag{108}$$

The operator $iV(-i)J + I$ is invertible and so is the right hand side of (108). Since $I + BJ + iQJ = J(I + JB + iJQ)J$, where J is a unitary self-adjoint operator in the space E, 0 is a regular point for the operator $I + BJ + iJQ$. At the same time 0 is a regular point for the operators $I + JB - iJQ = (BJ + iQJ + I)^*$ and $I + BJ - iQJ = (I + JB + iJQ)^*$. Let

$$\begin{aligned} \mathbb{Z} &= (I + BJ - iQJ)^{-1}, \quad \mathbb{Z} \in [E, E], \\ \mathbb{Z}^* &= (I + JB + iJQ)^{-1}, \quad \mathbb{Z}^* \in [E, E], \end{aligned} \tag{109}$$

and let $\Gamma = (I + JB + iJQ)^{-1}$. Clearly $\mathrm{Ker}\Gamma = 0$. We will show that for any $f \in E$, the equation

$$(\mathbb{A} + iI)g = Kf, \tag{110}$$

has a unique solution $g = \hat{R}_{-i}K\Gamma f$, where $\hat{R}_{-i} = \overline{(\mathbb{B}+iI)^{-1}}$ and $\mathbb{A} = \mathbb{B} + i\mathbb{C}$. Moreover,

$$\mathbb{A}\hat{R}_{-i}K\Gamma f = \mathbb{B}\,\hat{R}_{-i}K\Gamma f + iKJK^*\hat{R}_{-i}K\Gamma f,\ f \in E.$$

As shown above (see also [2])

$$\begin{aligned} K^*\hat{R}_{-i}\Gamma f &= V(-i)\Gamma f = (Q - iB)\Gamma f, \\ iKJK^*\hat{R}_{-i}K\Gamma f &= K(JB + iJQ)\Gamma f \\ &= K(I + JB + iJQ)(I + JB + iJQ)^{-1}f - K\Gamma f \\ &= Kf - K\Gamma f, \quad f \in E. \end{aligned}$$

Also,

$$\begin{aligned} (\mathbb{A} + iI)\hat{R}_{-i}K\Gamma f &= (\mathbb{B} + iI)\hat{R}_{-i}K\Gamma f + iKJK^*\hat{R}_{-i}K\Gamma f \\ &= Kf, \quad f \in E. \end{aligned}$$

If there exists a $g \in \mathfrak{H}_+$ such that $\mathbb{A}g = -ig$, then $g \in \mathfrak{N}_{-i}$. Since $\mathfrak{R}(\Gamma) = E$, we find that $\hat{R}_{-i}K\Gamma E = \mathfrak{N}_{-i}$. Therefore $g = \hat{R}_{-i}K\Gamma e$, $e \in E$, and $(\mathbb{A} + iI)\hat{R}_{-i}K\Gamma e = 0$, $Ke = 0$, $e = 0$, and $g = 0$. It follows that the equation $(\mathbb{A} + iI)g = Kf$ has a unique solution given by $g = \hat{R}_{-i}K\Gamma f$ and $(\mathbb{A} + iI)^{-1}KE = \mathfrak{N}_{-i}$.

Similarly, 0 is the regular point for the operator $I + JB - iJQ$ in E. Let

$$\Gamma_1 = (I + JB - iJQ)^{-1}. \tag{111}$$

In the same way as above, we can show that the equation $(\mathbb{A}^* - iI)gKf$, $f \in E$, has a unique solution of the form $g = \hat{R}_iK\Gamma_1 f$ and $(\mathbb{A}^* - iI)^{-1}KE = \mathfrak{N}_i$.

If $f_i \in \mathfrak{N}_i$, then $f_i = f_A + f_{\mathfrak{M}}$, where $f_A \in \mathfrak{D}(A)$, $f_{\mathfrak{M}} \in \mathfrak{M} = \mathfrak{N}'_i \oplus \mathfrak{N}'_{-i} \oplus \mathfrak{N}$. Therefore,

$$\begin{aligned} A^*f_i &= PAf_A + A^*f_{\mathfrak{M}} = iPf_i, \\ A^*f_{\mathfrak{M}} &= iPf_i - PAf_A, \end{aligned}$$

and

$$\begin{aligned} (\mathbb{A} + iI)f_i &= (A + iI)f_A + iPf_i - PAf_A + if_{\mathfrak{M}} \\ &\quad + \mathcal{R}_1^{-1}\left(S - \frac{i}{2}P^+_{\mathfrak{N}'_i} + \frac{i}{2}P^+_{\mathfrak{N}'_{-i}}\right)f_{\mathfrak{M}} + iKJK^*f_i, \\ &= (I - P)(A + iI)f_A + i(P - I)f_i \in E_\infty \subset \mathfrak{R}(K). \end{aligned}$$

This implies that

$$(\mathbb{A} + iI)f_i - 2if_i = (\mathbb{A} + iI)f_i.$$

That is $2if_i = (\mathbb{A} + iI)(f_i - f_{-i})$, $(f_{-i} \in \mathfrak{N}_{-i})$. Hence $(\mathbb{A} + iI)\mathfrak{H}_+ \subset \mathfrak{N}_i$. Since

$$(\mathbb{A} + iI)\mathfrak{D}(A) = (A + iI)\mathfrak{D}(A),$$

and $(A+iI)\mathfrak{D}(A)\oplus\mathfrak{N}_i = \mathfrak{H}$, we have $(\mathbb{A}+iI)\mathfrak{H}_+ \subset \mathfrak{H}$. Similarly, $(\mathbb{A}^*-iI)\mathfrak{H}_+ \subset \mathfrak{H}$. Therefore we can conclude that the operators $(\mathbb{A}+iI)^{-1}$ and $(\mathbb{A}^*-iI)^{-1}$ are $(-,\cdot)$-continuous (see [25]). Let

$$\begin{aligned} \mathfrak{D}(T) &= (\mathbb{A}+iI)^{-1}\mathfrak{H}, \\ \mathfrak{D}(T_1) &= (\mathbb{A}^*-iI)^{-1}\mathfrak{H}. \end{aligned} \tag{112}$$

It is easy to see that $\mathfrak{D}(T)$ and $\mathfrak{D}(T_1)$ are dense in $\mathfrak{H}$ and that the operators $(\mathbb{A}+iI)^{-1}\Big|_{\mathfrak{H}}$ and $(\mathbb{A}^*-iI)^{-1}\Big|_{\mathfrak{H}}$ are $(\cdot,\cdot)$-continuous.

Let us define

$$\begin{aligned} T &= \mathbb{A}\Big|_{\mathfrak{D}(T)}, \\ T_1 &= \mathbb{A}^*\Big|_{\mathfrak{D}(T_1)}. \end{aligned} \tag{113}$$

The points (i) and $(-i)$ are regular points for the operators T and T_1 respectively. This implies that $T_1 = T^*$.

Since T and T^* are quasi-kernels of operators $\mathbb{A}$ and $\mathbb{A}^*$ respectively, and $\mathrm{Re}\mathbb{A} = \mathbb{B}$ is a strong self-adjoint bi-extension of the operator A we find that $T \in \Lambda_A$ (the fact that PT and PT^* are closed follows from the $(+,\cdot)$-continuity of T and T^*).

STEP 5. Let us construct a linear stationary conservative dynamical system θ. Let $K \in [E, \mathfrak{H}_-]$ be the operator defined in the Step 4. It is easy to see that

$$\frac{1}{2i}(\mathbb{A}-\mathbb{A}^*) = KJK^*.$$

Therefore,

$$\theta = \begin{pmatrix} \mathbb{A} & K & J \\ \mathfrak{H}_+ \subset \mathfrak{H} \subset \mathfrak{H}_- & & E \end{pmatrix}$$

is a l.s.c.d.s. In particular, θ is a scattering system if $J = I$. Since $V_\theta(z)$ is a linear-fractional transformation of $W_\theta(z)$ then $V_\theta(z) = V(z)$ whenever z is in some neighborhood G_{-i} of the point $(-i)$. This completes the proof of the theorem.

Remark. It can be seen that when $J = I$ the invertibility condition for $I + iV(\lambda)J$ is satisfied automatically.

Theorem 10. *Let an operator-valued function $V(z)$ belong to the class $N(R)$. Then $V(z)$ can be realized by the scattering ($J = I$) system (dissipative operator colligation) θ of the form (30).*

The following theorem deals with the realization of two realizable operator-valued R-functions differing from each other only by the constant terms in the representation (48).

Theorem 11. *Let the operator-valued functions*

$$V_1(\lambda) = Q_1 + \int_{-\infty}^{+\infty} \left(\frac{1}{t-\lambda} - \frac{t}{1+t^2}\right) dG(t) \tag{114}$$

and

$$V_2(\lambda) = Q_2 + \int_{-\infty}^{+\infty} \left(\frac{1}{t-\lambda} - \frac{t}{1+t^2}\right) dG(t) \tag{115}$$

belong to the class $N(R)$. Then they can be realized by systems

$$\theta_1 = \begin{pmatrix} \mathbb{A}_1 & K_1 & J \\ \mathfrak{H}_+ \subset \mathfrak{H} \subset \mathfrak{H}_- & & E \end{pmatrix} \qquad (\mathbb{A}_1 \supset T_1) \tag{116}$$

and

$$\theta_2 = \begin{pmatrix} \mathbb{A}_2 & K_2 & J \\ \mathfrak{H}_+ \subset \mathfrak{H} \subset \mathfrak{H}_- & & E \end{pmatrix} \qquad (\mathbb{A}_2 \supset T_2) \tag{117}$$

respectively, so that the operators T_1 and T_2 acting on the Hilbert space $\mathfrak{H}$ are both extensions of the Hermitian operator A defined in this Hilbert space.

Proof. Applying Theorem 9 to the function $V_1(\lambda)$, we obtain a l.s.c.d.s. θ_1 of the type (116). The corresponding Hermitian operator A_1 constructed in the Steps 1 and 2 of the proof of Theorem 9 satisfies the formulas (72) and (73). The construction of A_1 doesn't involve the operator Q_1 from (114). It is easy to see that the corresponding rigged Hilbert space $\mathfrak{H}_+^{(1)} \subset \mathfrak{H}^{(1)} \subset \mathfrak{H}_-^{(1)}$ was built without the use of the operator Q_1 too.

Similarly, if we apply Theorem 9 to the function $V_2(\lambda)$ we get the corresponding Hermitian operator $A_2 = A_1$ and the same rigged Hilbert space. This occurs because the operator-functions $V_1(\lambda)$ and $V_2(\lambda)$ differ from each other only by the constant terms Q_1 and Q_2. Setting $A = A_1 = A_2$, we can conclude that T_1 and T_2 are both extensions of the Hermitian operator A.

A closed Hermitian operator A is called a *prime operator* [25] if there exists no reducing invariant subspace on which it induces a self-adjoint operator.

Definition. A l.s.c.d.s. θ of the form (30) is said to be a *prime system* if its Hermitian operator A is a prime operator.

Theorem 12. *Let the operator-valued function $V(z)$ belong to the class $N(R)$. Then it can be realized by the prime system θ of the form (30) with a preassigned direction operator J for which $I + iV(-i)J$ is invertible.*

Proof. Theorem 9 provides us with a possibility of realization for a given operator-valued function $V(z)$ from the class $N(R)$. Let us assume that its Hermitian operator A has a reducing invariant subspace $\mathfrak{H}^1 \subset \mathfrak{H}$ on which it generates the self-adjoint operator A_1. Then we can write the following $(\cdot)$-orthogonal decomposition

$$\mathfrak{H} = \mathfrak{H}^0 + \mathfrak{H}^1, \qquad A = A_0 \oplus A_1, \tag{118}$$

where A_0 is an operator induced by A on $\mathfrak{H}^0$.

Now let us consider an operator $T \supset A$ as in the definition of the system θ. We have

$$T = T_0 \oplus A_1, \tag{119}$$

where $T_0 \supset A_0$. Indeed, since A_1 is a self-adjoint operator it can not be extended any further. Clearly, $\overline{\mathfrak{D}(A_1)} = \mathfrak{H}^1$. Similarly,

$$T^* = T_0^* \oplus A_1, \tag{120}$$

where $T_0^* \supset A_0$. Furthermore,

$$\mathfrak{H}_+ = \mathfrak{H}_+^0 \oplus \mathfrak{H}_+^1 = \mathfrak{D}(A_0^*) \oplus \mathfrak{D}(A_1).$$

We now show that the same holds in the $(+)$-orthogonality sense. Indeed, if $f_0 \in \mathfrak{H}_+^0$, $f_1 \in \mathfrak{H}_+^1 = \mathfrak{D}(A_1)$ then

$$\begin{aligned}(f_0, f_1)_+ &= (f_0, f_1) + (A^* f_0, A^* f_1)\\ &= (f_0, f_1) + (A_0^* f_0, A_1 f_1)\\ &= 0 + 0 = 0.\end{aligned}$$

Consequently, we have

$$\begin{aligned}\mathfrak{H}_+ \subset \mathfrak{H} \subset \mathfrak{H}_- &= \mathfrak{H}_+^0 \oplus \mathfrak{H}_+^1 \subset \mathfrak{H}^0 \oplus \mathfrak{H}^1 \subset \mathfrak{H}_-^0 \oplus \mathfrak{H}_-^1\\ &= \mathfrak{H}_+^0 \oplus \mathfrak{D}(A_1) \subset \mathfrak{H}^0 \oplus \overline{\mathfrak{D}(A_1)} \subset \mathfrak{H}_-^0 \oplus \mathfrak{H}_-^1.\end{aligned}$$

Similarly, we obtain $\mathbb{A} = \mathbb{A}_0 \oplus A_1$ and $\mathbb{A}^* = A_0 \oplus A_1$. Therefore,

$$\begin{aligned}\frac{\mathbb{A} - \mathbb{A}^*}{2i} &= \frac{(\mathbb{A}_0 \oplus A_1) - (\mathbb{A}_0^* \oplus A_1)}{2i}\\ &= \frac{\mathbb{A}_0 - \mathbb{A}_0^*}{2i} \oplus \frac{A_1 - A_1}{2i}\\ &= \frac{\mathbb{A}_0 - \mathbb{A}_0^*}{2i} \oplus O,\end{aligned}$$

where O is the zero operator. This implies that

$$KJK^* = K_0 J K_0^* \oplus O.$$

Let P_+^0 be an orthoprojection operator of $\mathfrak{H}_+$ onto $\mathfrak{H}_+^0$ and set $K = K_0$. Now $K^* = K_0^* P_+^0$, since for all $f \in E$, $g \in \mathfrak{H}_+$ we have:

$$\begin{aligned}(Kf, g) &= (K_0 f, g) = (K_0 f, g_0 + g_1) = (K_0 f, g_0) + (K_0 f, g_1)\\ &= (K_0 f, g_0) = (f, K_0^* g_0) = (f, K_0^* P_+^0 g).\end{aligned}$$

Next, consider $e \in E$ and $x = x^0 + x^1$ in $\mathfrak{H}_+$ such that

$$(\mathbb{A} - \lambda I) P_+^0 x = Ke.$$

Then

$$\begin{aligned}(\mathbb{A}_0 \oplus A_1 - \lambda I) P_+^0 x &= K_0 e,\\ \mathbb{A}_0 x^0 - \lambda x^0 &= K_0 e,\\ (\mathbb{A} - \lambda I) x^0 &= K_0 e,\\ x^0 &= (\mathbb{A}_0 - \lambda I)^{-1} K_0 e.\end{aligned}$$

On the other hand, $x^0 = (\mathbb{A} - \lambda I)^{-1} Ke$. Therefore

$$(\mathbb{A} - \lambda I)^{-1} Ke = (\mathbb{A}_0 - \lambda I)^{-1} K_0 e,$$

and

$$K^* (\mathbb{A} - \lambda I)^{-1} Ke = K_0^* (\mathbb{A}_0 - \lambda I)^{-1} K_0 e.$$

This means that the transfer operator-functions of our system θ and of the system

$$\theta_0 = \begin{pmatrix} \mathbb{A}_0 & K_0 & J\\ \mathfrak{H}_+ \subset \mathfrak{H} \subset \mathfrak{H}_- & & E \end{pmatrix}$$

coincide. This proves the statement of the theorem.

4. Example

Let

$$Tx = \frac{1}{i}\frac{dx}{dt},$$

with

$$\mathfrak{D}(T) = \left\{ x(t) : x'(t) \in L^2_{[0,l]}, x(0) = 0 \right\},$$

be a differential operator in $\mathfrak{H} = L^2_{[0,l]}$ $(l > 0)$. Obviously,

$$T^* x = \frac{1}{i}\frac{dx}{dt},$$

with

$$\mathfrak{D}(T^*) = \left\{ x(t) : x'(t) \in L^2_{[0,l]}, x(l) = 0 \right\},$$

is the adjoint operator of T. Consider the Hermitian operator A (see also [1]) defined by

$$\begin{aligned} Ax &= \frac{1}{i}\frac{dx}{dt}, \\ \mathfrak{D}(A) &= \left\{x(t) : x'(t) \in L^2_{[0,l]}, x(0) = x(l) = 0\right\}, \end{aligned}$$

where its adjoint A^* is given by

$$\begin{aligned} A^*x &= \frac{1}{i}\frac{dx}{dt}, \\ \mathfrak{D}(A^*) &= \left\{x(t) : x'(t) \in L^2_{[0,l]}\right\}. \end{aligned}$$

Then $\mathfrak{H}_+ = \mathfrak{D}(A^*) = W_2^1$ is a Sobolev space with scalar product

$$(x,y)_+ = \int_0^l x(t)\overline{y(t)}\,dt + \int_0^l x'(t)\overline{y'(t)}\,dt.$$

We construct the rigged Hilbert space [9]

$$W_2^1 \subset L^2_{[0,l]} \subset (W_2^1)_-,$$

and consider the operators

$$\begin{aligned} \mathbb{A}x &= \frac{1}{i}\frac{dx}{dt} + ix(0)\left[\delta(x-l) - \delta(x)\right], \\ \mathbb{A}^*x &= \frac{1}{i}\frac{dx}{dt} + ix(l)\left[\delta(x-l) - \delta(x)\right], \end{aligned}$$

where $x(t) \in W_2^1$, $\delta(x)$, $\delta(x-l)$ are delta-functions in $(W_2^1)_-$. It is easy to see that

$$\mathbb{A} \supset T \supset A, \qquad \mathbb{A}^* \supset T^* \supset A,$$

and

$$\theta = \begin{pmatrix} \frac{1}{i}\frac{dx}{dt} + ix(0)[\delta(x-l) - \delta(x)] & K & -1 \\ W_1^2 \subset L^2_{[0,l]} \subset (W_2^1)_- & & \mathbb{C}^1 \end{pmatrix} \qquad (J=-1)$$

is a Brodskiĭ-Livšic rigged operator colligation where

$$\begin{aligned} Kc &= c \cdot \frac{1}{\sqrt{2}}[\delta(x-l) - \delta(x)], \quad (c \in \mathbb{C}^1) \\ K^*x &= \left(x, \frac{1}{\sqrt{2}}[\delta(x-l) - \delta(x)]\right) = \frac{1}{\sqrt{2}}[x(l) - x(0)], \end{aligned}$$

for $x(t) \in W_2^1$. Also

$$\frac{\mathbb{A} - \mathbb{A}^*}{2i} = -\left(\cdot, \frac{1}{\sqrt{2}}[\delta(x-l) - \delta(x)]\right)\frac{1}{\sqrt{2}}[\delta(x-l) - \delta(x)].$$

The characteristic function of this colligation is

$$W_\theta(\lambda) = I - 2iK^*(\mathbb{A} - \lambda I)^{-1}KJ = e^{i\lambda l}.$$

Consider the following R-function (hyperbolic tangent)

$$V(\lambda) = -i\tanh\left(\frac{i}{2}\lambda l\right).$$

Obviously this fucntion can be realized as follows

$$\begin{aligned} V(\lambda) = -i\tanh\left(\frac{i}{2}\lambda l\right) &= -i\frac{e^{\frac{i}{2}\lambda l} - e^{-\frac{i}{2}\lambda l}}{e^{\frac{i}{2}\lambda l} + e^{-\frac{i}{2}\lambda l}} = -i\frac{e^{i\lambda l} - 1}{e^{i\lambda l} + 1} \\ &= i\left[W_\theta(\lambda) + I\right]^{-1}\left[W_\theta(\lambda) - I\right]J. \qquad (J = -1) \end{aligned}$$

References

1. N.I. Akhiezer, I.M. Glazman, *Theory of linear operators in Hilbert spaces*, F. Ungar. Pub. Co., New York, 1966.
2. Yu.M.Arlinskiĭ, *On inverse problem of the theory of characteristic functions of unbounded operator colligations*, Dopovidi Akad. Nauk Ukrain. RSR **2** (1976), no. Ser. A, 105–109.
3. Yu.M.Arlinskiĭ, E.R.Tsekanovskiĭ, *Regular (∗)-extension of unbounded operators, characteristic operator-functions and realization problems of transfer mappings of linear systems*, Preprint, VINITI,-2867.-79 Dep. - 72 p.
4. ———, *The method of equipped spaces in the theory of extensions of Hermitian operators with a nondense domain of definition*, Sibirsk. Mat. Zh. **15** (1974), 243–261.
5. D.R. Arov, M.A. Nudelman, *Passive linear stationary dynamical scattering systems with continuous time*, Integr. Equat. Oper. Th. **24** (1996), 1–45.
6. S.V.Belyi, E.R.Tsekanovskiĭ, *Classes of operator-valued R-functions and their realization by conservative systems*, Dokl. Akad. Nauk SSR **321** (1991), no. 3, 441–445.
7. ———, *Realization and factorization problems for J-contractive operator-valued functions in half-plane and systems with unbounded operators*, Systems and Networks: Mathematical Theory and Applications **2** (1994), Akademie Verlag, 621–624.
8. H.Bart, I.Gohberg, M.A. Kaashoek, *Minimal factorizations of matrix and operator-functions. Operator theory: Advances and Applications*, Birkhäuser Verlag Basel, 1979.
9. Ju. M. Berezanskii, *Expansion in eigenfunctions of self-adjoint operators*, vol. 17, Transl. Math. Monographs, AMS, Providence RI, 1968.
10. M.S. Brodskiĭ, *Triangular and Jordan representations of linear operators*, Moscow, Nauka, 1969.
11. M.S. Brodskiĭ, M.S. Livšic, *Spectral analysis of non-selfadjoint operators and intermediate systems*, Uspekhi Matem. Nauk **XIII** (1958), no. 1 (79), 3–84.
12. J.A. Ball, Nir Cohen, *De Branges-Rovnyak operator models and systems theory: a survey*, In book: Operator Theory: Advances and Applications, Birkhäuser Verlag Basel **50** (1991), 93–136.
13. I. Gohberg, M.A. Kaashoek, A.C.M. Ran, *Factorizations of and extensions to J-unitary rational matrix-functions on the unit circle*, Integr. Equat. Oper. Th., **5** (1992), 262 – 300.
14. J.W. Helton, *Systems with infinite-dimensional state space: the Hilbert space approach*, Proc. IEEE **64** (1976), no. 1, 145 – 160.
15. I.S. Kač, M.G. Krein, *The R-functions – analytic functions mapping the upper half-plane into itself*, Supplement I to the Russian edition of F.V. Atkinson, Discrete and continuous boundary problems, (1968), Mir, Moscow (Russian) (English translation: Amer. Math. Soc. Transl. (2) 103 (1974), 1-18).
16. M.A. Krasnoselskii, *On self-adjoint extensions of Hermitian operators*, Ukrain. Mat. Zh. **1** (1949), 21 – 38.

17. M.S. Livšic, *On spectral decomposition of linear nonselfadjoint operators*, Math. Sbornik **34** (1954), no. 76, 145–198.
18. ———, *Operators, oscillations, waves*, Moscow Nauka, 1966.
19. Ju.L. Šmuljan, *Extended resolvents and extended spectral functions of Hermitian operator*, Math. USSR Sbornick **13** (1971), no. 3, 435–450.
20. ———, *On operator R-functions*, Sibirsk. Mat. Zh. **12** (1971), no. 2, 442–452.
21. B. Sz.-Nagy, C. Foias, *Harmonic analysis of operators on Hilbert space*, North-Holland Pub. Co., Amsterdam, 1970.
22. E.R.Tsekanovskiĭ, *Generalized self-adjoint extensions of symmetric operators*, Dokl. Akad. Nauk SSR **178** (1968), 1267–1270.
23. ———, *On the description and uniqueness of the generalized extensions of quasi-Hermitian operators*, Functional Anal. Appl. **3** (1969), 79–80.
24. ———, *Analytical properties of the resolvent matrix-valued functions and inverse problem*, Abstracts of the All Union Conference on Complex Analysis, Kharkov, FTINT **3** (1971), 233–235.
25. E.R. Tsekanovskiĭ, Ju.L. Šmuljan, *The Theory of bi-extensions of operators on rigged Hilbert spaces. Unbounded operator colligations and characteristic functions.*, Russian Math. Surveys **32** (1977), no. 5, 69–124.
26. V.E. Tsekanovskiĭ, E.R. Tsekanovskiĭ, *Stieltjes operator-functions with the gaps and their realization by conservative systems*, Proceedings of the International symposium MTNS-91 **1** (1992), 37–43.
27. G. Weis, *The representation of regular linear systems on Hilbert spaces*, International series of Numerical Mathematics **91** (1989), 401–415.

Department of Mathematics
Troy State University
Troy, AL 36082
E-mail address: sbelyi@trojan.troyst.edu

Department of Mathematics
University of Missouri-Columbia
Columbia, MO 65211
E-mail address: tsekanov@math.missouri.edu

1991 Mathematics Subject Classification.
Primary 47A10, 47B44; Secondary 46E20, 46F05

Operator Theory
Advances and Applications, Vol. 98

TIMAN'S TYPE RESULT ON APPROXIMATION BY ALGEBRAIC POLYNOMIALS

YURI BRUDNYI

The paper presents two approximation results of the Jackson-Timan type, which were included in my lecture at the All-Union Conference in Approximation Theory (Dniepropetrovsk, June 26-28, 1990).[1] The conference was dedicated to the seventieth birthday of Professor Alexander Timan (26.6.1920-13.8.1988). Unfortunately, proceedings of the conference never appeared because of the well-known events in the former Soviet Union, and only the abstracts were published (see [B3]). Here we present the approximation method which has been used in the proof of these results. In fact, it can be applied to many other approximation problems of such a kind.

In view of the character of this volume, let me give some details regarding Timan's life, which has many common features with Glazman's life. Professor Timan belonged to the same unfortunate generation as Professor Glazman, with whom he was closely acquainted. More precisely, they were included in the small part of that generation which was lucky enough to survive.

Professor Alexander Timan was born in the small Ukrainian town Zolotonosha. At the age of sixteen Alexander Timan entered the Dniepropetrovsk State University. He started his creative work under the strong influence of academicians Kolmogorov and Kagan, who had positions at the university. After graduating in 1941, Alexander Timan passed the entrance examinations for postdoctoral studies at the Moscow State University. The Second World War drastically changed his plans. From September 1941 until the end of the war Alexander Timan was an artillery officer at the front line. As Glazman, he took part in many heavy battles, including the Moscow military operation of 1941 and the capture of Berlin in May 1945.

After returning to Dniepropetrovsk he started working at the university. In 1952, he became a professor and head of the Chair of Function Theory and Geometry. His mathematical achievements during that fruitful period of his scientific career were summarized in his classical book "Approximation Theory of Real Functions" (the third edition was published recently, see [T2]).

In 1961 high ranking party officials put pressure on the university administration to dismiss him from his university position. The official reason given was some dispute with another colleague. But the real reason was his Jewish origins with which he strongly identified. This event had an extremely negative effect on his mathematical creativity and on

[1]Professor J. Gopengauz kindly presented my talk at the Conference.

the subsequent scientific careers of his Jewish students. (One of them lost his job almost immediately, after trying to defend his teacher.)

For the rest of his life, Alexander Timan worked as a professor at the Dniepropetrovsk Institute of Chemical Technology.

To formulate the basic results we let $\mathcal{P}_n$ denote the space of algebraic polynomials of degree n. Then we set

$$E_n(f) := \text{dist}_{C[-1,1]}(f; \mathcal{P}_n).$$

The classical Jackson theorem states

$$E_n(f) \leq c(k)n^{-k}\omega(f^{(k)}; n^{-1})$$

for every $f \in C^k[-1,1]$ and $n \geq$ k. It is known that the result cannot be improved. Moreover, one can readily check that

$$\varlimsup_{n\to\infty} \frac{n^k E_n(f)}{\omega(f^{(k)}; n^{-1})} > 0$$

for almost all (in the Baire category sense) functions f from $C^k[-1,1]$. Timan's aforementioned theorem, therefore, was very striking for that time. Let me recall its formulation.

Theorem 1 [T1]. *There exists a sequence of linear operators $T_n : C^k[-1,1] \to \mathcal{P}_n (n \geq k)$ such that*

$$(1) \qquad | (f - T_n f)(x) | \leq c(k)\Delta_n(x)^k \omega(f^{(n)}; \Delta_n(x)$$

where

$$\Delta_n(x) := \frac{\sqrt{1-x^2}}{n} + \frac{1}{n^2}$$

The theorem demonstrated the phenomen of "the better than best" approximation to its full extent.

Below we shall present a relatively simple proof of the theorem which contains as consequences a few generalizations of Timan's aforementioned result ([Dz] and [F]) for the case $r = 2, m = 0$, [Tr] for $r = 1, m \in \{0, \ldots, k\}$, [B1] for arbitrary k and r but with $m = 0$). All of them were proved by other more elaborate methods.

To formulate the result we let $\omega_r(f; \cdot)$ denote r -modulus of continuity of the function $f \in C[-1,1]$, i.e., the quantity

$$(2) \qquad \omega_r(f;t) := \sup_{0\leq h\leq t} \sup_{-1\leq x\leq 1-rh} | \Delta_k^r(f;x) |, \quad (t \leq 2/k).$$

Theorem 2. *There exists a sequence of linear operators $T_n : C^k[-1,1] \to \mathcal{P}_n (n \geq k+r-1)$ such that*

$$(3) \qquad | (f - T_n f)^{(m)}(x) | \leq C(K,r)\Delta_n(x)^{k-m}\omega_r(f^{(m)}; \Delta_n(x)$$

where $m = 0, 1, \ldots, k-1$.

Proof. Let $\varphi \in C_0^\infty(\mathbb{R})$ and satisfy

(a) $supp\varphi \subset [-k-r, k=r]$;
(b) $\varphi(x) = 1$ for $|x| \leq k+r-1$;
(c) $\varphi(-x) = \varphi(x)$.

If g is an even 2π-periodic continuous function, then $g * K_n$, where

$$K_n(u) := n^{-1}\widehat{\varphi}(nu),$$

is an even trigonometrical polynomial of degree $n(k+r)$ due to (a) and (c). Therefore

$$\varphi_n(f;x) := \int_{\mathbb{R}} f\left[\cos(t+\frac{u}{n})\right]\widehat{\varphi}(u)du(f \in C[-1,1]) \tag{4}$$

is an algebraic polynomial in $x:=$ cost of degree $n(k+r)$.

Besides, from (b) and (c) it follows that

$$\int_{\mathbb{R}} \widehat{\varphi}(u)p(\frac{u}{n})du = \frac{1}{n}p(0) \quad (n \in \mathbb{N}) \tag{5}$$

for every trigonometrical polynomial p of degree $\leq n(r+k-1)$.

Now verify that

$$|\,[f-\varphi_n(f)]^{(m)}(x)\,| \leq c(K,r)\Delta_n(x)^{p-m} \parallel f^{(p)} \parallel_C \tag{6}$$

where $f \in C^p[-1,1], m = 0,1,\ldots,p,$ and $p \leq r+k-1$.

Applying the Tailor expansion to $f(y) - f(x)$ with

$$x := \cos t \text{ and } y := \cos(t+\frac{u}{n})$$

we get

$$(f-\varphi_n)(x) = \sum_{i=1}^{p-1}\frac{f^{(i)}(x)}{i!}\int_{\mathbb{R}}(y-x)^i\widehat{\varphi}(u)du + \int_{\mathbb{R}} R(x,y)\widehat{\varphi}(u)du,$$

where we set

$$R(x,y) := \int_x^y \frac{(y-s)^{p-1}}{(p-1)!}f^{(k)}(s)ds.$$

Since $(y-x)^j = p_j(\frac{u}{n})$ where p_j is a trigonometrical polynomial of degree $\leq p-1 \leq r+k-1$ and $p_j(0) = 0$ for $j \geq 1$, the first sum on the right equals zero. So,

$$(f-\varphi_n)(x) = \int_{\mathbb{R}} R(x,y)\widehat{\varphi}(u)du. \tag{7}$$

Besides, $\widehat{\varphi}$ belongs to the Schwartz class $\mathcal{S}(\mathbb{R})$ and

$$| y - x | \leq \Delta_n(x)(1+u^2).$$

Together with (7) this leads to inequality (6) with $m = 0$ and $c = \int_{\mathbb{R}} (1+n^2)^p \mid \widehat{\varphi}(u) \mid du$.

We now continue with the proof of (6) by induction in p. By the supposition of induction we get

$$| [f' - \varphi_n(f')]^{(m)}(x) | \leq c\Delta_n(x)^{p-m-1} \| f^{(p)} \|_C \tag{8}$$

for $m = 0, 1, \dots, p-1$.

We then write

$$[f - \varphi_n(f)]^{(m)}(x) = [f' - \varphi_n(f')]^{(m-1)}(x) + Q^{(m-1)}(x),$$

where $m \leq p$ and

$$Q := \varphi_n(f)' - \varphi_n(f')$$

is a polynomial of degree $(k+r)n$. According to (8) the first term on the right is controlled by $\Delta_n(x)^{p-m} \| f^{(p)} \|_C$. So it remains to evaluate the second term by the same majorant. But according to Dziadyk's inequality (see, e.g., [T2] sec. 4.8.72),

$$| Q^{(m-1)}(x) | \leq c(k,r)\Delta_n(x)^{p-m} \| f^{(p)} \|_c$$

if the polynomial Q satisfies

$$| Q(x) | \leq c(k,r)\Delta_n(x)^{p-1} \| f^{(p)} \|_c . \tag{9}$$

So the proof of (6) will be complete if we prove the latter inequality. By definition of Q we get

$$| Q(x) \leq | [f' - \varphi_n(f')](x) | + | [f - \varphi_n(f)]'(x) |$$

and the first term on the right is majorized by $\Delta_n(x)^{p-1} \| f^{(p)} \|_c$ due to the supposition of induction. According to (7) we get

$$[f - \varphi_n(f)]'(x) = \int_{\mathbb{R}} R'_s(x,y)\widehat{\varphi}(u)du + \frac{n}{\sqrt{1-x^2}} \int_{\mathbb{R}} R(x,y)\widehat{\varphi}'(u)du.$$

Since $R'_x(x,y) = -\frac{f^{(p)}(x)}{(p-1)!}(y-x)^{p-1}$, the first integral equals zero. Thus the right side is majorized by $\frac{n}{\sqrt{1-x^2}}\Delta_n(x)^p \| f^{(p)} \|_c$ and this means that (9) holds for all $x \in I_n := \{x; n\sqrt{1-x^2} \geq 1\}$. Applying the classical Chebyshev inequality we can extend (9) with some larger absolute constant on the whole of interval [-1,1].

The proof of (6) is complete.

Returning to the proof of the theorem we set for $N \geq r+k$

$$T_N(f) := \varphi_n(f)$$

where $n = n(N) := \sup\{n \in \mathbb{N}; (r+k)n \leq N\}$.

In the remaining case $N = r+k-1$ we set

$$T_N(f;x) := \sum_{i=0}^{p-1} \frac{f^{(i)}(-1)}{i!}(x+1)^i.$$

Then from inequality (6) in case $N \geq r+k$ and from the Taylor formula in the remaining case we get

$$| (f - T_N f)^{(m)}(x) | \leq c(k,r)\Delta_n(x)^{p-m} \| f^{(p)} \|_C \tag{10}$$

for $m = 0, 1, \ldots, p$ and $p \leq k+r$.

In addition, T_N is a linear operator from $C^p[-1,1]$ into $\mathcal{P}_N$ $(N \geq k+r-1)$

Now let f be an arbitrary function from $C^k[-1,1]$. We represent f as $f_0 + f_1$ where $f_0 \in C^k[-1,1]$ and $f_1 \in C^{k+r}[-1,1]$. Applying inequality (10) to the summands, with $p = k$ and $k+r$ respectively, we get

$$\begin{aligned} | 1(f - T_n f)^{(m)}(x) | &\leq | (f_0 - T_n f_0)^{(m)}(x) | + 1(f_1 - T_n f_1)^{(m)}(x) | \\ &\leq C(K,r)\Delta_n(x)^{k-m}\{\| f_0^{(k)} \|_C + \Delta_n(x)^r \| f_1^{(k+r)} \|_C\}. \end{aligned}$$

Here $m = 0, 1, \ldots, k$.

It remains to apply the following well-known inequality (see [B2] for the basic case $k = 0$):

$$\inf_{f = f_0 + f_1} \{\| f_0^{(k)} \|_C + t^r \| f_1^{(k+r)} \|_C\} \leq C(k,r)\omega_r(f^{k)};t). \tag{11}$$

Together with the previous inequality, this proves the desired result.

The possibility of a further improvement of Timan's theorem was investigated by I. Gopengauz [G1]. He proved, in particular, that in the Timan-Trigub result (i.e., in Theorem 2 with $r = 1$) the quantity $\Delta_n(x)$ can be replaced by $\frac{\sqrt{1-x^2}}{n}$ (the special case $m = 0$ was proved independently and simultaneously by S. Telyakowskii [Te]). Surprisingly, the similar result is incorrect, generally speaking, for higher moduli of continuity. Combining the approximation method of this paper with the Gopengauz approach, we get the following sharp result in a simple proof.

Theorem 3. *There exists a sequence of linear operators $T_n : C^k[-1,1] \to \mathcal{P}_n$ ($n \geq 3k+3$) such that*

$$|(f - T_n f)^{(m)}(x)| \leq c(k)\left(\frac{\sqrt{1-x^2}}{n}\right)^{k-m} \omega_r\left(f(k); \frac{\sqrt{1-x^2}}{n}\right) \tag{12}$$

for $m = 0, 1, \ldots, \min\{k, k-r+2\}$.

Proof. First, as in the proof of Theorem 2, it is enough to construct a linear operator $\widehat{T}_n : C^k \to \mathcal{P}_n$ ($n \geq 3k+3$) such that

$$|(f - \widehat{T}_n f)^{(m)}(x)| \leq C(K)\left(\frac{\sqrt{1-x^2}}{n}\right)^{p-m} \| f^{(p)} \|_C \tag{13}$$

where $p = k$ and $k + r$ and m satisfies

$$0 \leq m \leq k - r + 2. \tag{14}$$

To this end we set

$$f_n := f - \varphi_n(f)$$

where φ_n is defined by (4).

Using Gopingauz's method, we consider the linear operator $G_n : C^k \to \mathcal{P}_{2K+1}$, sending f to the polynomial interpolating simultaneously g_n and their k derivatives at the points ± 1. Then

$$G_n(f;x) = \sum_{i=0}^{k}(1-x^2)A_i(x)$$

where

$$A_i(x) := f_n^{(i)}(1)(1+x)^{1+k-i} + f_n^{(i)}(-1)(1-x)^{1+k-i}.$$

According to inequality (6),

$$|A_i(x)| \leq c(k,r)n^{-2(p-i)} \| f^{(p)} \|_c \tag{15}$$

for $p \leq k + r$.

Now let t_i be a polynomial of degreee $\leq n$ such that

$$|(f - x^2)^i - (1-x^2)^{k+1}t_i(x)| \leq \frac{C(k,i)}{(n+1)^{2i}} \ (n \geq 0) \tag{16}$$

for $|x| \leq 1$ (see [Tr], Lemma 2''). Then we can represent G_n as follows:

$$G_n = P_n + Q_n,$$

where

$$\begin{aligned} P_n(f;x) &:= \sum_{i=0}^{k} \left[(1-x^2)^i - (1-x^2)^{k+1} t_i(x) \right] A_i(x), \\ Q_n(f;x) &:= (1-x^2)^{k+1} \sum_{i=0}^{k} A_i(x) t_i(x). \end{aligned}$$

Finally, we set

$$\widehat{T}_n := \varphi_n + P_n.$$

Applying inequality (6), (15) and (16) we get

$$| (f - \widehat{T}_n f)^{(m)}(x) | \leq C(k,r) \Delta_n(x)^p \cdot \| f^{(p)} \|_C \tag{17}$$

where $m \in \{0, 1, \ldots, p\}$ and $p = k, k+s$.

Besides, the degree of $\widehat{T}_n f$ (equals n) is more than or equal to $\max\{3k+3, k+r-1\} = 3k+3$ (see (14)) and

$$(\widehat{T}_n f)^{(m)}(\pm) = f^{(m)}(\pm 1). \tag{18}$$

$m = 0, 1, \ldots, k$, as well.

From this and from (17) with $p = k$ we obtain inequality (13) with $p = k$ (see [G], 169-170 for details).

Now let $p = k + r$ and m satisfy (14). We shall now prove (17) with these p and m. Due to the identity

$$(f - \widehat{T}_n f)^{(m)} = \sum_{j=0}^{\infty} R_{2^j n}^{(m)},$$

where $R_n := \widehat{T}_{2n} f - \widehat{T}_n f$, it suffices to prove that

$$| R_n^{(m)}(x) | \leq C(K,r) \left(\frac{\sqrt{1-x^2}}{n} \right)^{k+r-m} \| f^{(i+r)} \|_C \tag{19}$$

where $m = 0, 1, \ldots, k - r + 2$.

Applying (17) with $p = k + r$ and then Dziadyk's inequality (see, e.g., [T2], sec. 4.8.72), we obtain

(20) $$| R_n^{(m)}(x) | \leq c(K,r)\Delta_n(x)^{K+r-m} \| f^{(k+r)} \|_C .$$

Hence inequality (19) follows from inequality (20) for $n\sqrt{1-x^2} \geq 1$.

Now let $n\sqrt{1-x^2} < 1$. According to (18) the polynomial R_n has zeroes of multiplicity $k+1$ at ± 1. Therefore

$$| R_n^{(m)}(x) | \leq \frac{(1-|x|)^{k+1-m}}{(k+1-m)!} \max_{|x|\leq|\zeta|\leq 1} |R^{(k+1}(\zeta)|.$$

Using this and inequality (20) with $m = k+1$ we obtain

$$| R_n^{(m)}(x) | \leq c(k,r)(1-x^2)^{k+1-m}\Delta_n(x)^{r-1} \| f^{(k+r)} \|_C .$$

But $\Delta_n(x) \leq 2/n^2$ and

$$(1-x^2)^{k+1-m} \leq \frac{(\sqrt{1-x^2})^{k+r-m}}{n^{k+2-r-m}}$$

for $n\sqrt{1-x^2} \leq 1$. (It is worth pointing out that this is the only part of the proof where the restriction (14) has been used.)

Together with the previous inequality this gives

$$| R_n^{(m)}(x) | \leq C(k,r)\left(\frac{\sqrt{1-x^2}}{n}\right)^{k+s-m} \| f^{(k+s)} \|_C .$$

This complete the proof of inequality (19) and the theorem.

Remark 1. Thereom 3 does not hold without restriction (14). To see this let us suppose that inequality (12) holds for all $m = 0, 1, \ldots, m_0$ where

$$m_0 > \min\{k, k-r+r\}.$$

Applying (12) to the function $f_N(x) := x^N$ and to a fixed $n = n_0$ we get

(21) $$| (f_N - Q)^{(m)}(x) | \leq C(\sqrt{1-x^2})^{k+r-m} \| f_N^{(k)} \|_C$$

for $m = 0, 1, \ldots, m_0$, where $c = c(k, r, n_0)$ and Q is a polynomial of degree n_0.

From this inequality with $m = 0$, it follows that

$$| Q(x) | = O(N^k), \qquad | x | \leq 1$$

and the same estimate holds for arbitrary $Q^{(m)}$ due to the Markov inequality. Now in (21) taking $m = k$ if $r = 1$ and $m = (k - r + 3)_+$ if $r \geq 2$ we get $m_0 \leq m$ and therefore,

$$\mid (f_N - Q)^{(m)}(x) \mid = O\left((1-x)^{\frac{k+r-m}{2}}\right)$$

as $x \to 1$. So the polynomial $f_N - Q$ has a zero at 1 of multiplicity $\geq k+1$. Hence we have

$$N^{k+1} \approx \mid f_N^{(k+1)}(1) \mid = \mid Q^{(k+1)}(1) \mid = O(N^k),$$

which leads to a contradiction as $N \to \infty$.

Remark 2. After reading the text of the lecture (in 1989) Professor Trigub called my attention to the paper [D] of R. Dahlhaus, which contains a more elaborate proof of Theorem 3 under the restriction $n \geq 3k+4$. In this paper the reader can find references to the previous results of R. DeVore ($k = 0, r = 2$). H. Gonska and E. Hineman ($r \leq k+2, m = 0$ and $r \leq k, 0 \leq m \leq k - r$) and Yu (a counter example to the case $r \geq k+3$).

It is worth mentioning that Theorem 3 with $n \in \{k + r - 1, \ldots, 3k + 2\}$ is incorrect for some values of n (for instance, for $n = k + r - 1$ and $r < k + 2$). Nevertheless, it holds for these values of n and $m = 0$ (see [G2]).

REFERENCES

[B1] Yu. A. Brudnyi. Generalization of a theorem of A.F. Timan. *Soviet Math. Dokl.* **4** (1963), 244-247.

[B2] Yu. A. Brudnyi. Approximation of functions by algebraic polynomials. *in* "Investigations in the contemporary problems of constructive functions theory" (Proceedings of the conference, Baku, October 8-13, 1962), Izdat. AN Azerbajnskoy SSR, Baku, 1968, 40-45 [Russian].

[B3] Yu. A. Brudnyi. The A.F. Timan works on algebraic polynomial approximation *in* Materials of the All-Union Conference in Approximation Theory (Dniepropetrovsk, June 26-29, 1990). Dniepropetrovsk, 1991 13-17 [Russian].

[Da] R. Dahlhaus. Pointwise approximation by algebraic polynomials. *Journal of Approx. Theory*, **57** (1989), 274-277.

[Dz] V.K. Dziadyk. On approximation of function by ordinary polynomials on finite interval of the real line. Jzv. AN SSR, ser. matem. **22** (1958), 351-356 [Russian].

[F] G. Freud. `Uber die approximation reelen stetiger Functionen durch gewohnlicke Polinome. *Math. Ann.* **137** (1959), 17-25.

[G1] I.E. Gopengauz. A theorem of A.F. Timan on the approximation of functions by polynomials on a finite segment. *Mat. Zametki,* **1** (1967), 163-172 [Russian].

[G2] I.E. Gopengauz. Pointwise estimates of the hermitian interpolation. *Journal of Approx. Theory* **77** (1994), 31-41.

[Te] S.A. Telyakovskii. Two theorems on approximation of functions by algebraic polynomials, *in* "American Mathematical Society Translations, Series 2", **77**, 163-177, AMS, Providence, RI, 1966.

[T1] A.F. Timan. Improvement of the Jackson theorem on best approximation of continuous functions by polynomials on a finite interval. Soviet Doklady, **78** (1951), 17-20 [Russian].

[T2] A.F. Timan. Theory of Approximation of Real Functions. Dover Publishers Inc., New-York, 1994.

[Tr] R.M. Trigub. Approximation of functions by algebraic polynomials with integer coefficients. Jzv. AN SSSR. **26** (1962), 261-280.

Department of Mathematics
Technion—Israel Institute of Technology
32000 Haifa, Israel

1991 *Mathematics Subject Classification.* 41A10

Operator Theory
Advances and Applications, Vol. 98

ASYMPTOTIC FORMULAS FOR SPECTRAL AND WEYL FUNCTIONS OF STURM-LIOUVILLE OPERATORS WITH SMOOTH COEFFICIENTS

ANNE BOUTET DE MONVEL, VLADIMIR MARCHENKO

The exact asymptotic formulas are proved for the Weyl functions and spectral functions of Sturm-Liouville operators with smooth coefficients

Let l be the differential operation

$$l = -\frac{d^2}{dx^2} + q(x) \qquad (-\infty < x < \infty)$$

where q is a real continuous potential. For $\tau = \tan\alpha$ we denote $L_\tau^\pm$ a selfadjoint extension in $L_2(\mathbb{R}^\pm)$ of the symmetric operator defined by l, whose domain contains the set of smooth functions with compact support in $\mathbb{R}^\pm$, satisfying the boundary condition

$$y(0)\sin\alpha - y'(0)\cos\alpha = 0.$$

Further on we shall deal only with the boundary conditions $y'(0) = 0$ or $y(0) = 0$ and corresponding operators $L_0^\pm$ or $L_\infty^\pm$ ($\tau = 0$ or ∞).

We denote by $c(\lambda, x), s(\lambda, x)$ the solutions of the equation

$$l[y] = \lambda^2 y \tag{1}$$

which satisfy the initial data

$$c(\lambda,0) = s'(\lambda,0) = 1, \quad c'(\lambda,0) = s(\lambda,0) = 0.$$

We will also denote by $\chi^+(x)$, $\chi^-(x)$ the indicators of the positive ($\mathbb{R}^+$) and negative ($\mathbb{R}^-$) semiaxes.

The formulas of expansion in eigenfunctions of operators $L_0^\pm$, $L_\infty^\pm$ and the related Parseval equalities have the form

$$\chi^\pm(x)f(x) = \chi^\pm(x)\int_{-\infty}^{\infty} c^\pm(\sqrt{\mu}, f)c(\sqrt{\mu}, x)\, d\rho_0^\pm(\mu)$$

$$= \chi^\pm(x)\int_{-\infty}^{\infty} s^\pm(\sqrt{\mu}, f)s(\sqrt{\mu}, x)\, d\rho_\infty^\pm(\mu),$$

$$\int_{\mathbb{R}^\pm}(f(x))^2\,dx = \int_{-\infty}^{\infty} c^\pm(\sqrt{\mu}, f)^2\, d\rho_0^\pm(\mu) = \int_{-\infty}^{\infty} s^\pm(\sqrt{\mu}, f)^2\, d\rho_\infty^\pm(\mu), \tag{2}$$

with

$$c^\pm(\sqrt{\mu}, f) = \int_{\mathbb{R}^\pm} f(x)c(\sqrt{\mu}, x)\,dx; \quad s^\pm(\sqrt{\mu}, f) = \int_{\mathbb{R}^\pm} f(x)s(\sqrt{\mu}, x)\,dx$$

where $\rho_0^\pm(\mu)$ and $\rho_\infty^\pm(\mu)$ are the spectral functions of operators $L_0^\pm$ and $L_\infty^\pm$. These are nondecreasing functions, and we will normalize them by the condition that they be right semicontinuous ($\rho(\mu) = \rho(\mu + 0)$) and $\rho(+0) = 0$.

The spectral functions satisfy the equalities

$$\lim_{\lambda\to+\infty}\{\rho_0^\pm((\lambda + a)^2) - \rho_0^\pm(\lambda^2)\} = \lim_{\lambda\to+\infty}\lambda^{-2}\{\rho_\infty^\pm((\lambda + a)^2) - \rho_\infty^\pm(\lambda^2)\}$$

$$= \frac{2}{\pi}a \tag{3}$$

as $\mu \to +\infty$, they rapidly tend to the bounded limits

$$\lim_{\mu\to-\infty} e^{a\sqrt{|\mu|}}\{\rho_0^\pm(\mu) - \rho_0^\pm(-\infty)\} = \lim_{\mu\to-\infty} e^{a\sqrt{|\mu|}}\{\rho_\infty^\pm(\mu) - \rho_\infty^\pm(-\infty)\} = 0 \tag{4}$$

as $\mu \to -\infty$ ([1]), and according to [2]

$$\lim_{\mu\to+\infty}\{\rho_0^\pm(\mu) - \frac{2}{\pi}\sqrt{\mu} - \rho_0^\pm(-\infty)\} = 0.$$

According to the Weyl theorem the equation

$$l[\psi] = z\psi \tag{5}$$

has the solution for all nonreal z

$$\psi^\pm(z, x) = c(\sqrt{z}, x) + m_\infty^\pm(z)\, s(\sqrt{z}, x) \in L_2(\mathbb{R}^\pm).$$

This is called the Weyl solution of equation (5). The functions $m_\infty^\pm(z)$ and functions

$$m_0^\pm = -\frac{1}{m_\infty^\pm(z)}$$

are called Weyl functions of the operators $L_\infty^\pm$ and $L_0^\pm$. They are holomorphic outside the real axis, take conjugate values at conjugate points ($m_\infty^\pm(\overline{z}) = \overline{m_\infty^\pm(z)}$) and satisfy the inequalities

$$\pm\operatorname{Im} z\ \operatorname{Im} m_\infty^\pm(z) > 0 \quad (\operatorname{Im} z \neq 0).$$

The spectral functions are related to the Weyl functions by the equalities

$$\rho_\tau^\pm(\beta) - \rho_\tau^\pm(\alpha) = \pm\frac{1}{\pi}\lim_{\varepsilon\to+0}\int_\alpha^\beta \operatorname{Im} m_\tau^\pm(t+i\varepsilon)\,dt, \quad (\tau = 0,\infty) \tag{6}$$

$$m_0^\pm(z) = \int_{-\infty}^{\infty} \frac{d\rho_0^\pm(t)}{t-z}, \tag{7}$$

$$m_\infty^\pm(z) = \pm\left[a^\pm + \int_{-\infty}^{\infty}\left\{\frac{1}{t-z} - \frac{t}{1+t^2}\right\} d\rho_\infty^\pm(t)\right], \tag{8}$$

where α, β are any points of continuity of the spectral functions and $a^\pm$ are real numbers.

If the potential is semibounded, i.e.

$$\inf_{x\in\mathbb{R}} q(s) > -\infty,$$

then equation (5) has only one linearly independent solution in the space $L_2(\mathbb{R}^+)$ (resp. $L_2(\mathbb{R}^-)$). In this case the Weyl functions are unique and

$$m_0^\pm(z) = \frac{-\varphi_\pm(z,0)}{\varphi'_\pm(z,0)}, \quad m_\infty^\pm(z) = \frac{\varphi'_\pm(z,0)}{\varphi_\pm(z,0)}, \tag{9}$$

where $\varphi_\pm(z,x)$ are arbitrary nonzero solutions of equation (5), which belong respectively to the spaces $L_2(\mathbb{R}^+)$ and $L_2(\mathbb{R}^-)$.

Let us note that the functions $\varphi_\pm(z,-x)$ satisfy the equation

$$\tilde{l}[\varphi_\pm] = z\varphi_\pm, \tag{10}$$

with

$$\tilde{l} = -\frac{d^2}{dx^2} + q(-x)$$

and they belong to $L_2(\mathbb{R}^\pm)$:

$$\varphi_+(z,-x) \in L_2(\mathbb{R}^-), \quad \varphi_-(z,-x) \in L_2(\mathbb{R}^+).$$

From this and (9) it follows that

$$\tilde{m}_0^\pm(z) = \frac{\varphi_\mp(z,0)}{\varphi'_\mp(z,0)}, \quad \tilde{m}_\infty^\pm(z) = \frac{-\varphi'_\mp(z,0)}{\varphi_\mp(z,0)},$$

that is

$$\tilde{m}_\tau^\pm(z) = -m_\tau^\mp(z), \qquad (\tau = 0,\infty) \tag{11}$$

where $\tilde{m}_\tau^\pm(z)$ are the Weyl functions of equation (10).

The behaviour of spectral functions as $\mu \to +\infty$ and of the Weyl functions as $|z| \to \infty$ depends on the smoothness of the potential $q(x)$. In this paper we shall obtain sharp asymptotic formulas which take this dependence into account. In a forthcoming paper we shall use these results to solve the nonlinear Cauchy problems with nondecaying initial data.

Let us start with compactly supported potentials. If $q(x) = 0$ when $x > A$, then equation (1) has a solution $e(\lambda, x)$ which coincides with $e^{i\lambda x}$ for $x > A$. This solution is called the Jost solution and it can be represented in the form

$$e(\lambda, x) = e^{i\lambda x} + \int_x^\infty K(x,t)e^{i\lambda t}\,dt \tag{12}$$

where $K(x,t)$ is a real differentiable function, equal to zero when $t + x \geq 2A$. It is evident that the Jost solution decreases exponentially when $\operatorname{Im}\lambda > 0$, $x \to +\infty$ from which it follows according to (9) that

$$m_0^+(\lambda^2) = -\frac{e(\lambda,0)}{e'(\lambda,0)}, \quad m_\infty^+(\lambda^2) = \frac{e'(\lambda,0)}{e(\lambda,0)}, \quad (\operatorname{Im}\lambda > 0). \tag{13}$$

The functions $e(\lambda,0)$, $e'(\lambda,0)$ are entire and have only finite number of simple zeros on the imaginary semiaxis in the closed upper half-plane $\operatorname{Im}\lambda \geq 0$. Therefore the functions $m_0^+(z)$, $m_\infty^+(z)$ have continuous limits on the positive semiaxis:

$$m_0^+(\tau^2 + i0) = -\frac{e(\tau,0)}{e'(\tau,0)}, \quad m_\infty^+(\tau^2 + i0) = \frac{e'(\tau,0)}{e(\tau,0)}, \quad (0 < \tau < \infty)$$

and on the negative semiaxis they take on real values and have simple poles at the points $-\kappa_l^2$ $(-\nu_l^2)$, where $i\kappa_l$ (resp. $i\nu_l$) are the zeros of the function $e(\lambda,0)$ (resp. $e'(\lambda,0)$), which lie in the upper half-plane. It follows from this and from (6) that the spectral functions $\rho_0^+(\mu)$, $\rho_\infty^+(\mu)$ are absolutely continuous on the positive semiaxes and

$$\left.\frac{d\rho_0^+(\mu)}{d\mu}\right|_{\mu=\tau^2} = -\frac{1}{\pi}\operatorname{Im}\frac{e(\tau,0)}{e'(\tau,0)}; \quad \left.\frac{d\rho_\infty^+(\mu)}{d\mu}\right|_{\mu=\tau^2} = \frac{1}{\pi}\operatorname{Im}\frac{e'(\tau,0)}{e(\tau,0)}, \tag{14}$$

and they are piecewise constant on the negative semiaxis with jumps

$$\rho_0^+(-\nu_l^2) - \rho_0^+(-\nu_l^2 - 0) = 2i\nu_l\,\frac{e(i\nu_l,0)}{\dot{e}'(i\nu_l,0)}, \tag{15}$$

$$\rho_\infty^+(-\kappa_l^2) - \rho_\infty^+(-\kappa_l^2 - 0) = 2ik_l\,\frac{e'(i\kappa_l,0)}{\dot{e}(i\kappa_l,0)}, \tag{15'}$$

at the points $-\kappa_l^2$ $(-\nu_l^2)$ ("dot" denotes derivatives with respect to λ).

It follows from (12) that in the closed upper half-plane $\operatorname{Im}\lambda \geq 0$ we have

$$\frac{e'(\lambda,x)}{e(\lambda,x)} = i\lambda + \alpha(\lambda,x), \tag{16}$$

where $\alpha(\lambda,x) \to 0$ when $|\lambda| \to \infty$, $\operatorname{Im}\lambda \geq 0$, uniformly with respect to $x \in (0,\infty)$.

Hence

$$\frac{e(\lambda,x)}{e'(\lambda,x)} = (i\lambda + \alpha(\lambda,x))^{-1} = (i\lambda)^{-1}\left(1 - \frac{\beta(\lambda,x)}{i\lambda}\right),$$

where

$$\beta(\lambda,x) = -i\lambda\sum_{k=1}^\infty \left(\frac{i\alpha(\lambda,x)}{\lambda}\right)^k; \quad \beta(\lambda,x) = \alpha(\lambda,x)\left(1 - \frac{\beta(\lambda,x)}{i\lambda}\right),$$

from which it follows that

$$-\frac{e(\lambda,x)}{e'(\lambda,x)} = \frac{1}{\lambda^2}\left(i\lambda - \beta(\lambda,x)\right) \tag{17}$$

and

$$\lim_{|\lambda|\to\infty} \beta(\lambda,x) = 0 \qquad (\operatorname{Im}\lambda \geq 0)$$

where the second limit is uniform with respect to $x \in (0,\infty)$.

Suppose now that the potential $q(x)$ has $n \geq 0$ continuous derivatives and vanish outside of a compact set. In this case it is known that

$$\alpha(\lambda,x) = \sum_{l\leq n} \frac{\alpha_l(x)}{(2i\lambda)^l} + \frac{\alpha_n(\lambda,x)}{(2i\lambda)^n}, \tag{18}$$

where the functions $\alpha_l(x)$ are given by the recurrence relations

$$\alpha_l(x) = 0 \quad l \leq 0; \qquad \alpha_1(x) = q(x)$$

$$\alpha_{l+1}(x) = -\alpha_l'(x) - \sum_{j\geq 1} \alpha_{l-j}(x)\alpha_j(x), \qquad (l \geq 1) \tag{18'}$$

and the function $\alpha_n(\lambda,x)$ tends to zero when $|\lambda| \to \infty$, $\operatorname{Im}\lambda \geq 0$ uniformly with respect to $x \in (0,\infty)$ (see for example [3]). It follows from this that

$$\beta(\lambda,x) = \sum_{l\leq n} \frac{\beta_l(x)}{(2i\lambda)^l} + \frac{\beta_n(\lambda,x)}{(2i\lambda)^n}, \tag{19}$$

where

$$\beta_l(x) = \alpha_l(x) - 2\sum_{j\geq 1} \alpha_{l-j-1}(x)\beta_j(x), \tag{19'}$$

and the function $\beta_n(\lambda,x)$ tends to zero when $|\lambda| \to \infty$, $\operatorname{Im}\lambda \geq 0$ uniformly with respect to $x \in (0,\infty)$.

The functions $\beta_l(x)$ are expressed also directly in terms of the potential by means of the recurrence formulas:

$$\beta_l(x) = 0 \quad l \leq 0; \qquad \beta_1(x) = q(x),$$

$$\beta_{l+1}(x) = -\beta_l'(x) + \sum_{j\geq 1} \left[\left(\beta_{l-j}(x) + 4q(x)\beta_{l-2-j}(x)\right)\beta_j(x) - 4q(x)\beta_{l-1}(x)\right]$$

Let us denote by $\sqrt{z}$ the branch which takes on positive values on the upper side of the cut traced along the positive semiaxis.

Lemma. *Let q be n-times differentiable with compact support. Then*

1. *The Weyl functions can be expressed in the form*

$$m_\infty^\pm(z) = \pm i\sqrt{z}\left\{1 + 2\sum_{l\leq n} \frac{\alpha_l(0)}{(\pm 2i\sqrt{z})^{l+1}} + \frac{\alpha_n^\pm(z)}{(\pm 2i\sqrt{z})^{n+1}}\right\} \tag{20}$$

$$m_0^\pm(z) = \pm\frac{i}{\sqrt{z}}\left\{1 - 2\sum_{l\leq n} \frac{\beta_l(0)}{(\pm 2i\sqrt{z})^{l+1}} + \frac{\beta_n^\pm(z)}{(\pm 2i\sqrt{z})^{n+1}}\right\}, \tag{20'}$$

where

$$\lim_{|z|\to\infty} \alpha_n^{\pm}(z) = \lim_{|z|\to\infty} \beta_n^{\pm}(z) = 0.$$

2. *For all integer $k \geq 0$ the spectral functions satisfy the asymptotic equalities:*

$$\int_{-\infty}^{\mu} \frac{(\mu-t)^k}{k!}\, d\rho_\infty^{\pm}(t)$$

$$= \mu^{k+\frac{3}{2}} \left\{ A(k,3) + 2\sum_{l\leq n} A(k,2-l)\frac{\alpha_l(0)}{(\pm 2\sqrt{\mu})^{l+1}} + \frac{\varepsilon_\infty^{\pm}(\mu)}{(\pm 2\sqrt{\mu})^{n+1}} \right\}, \tag{21}$$

$$\int_{-\infty}^{\mu} \frac{(\mu-t)^k}{k!}\, d\rho_0^{\pm}(t) =$$

$$= -\mu^{k+\frac{1}{2}} \left\{ A(k,1) - 2\sum_{l\leq n} A(k,-l)\frac{\beta_l(0)}{(\pm 2\sqrt{\mu})^{l+1}} + \frac{\varepsilon_0^{\pm}(\mu)}{(\pm 2\sqrt{\mu})^{n+1}} \right\} \tag{21'}$$

where

$$A(k,s) = \frac{(-1)^s \sin\frac{\pi s}{2}}{\pi \prod_{p=0}^{k} (p+\frac{s}{2})},$$

and $\lim_{\mu\to+\infty} \varepsilon_0^{\pm}(\mu) = \lim_{\mu\to+\infty} \varepsilon_\infty^{\pm}(\mu) = 0.$

Proof. If the conditions of the lemma hold, we can use the formulas (13) and (16), (17), (18), (19) from which it follows that when $\operatorname{Im}\lambda \geq 0$

$$m_\infty^{+}(\lambda^2) = i\lambda + \sum_{l\leq n} \frac{\alpha_l(0)}{(2i\lambda)^l} - \frac{\alpha_n(\lambda,0)}{(2i\lambda)^n} = \frac{e'(\lambda,0)}{e(\lambda,0)}$$

$$m_0^{+}(\lambda^2) = \frac{1}{\lambda^2}\left\{ i\lambda - \sum_{l\leq n} \frac{\beta_l(0)}{(2i\lambda)^l} - \frac{\beta_n(\lambda,0)}{(2i\lambda)^n} \right\} = -\frac{e(\lambda,0)}{e'(\lambda,0)}\,.$$

The formulas (20), (20′) for the functions $m_0^{+}(z)$, $m_\infty^{+}(z)$ are direct consequences of these equalities and of the definition of the function $\sqrt{z}$. Since the potential $\tilde{q}(x) = q(-x)$ also satisfies the conditions of the lemma, then the Weyl functions $\tilde{m}_0^{+}(z)$, $\tilde{m}_\infty^{+}(z)$ of equation (10) can also be expressed by (20), (20′) with the coefficients $\tilde{\alpha}_l(0)$, $\tilde{\beta}_l(0)$ defined by the recurrence formulas (18′), (19′) where $q(x)$ must be replaced by $\tilde{q}(x) = q(-x)$.

It follows by induction from these recurrence formulas that

$$\tilde{\alpha}_l(x) = (-1)^{l+1}\alpha_l(-x), \quad \tilde{\beta}_l(x) = (-1)^{l+1}\beta_l(-x).$$

Therefore

$$\tilde{\alpha}_l(0) = (-1)^{l+1}\alpha_l(0), \quad \tilde{\beta}_l(0) = (-1)^{l+1}\beta_l(0)$$

from which according to (11) it follows that the representations (20), (20′) for the functions $m_0^{-}(z)$, $m_\infty^{-}(z)$ are obtained from the representations of the functions $m_0^{+}(z)$, $m_\infty^{+}(z)$ by exchanging $\sqrt{z}$ and $-\sqrt{z}$.

The proofs of the equalities (21), (21′) are similar for all spectral functions and we shall restrict ourselves with the consideration of one of them. If the conditions of the lemma hold the spectral functions are absolutely continuous on the positive semiaxis. Hence

$$\int_0^{\mu} (\mu - t)^k \, d\rho_\infty^+(t) = \int_0^{\mu} (\mu - t)^k \rho_\infty^+(t)' \, dt = \int_0^{\sqrt{\mu}} (\mu - \tau^2)^k \rho_\infty^+(\tau^2)' 2\tau \, d\tau$$

and since for real τ

$$\overline{e(\tau, 0)} = e(-\tau, 0), \quad \overline{e'(\tau, 0)} = e'(-\tau, 0),$$

then according to (14)

$$\int_0^{\mu} (\mu - t)^k \, d\rho_\infty^+(t) = \frac{1}{2\pi i} \int_0^{\sqrt{\mu}} (\mu - \tau^2)^k \, 2\tau \left\{ \frac{e'(\tau, 0)}{e(\tau, 0)} - \frac{e'(-\tau, 0)}{e(-\tau, 0)} \right\} d\tau$$

$$= \frac{1}{2\pi i} \int_{-\sqrt{\mu}}^{\sqrt{\mu}} (\mu - \tau^2)^k \, 2\tau \frac{e'(\tau, 0)}{e(\tau, 0)} \, d\tau.$$

In the upper half-plane the entire function $e(\tau, 0)$ has a finite number of simple zeros $i\kappa_1, \ldots, i\kappa_m$, and according to the Cauchy theorem when $\mu > \max\{\kappa_l\}$:

$$\frac{1}{2\pi i} \Bigl\{ \int_{-\sqrt{\mu}}^{\sqrt{\mu}} (\mu - \tau^2)^k \, 2\tau \frac{e'(\tau, 0)}{e(\tau, 0)} \, d\tau + \int_{C(\mu)} (\mu - \tau^2)^k \, 2\tau \frac{e'(\tau, 0)}{e(\tau, 0)} \, d\tau \Bigr\} =$$

$$= \sum_{j=1}^{m} (\mu + \kappa_j^2)^k \, 2i\kappa_j \frac{e'(i\kappa_j, 0)}{\dot{e}(i\kappa_j, 0)}$$

where $C(\mu)$ is the semicircle $\tau = \sqrt{\mu} e^{i\varphi}$ $(0 \leq \varphi \leq \pi)$ oriented according to increasing φ. But according to (15′)

$$\sum_{j=1}^{m} (\mu + \kappa_j^2)^k \, 2i\kappa_j \frac{e'(i\kappa_j, 0)}{\dot{e}(i\kappa_j, 0)} = -\int_{-\infty}^{0} (\mu - t)^k \, d\rho_\infty^+(t)$$

from which, according to (16) and (18), it follows that

$$\int_{-\infty}^{\mu} (\mu - t)^k \, d\rho_\infty^+(t) = -\frac{1}{\pi i} \int_{C(\mu)} (\mu - \tau^2)^k \, \tau \frac{e'(\tau, 0)}{e(\tau, 0)} \, d\tau$$

$$= \frac{\mu^{k+\frac{3}{2}}}{\pi} \int_0^{\pi} (1 - e^{2i\varphi})^k (ie^{i\varphi})^2 \times$$

$$\times \left\{ ie^{i\varphi} + 2 \sum_{l \leq n} \frac{\alpha_l(0)}{(2\sqrt{\mu})^{l+1}} (ie^{i\varphi})^{-l} + \frac{2\alpha_n(\sqrt{\mu} e^{i\varphi}, 0)}{(2\sqrt{\mu})^{n+1}} (ie^{i\varphi})^{-n} \right\} d\varphi$$

i.e.

$$\int_{-\infty}^{\mu} \frac{(\mu-t)^k}{k!}\, d\rho_{\infty}^{+}(t) = \mu^{k+\frac{3}{2}} \left\{ A(k,3) + 2\sum_{l\leq n} A(k,2-l)\frac{\alpha_l(0)}{(2\sqrt{\mu})^{l+1}} + \frac{\varepsilon_{\infty}^{+}(\mu)}{(2\sqrt{\mu})^{n+1}} \right\}$$

where

$$A(k,s) = \frac{1}{\pi k!}\int_0^{\pi} (1-e^{2i\varphi})^k (ie^{i\varphi})^s d\varphi,$$

$$\varepsilon_{\infty}^{+}(\mu) = \frac{2}{\pi k!}\int_0^{\pi} (1-e^{2i\varphi})^k (ie^{i\varphi})^{2-n} \alpha_n(\sqrt{\mu}e^{i\varphi},0) d\varphi$$

and $\lim\limits_{\mu\to+\infty} \varepsilon_{\infty}^{+}(\mu) = 0$ because in the upper half-plane $\operatorname{Im}\lambda \geq 0$ the function $\alpha_n(\lambda,0)$ tends to zero as $|\lambda| \to \infty$. By substituting $\varphi = \alpha + \frac{\pi}{2}$ the right part of the formula for coefficients $A(k,s)$ can be reduced to the integral:

$$\begin{aligned} A(k,s) &= \frac{2^k(-1)^s}{\pi k!}\int_{-\frac{\pi}{2}}^{\frac{\pi}{2}} (\cos\alpha)^k e^{i(k+s)\alpha}\, d\alpha \\ &= \frac{2^{k+1}(-1)^s}{\pi k!}\int_0^{\frac{\pi}{2}} (\cos\alpha)^k \cos(k+s)\alpha\, d\alpha, \end{aligned}$$

which can be expressed in terms of gamma-function:

$$\begin{aligned} \int_0^{\frac{\pi}{2}} (\cos\alpha)^k \cos(k+s)\alpha\, d\alpha &= \frac{\pi\Gamma(k+2)}{(k+1)2^{k+1}\Gamma(k+1+\frac{s}{2})\Gamma(1-\frac{s}{2})} \\ &= \frac{k!\sin\frac{\pi s}{2}}{2^{k+1}\prod\limits_{p=0}^{k}(p+\frac{s}{2})}. \end{aligned}$$

Hence

$$A(k,s) = \frac{(-1)^s \sin\frac{\pi s}{2}}{\pi\prod\limits_{p=0}^{k}(p+\frac{s}{2})}. \quad \square$$

The results of this lemma can be extended to arbitrary $n \geq 0$ times continuously differentiable potentials.

Theorem. *Let $q(x)$ be a real potential with $n \geq 0$ continuous derivatives. Then*

1. *for any $\varepsilon > 0$ the equalities* (20), (20′) *are satisfied in angular domains*

$$\varepsilon \leq \arg z \leq \pi - \varepsilon, \quad \pi + \varepsilon \leq \arg z \leq 2\pi - \varepsilon,$$

2. *the spectral functions satisfy the asymptotic equalities* (21), (21′) *for all $k \geq n$.*

The proof of this theorem is based on the Tauberian theorems of [5]. Here is the statement of the special case we need.

Tauberian Theorem. *Let $\tau_1(\mu)$, $\tau_2(\mu)$ be nondecreasing functions satisfying the conditions:*

1. *on the set of infinitely differentiable functions $f(t)$ which are equal to zero outside of the segment $[-2h, 2h]$ the equalities*

$$\int_{-\infty}^{\infty} E_f(\mu)\, d\tau_1(\mu) = \int_{-\infty}^{\infty} E_f(\mu)\, d\tau_2(\mu) + \int_{-\infty}^{\infty} f(t)G(t)\, dt$$

hold where $G(t)$ is some infinitely differentiable function and

$$E_f(\mu) = \int_{-\infty}^{\infty} f(t)e^{-i\mu t} dt;$$

2. *for some integer $m \geq 0$ and all $x \in \mathbb{R}$ the equality*

$$\overline{\lim_{|\mu|\to\infty}} \left| \frac{\tau_2(\mu + x) - \tau_2(mu)}{\mu^m} \right| \leq B|x| \qquad B = const. < \infty$$

is fulfilled.

Then for all integer $k \geq 0$

$$\overline{\lim_{N\to\infty}} N^{k-m} \left| \int_{-N}^{N} \left(1 - \left(\tfrac{t}{N}\right)^2\right)^k d(\tau_1(t) - \tau_2(t)) + \right.$$

$$\left. + \sum_{0\leq l \leq k-m} \frac{(-i)^l G^{(l)}(0) P^{(l)}(0)}{l! N^l} \right| \leq h^{-(k+1)} C_k, \tag{22}$$

where

$$P(t) = (1 - t^2)^k, \quad C_k = 10^8 B \left\{ \int_{-1}^{1} |P^{(k+1)}(t)| dt + 2^{k+1} k! \right\}.$$

Let us consider two Sturm-Liouville operations l_1, l_2 with the potentials $q_1(x)$, $q_2(x)$ which are coinciding on the segment $[-h, h]$. It is obvious that the solutions $s_1(\lambda, x)$, $s_2(\lambda, x)$ of the corresponding equations (1) are coinciding on this segment too. Therefore for all functions $g(x) \in L_2(\mathbb{R}^\pm)$ which are equal to zero when $|x| > h$ the equalities $s_1^\pm(\lambda, g) = s_2^\pm(\lambda, g)$ are fulfilled from which according to Parseval equalities (2) it follows that

$$\int_{\mathbb{R}^\pm} g(x)^2 dx = \int_{-\infty}^{\infty} s_1^\pm(\sqrt{\mu}, g)^2 d\rho_\infty^\pm(\mu, 1) = \int_{-\infty}^{\infty} s_1^\pm(\sqrt{\mu}, g)^2 d\rho_\infty^\pm(\mu, 2), \tag{23}$$

where $\rho_\infty^\pm(\mu, i)$ $(i = 1, 2)$ are the spectral functions of the operators $L_\infty^\pm(i)$ which are generated by the operations l_1, l_2. It follows from the existence of the transform operators and from Paley-Wiener theorem that the set of functions $s_1^\pm(\lambda, g)$ where $g(x) \in L_2(\mathbb{R}^\pm)$ and $g(x) = 0$ when $|x| > h$ coincides with the set of functions of the form $\lambda^{-1}\psi(\lambda)$ where $\psi(\lambda)$ is an arbitrary odd function of exponential type $\leq h$ which is square integrable on the positive semiaxis (see for example [3],[4]). Let us assume

$$p(\lambda) = \prod_{k=1}^{\infty} \left(\frac{\sin 2^{-k}\lambda}{2^{-k}\lambda} \right)$$

and consider the functions

$$\varphi_\varepsilon(\lambda) = \sqrt{2}\,\frac{\sin\frac{\lambda t}{2}}{\lambda}\,p(\varepsilon\lambda).$$

As $\sin\frac{\lambda t}{2}\,p(\varepsilon\lambda)$ is an odd function of exponential type $\frac{|t|}{2}+\varepsilon$ and it belongs to the space $L_2(-\infty,\infty)$ then according to the previous $\varphi_\varepsilon(\lambda) = s_1^\pm(\lambda, g_\varepsilon)$ where $g_\varepsilon \in L_2(\mathbb{R}^\pm)$ and $g_\varepsilon(x) = 0$ when $|x| > \frac{|t|}{2}+\varepsilon$. Therefore if $\frac{|t|}{2}+\varepsilon \leq h$ then according to (23)

$$\int\limits_{-\infty}^{\infty} \varphi_\varepsilon(\sqrt{\mu})^2\, d\rho_\infty^\pm(\mu,1) = \int\limits_{-\infty}^{\infty} \varphi_\varepsilon(\sqrt{\mu})^2\, d\rho_\infty^\pm(\mu,2), \tag{24}$$

i.e.

$$\int\limits_{-\infty}^{\infty} \frac{1-\cos\sqrt{\mu}t}{\mu}\,p(\varepsilon\sqrt{\mu})^2\, d\rho_\infty^\pm(\mu,1) = \int\limits_{-\infty}^{\infty} \frac{1-\cos\sqrt{\mu}t}{\mu}\,p(\varepsilon\sqrt{\mu})^2\, d\rho_\infty^\pm(\mu,2).$$

Let $f(t)$ is an arbitrary infinitely differentiable function which is equal to zero when $|t| > 2(h-\delta)$ and $\varepsilon \in (0,\delta)$. By multiplying both parts of (24) by $f''(t)$ we obtain after integration the equality

$$\int\limits_{-\infty}^{\infty} c_f(\sqrt{\mu})p(\varepsilon\sqrt{\mu})^2\, d\rho_\infty^\pm(\mu,1) = \int\limits_{-\infty}^{\infty} c_f(\sqrt{\mu})p(\varepsilon\sqrt{\mu})^2\, d\rho_\infty^\pm(\mu,2),$$

where

$$c_f(\lambda) = \int\limits_{-\infty}^{\infty} f(x)\cos\lambda x\, dx.$$

Here the integrals and limits for $\varepsilon \to 0$ commute because $\lim\limits_{\mu\to+\infty} \mu^n c_f(\sqrt{\mu}) = 0$ for all $n > 0$ and the spectral functions satisfy the conditions (3), (4). Therefore

$$\int\limits_{-\infty}^{\infty} c_f(\sqrt{\mu})\, d\rho_\infty^\pm(\mu,1) = \int\limits_{-\infty}^{\infty} c_f(\sqrt{\mu})\, d\rho_\infty^\pm(\mu,2),$$

i.e.

$$\int\limits_{0}^{\infty} c_f(\lambda)\, d(\rho_\infty^\pm(\lambda^2,1) - \rho_\infty^\pm(\lambda^2,2)) = \int\limits_{-\infty}^{\infty} f(t)G_\infty^\pm(t)dt$$

where the even functions

$$G_\infty^\pm(t) = \int\limits_{-\infty}^{+0} \cos\sqrt{\mu}t\, d(\rho_\infty^\pm(\mu,1) - \rho_\infty^\pm(\mu,2)) \tag{25}$$

are infinitely differentiable by virtue of (4).

Setting

$$\tau_\infty^\pm(\lambda,i) = \frac{\lambda}{2|\lambda|}\,\rho_\infty^\pm(\lambda^2,i) \quad (i=1,2) \tag{26}$$

we can replace the last equality to the equivalent one

$$\int_{-\infty}^{\infty} c_f(\lambda)\, d\tau_\infty^\pm(\lambda,1) = \int_{-\infty}^{\infty} c_f(\lambda)\, d\tau_\infty^\pm(\lambda,2) + \int_{-\infty}^{\infty} f(t) G_\infty^\pm(t)\, dt.$$

Finally since

$$E_f(\lambda) = \int_{-\infty}^{\infty} f(t) e^{-i\lambda t}\, dt = c_f(\lambda) + i \int_{-\infty}^{\infty} f(t) \sin \lambda t\, dt$$

and the functions $\tau_\infty^\pm(\lambda,i)$ are odd then

$$\int_{-\infty}^{\infty} E_f(\lambda)\, d\tau_\infty^\pm(\lambda,i) = \int_{-\infty}^{\infty} c_f(\lambda)\, d\tau_\infty^\pm(\lambda,i).$$

Since δ is arbitrary it follows from this that for all infinitely differentiable functions $f(t)$ which are equal to zero when $|t| \geq 2h$ the equalities

$$\int_{-\infty}^{\infty} E_f(\lambda)\, d\tau_\infty^\pm(\lambda,1) = \int_{-\infty}^{\infty} E_f(\lambda)\, d\tau_\infty^\pm(\lambda,2) + \int_{-\infty}^{\infty} f(t) G_\infty^\pm(t)\, dt$$

are satisfied. Besides according to (3), (26)

$$\lim_{|\lambda|\to\infty} \left| \frac{\tau_\infty^\pm(\lambda - \tau_\infty^\pm(\lambda)}{\lambda^2} \right| = \frac{1}{\pi}|x|$$

Hence the functions $\tau_\infty^\pm(\lambda,1)$, $\tau_\infty^\pm(\lambda,2)$ satisfy all conditions of the Tauberian theorem with $B = \frac{1}{\pi}$ and $m = 2$, $G(t) = G_\infty^\pm(t)$. It can be proved similarly that the functions

$$\tau_0^\pm(\lambda,i) = \frac{\lambda}{2|\lambda|}\, \rho_0^\pm(\lambda^2,i) \quad (i = 1,2) \tag{26'}$$

satisfy all conditions of this theorem as well with $B = \frac{1}{\pi}$, $m = 0$, $G(t) = G_0^\pm(t)$ where

$$G_0^\pm(t) = -\int_{-\infty}^{+0} \cos \sqrt{\mu} t\, d(\rho_0^\pm(\mu,1) - \rho_0^\pm(\mu,2)). \tag{25'}$$

Therefore, for corresponding $m = m(p)$, the functions

$$\tau_1(\mu) = \tau_p^\pm(\mu,1), \quad \tau_2(\mu) = \tau_p^\pm(\mu,2) \quad (p = 0,\infty)$$

satisfy the inequalities (22). According to (26), (26′)

$$\int_{-N}^{N}\left(1-\frac{t^2}{N^2}\right)^k d(\tau_1(t)-\tau_2(t)) =$$

$$= N^{-2k}\int_0^N (N^2-t^2)^k d(\rho_p^{\pm}(t^2,1)-\rho_p^{\pm}(t^2,2))$$

$$= N^{-2k}\int_0^{N^2} (N^2-\xi)^k d(\rho_p^{\pm}(\xi,1)-\rho_p^{\pm}(\xi,2)) \quad (p=0,\infty) \tag{27}$$

and according to (25), (25′) the derivatives of odd orders of the functions $G_p^{\pm}(t)$, $P(t)$ are equal to zero when $t=0$ and the derivatives of the even order $l=2s$ are equal to

$$G_p^{\pm}(t)^{(2s)}\Big|_{t=0} = -(-1)^s\int_{-\infty}^{+0}\xi^s d(\rho_p^{\pm}(\xi,1)-\rho_p^{\pm}(\xi,2))$$

$$P^{(2s)}(t)\Big|_{t=0} = \left(\frac{d}{dt}\right)^{2s}(1-t^2)^k\Big|_{t=0} = (-1)^s(2s!)C_k^s.$$

Hence

$$\sum_{0\le l\le k-m}\frac{(-i)^l G^{(l)}(0)P^{(l)}(0)}{l!N^l} =$$

$$= -\int_{-\infty}^{+0}\left(\sum_{0\le 2s\le k-m}(-1)^s C_k^s N^{-2s}\xi^s\right) d(\rho_p^{\pm}(\xi,1)-\rho_p^{\pm}(\xi,2))$$

$$= -\int_{-\infty}^{+0}\left\{\left(1-\frac{\xi}{N^2}\right)^k - \sum_{\frac{k-m}{2}<s\le k}(-1)^s C_k^s N^{-2s}\xi^s\right\} d(\rho_p^{\pm}(\xi,1)-\rho_p^{\pm}(\xi,2))$$

$$= -N^{-2k}\int_{-\infty}^{+0}(N^2-\xi)^k\, d(\rho_p^{\pm}(\xi,1)-\rho_p^{\pm}(\xi,2)) + N^{-(k+1-m)}\delta(N)$$

where $\overline{\lim}_{N\to+\infty}\delta(N)<\infty$ and it follows from (22), (27) that

$$\overline{\lim_{N\to+\infty}} N^{-(k+m(p))}\left|\int_{-\infty}^{N^2}(N^2-\xi)^k\, d(\rho_p^{\pm}(\xi,1)-\rho_p^{\pm}(\xi,2))\right| \le h^{-(k+1)}C_k$$

where $m(0)=0$ and $m(\infty)=2$. So if the potentials $q_1(x)$ and $q_2(x)$ coincide on the segment $[-h,h]$ then

$$\overline{\lim_{\mu\to+\infty}} \mu^{-\frac{k+m(p)}{2}}\left|\int_{-\infty}^{\mu}\frac{(\mu-\xi)^k}{k!}\, d(\rho_p^{\pm}(\xi,1)-\rho_p^{\pm}(\xi,2))\right| \le \frac{h^{-(k+1)}C_k}{k!}. \tag{28}$$

Proof of the theorem. Let $q(x)$ be a real potential with $n \geq 0$ continuous derivatives. Let $\varphi(x)$ be an arbitrary infinitely differentiable function equal to 1 on the segment $[-1,1]$ and to 0 outside of $[-2,2]$. Let us consider the set of potentials

$$q(x,h) = q(x)\varphi(\tfrac{x}{h}) \tag{29}$$

and the set of corresponding spectral functions $\rho_p^{\pm}(\xi,h)$ $(p = 0,\infty)$. It is obvious that the potential $q(x,h)$ satisfies the conditions of the lemma and coincides with the potential $q(x)$ on $[-h,h]$. Therefore the spectral functions $\rho_p^{\pm}(\xi,h)$ satisfy the asymptotic equalities (21), (21′) and because $q(x) = q(x,h)$ when $|x| \leq h$ the coefficients $\alpha_l(0)$, $\beta_l(0)$ in these equalities are independent of h and can be found from the potential $q(x)$ by the recurrence formulas (18′), (19′).

Hence

$$\int_{-\infty}^{\mu} \frac{(\mu-t)^k}{k!}\, d\rho_\infty^{\pm}(t) =$$

$$= \mu^{k+\frac{3}{2}} \left\{ A(k,3) + 2\sum_{l\leq n} A(k,2-l)\frac{\alpha_l(0)}{(\pm 2\sqrt{\mu})^{l+1}} + \frac{\varepsilon_\infty^{\pm}(\mu,h)}{(\pm 2\sqrt{\mu})^{n+1}} \right\}$$

$$+ \int_{-\infty}^{\mu} \frac{(\mu-t)^k}{k!}\, d(\rho_\infty^{\pm}(t) - \rho_\infty^{\pm}(t,h))$$

i.e.

$$R_\infty^{\pm}(\mu) = \mu^{-\frac{2k-n+2}{2}} \times$$

$$\times \left[\int_{-\infty}^{\mu} \frac{(\mu-t)^k}{k!}\, d\rho_\infty^{\pm}(t) - \mu^{k+\frac{3}{2}} \left\{ A(k,3) + 2\sum_{l\leq n} A(k,2-l)\frac{\alpha_l(0)}{(\pm 2\sqrt{\mu})^{l+1}} \right\} \right]$$

$$= \frac{\varepsilon_\infty^{\pm}(\mu,h)}{(\pm 2)^{n+1}} + \mu^{-\frac{2k-n+2}{2}} \int_{-\infty}^{\mu} \frac{(\mu-t)^k}{k!}\, d(\rho_\infty^{\pm}(t) - \rho_\infty^{\pm}(t,h)).$$

As the potentials $q(x,h)$ satisfy the conditions of the lemma then

$$\lim_{\mu\to+\infty} \varepsilon_\infty^{\pm}(\mu,h) = 0$$

for all $h > 0$ and since the potentials $q(x)$ and $q(x,h)$ coincide on the segment $[-h,h]$, it follows from (28) that

$$\overline{\lim_{\mu\to+\infty}}\, \mu^{-\frac{2k-n+2}{2}} \left| \int_{-\infty}^{\mu} \frac{(\mu-t)^k}{k!}\, d(\rho_\infty^{\pm}(t) - \rho_\infty^{\pm}(t,h)) \right| \leq \frac{h^{-(k+1)}C_k}{k!}$$

if $k \geq n$. Therefore for all $h > 0$ and $k \geq n$

$$\overline{\lim_{\mu\to+\infty}}\, |R_\infty^{\pm}(\mu)| \leq \frac{h^{-(k+1)}C_k}{k!}\,.$$

Now the left-hand side of this inequality does not depend on h and hence

$$\lim_{\mu\to+\infty} R_{\infty}^{\pm}(\mu) = 0.$$

Thus the spectral functions $\rho_{\infty}^{\pm}(\mu)$ of the operators $L_{\infty}^{\pm}$ with $n \geq 0$ times continuously differentiable potentials satisfy the asymptotic equalities (21) for all $k \geq n$.

The proof for the spectral functions $\rho_{0}^{\pm}(\mu)$ is similar.

Turning to the proof of equalities (20), (20′) we shall note that (3) and (4) imply the rough estimates:

$$\lim_{\mu\to-\infty} e^{x\sqrt{|\mu|}} \int_{-\infty}^{\mu} (\mu - t)^k \, d\rho_p^{\pm}(t) = 0$$

$$\lim_{\mu\to+\infty} \mu^{-\frac{2k+m(p)+1}{2}} \int_{-\infty}^{\mu} (\mu - t)^k \, d\rho_p^{\pm}(t) < \infty$$

($p = 0, \infty$, $m(0) = 0$, $m(\infty) = 2$). These estimates permit us to make any number of partial integrations in the right parts of the formulas (7), (8). As consequence we obtain the equalities

$$\pm\, m_0^{\pm}(z) = (k+1) \int_{-\infty}^{\infty} (t-z)^{-(k+2)} \left\{ \int_{-\infty}^{t} (t-\xi)^k \, d\rho_0^{\pm}(\xi) \right\} dt,$$

$$\pm\, m_{\infty}^{\pm}(z) = a^{\pm} + (k+1) \int_{-\infty}^{\infty} \left\{ (t-z)^{-(k+2)} - \tfrac{1}{2}(t+i)^{-(k+2)} - \tfrac{1}{2}(t-i)^{-(k+2)} \right\}$$
$$\times \left\{ \int_{-\infty}^{t} (t-\xi)^k \, d\rho_{\infty}^{\pm}(\xi) \right\} dt$$

from which it follows that

$$\pm \left(m_0^{\pm}(z) - m_0^{\pm}(z,h) \right) =$$
$$= (k+1) \int_{-\infty}^{\infty} (t-z)^{-(k+2)} \left\{ \int_{-\infty}^{t} (t-\xi)^k \, d(\rho_0^{\pm}(\xi) - \rho_0^{\pm}(\xi,h)) \right\} dt, \tag{30}$$

$$\pm \left(m_{\infty}^{\pm}(z) - m_{\infty}^{\pm}(z,h) \right) = a^{\pm} - a^{\pm}(h) + \tag{30′}$$
$$+ (k+1) \int_{-\infty}^{\infty} \left\{ (t-z)^{-(k+2)} - \tfrac{1}{2}(t+i)^{-(k+2)} - \tfrac{1}{2}(t-i)^{-(k+2)} \right\} \times$$
$$\times \left\{ \int_{-\infty}^{t} (t-\xi)^k \, d(\rho_{\infty}^{\pm}(\xi) - \rho_{\infty}^{\pm}(\xi,h)) \right\} dt$$

where $m_p^{\pm}(z,h)$, $\rho_p^{\pm}(t,h)$ ($p = 0, \infty$) denote Weyl functions and spectral functions of the operators with the potentials $q(x,h)$ defined by the formula (29).

Since the potentials $q(x,h)$ satisfy the conditions of the lemma, then the Weyl functions $m_p^{\pm}(z,h)$ can be represented in the form (20), (20′) in the whole z-plane with a cut along the positive semiaxes. Therefore in the angular domains $\varepsilon \le \arg z \le \pi - \varepsilon$, $\pi + \varepsilon \le \arg z \le 2\pi - \varepsilon$ the functions $m_p^{\pm}(z)$ can be represented in a similar form if they satisfy in these domains the equalities

$$\lim_{|z|\to\infty} |z|^{\frac{n+2-m(p)}{2}} \, |\, m_p^{\pm}(z) - m_p^{\pm}(z,h)\,| = 0.$$

Since the potentials $q(x)$ and $q(x,h)$ coincide on the segment $[-h,h]$ it follows from the already proved equalities (21), (21′) that

$$\int_{-\infty}^{t} (t-\xi)^n \, d(\rho_p^{\pm}(\xi) - \rho_p^{\pm}(\xi,h)) = (1+|t|)^{\frac{n+m(p)}{2}} \delta_p^{\pm}(t),$$

where

$$\lim_{|t|\to\infty} \delta_p^{\pm}(t) = 0$$

if the potential $q(x)$ has $n \ge 0$ continuous derivatives. From this, according to (30) and (30′), it follows that

$$\pm\left(m_0^{\pm}(z) - m_0^{\pm}(z,h)\right) = (n+1) \int_{-\infty}^{\infty} (t-z)^{-(n+2)} (1+|t|)^{\frac{n}{2}} \delta_0^{\pm}(t)\, dt,$$

$$\begin{aligned} \pm\left(m_\infty^{\pm}(z) - m_\infty^{\pm}(z,h)\right) &= a^{\pm} - a^{\pm}(h) + \\ &+ (n+1) \int_{-\infty}^{\infty} \left\{(t-z)^{-(n+2)} - \tfrac{1}{2}(t+i)^{-(n+2)} - \tfrac{1}{2}(t-i)^{-(n+2)}\right\} \times \\ &\times (1+|t|)^{\frac{n+2}{2}} \delta_\infty^{\pm}(t)\, dt \end{aligned}$$

where all three integrals are absolutely convergent if $n \ge 1$. Therefore if $n \ge 1$ we have

$$\pm\left(m_\infty^{\pm}(z) - m_\infty^{\pm}(z,h)\right) = C + (n+1) \int_{-\infty}^{\infty} (t-z)^{-(n+2)} (1+|t|)^{\frac{n}{2}} \delta_\infty^{\pm}(t)\, dt \tag{31}$$

where $C = C(n,h)$ are constants. After substitution $t = |z|\tau$ the integrals in the right-hand sides of these formulas are reduced to the form

$$\int_{-\infty}^{\infty} (t-z)^{-(n+2)} (1+|t|)^{\frac{n+m(p)}{2}} \delta_p^{\pm}(t)\, dt = |z|^{-\frac{n+2-m(p)}{2}} \varepsilon_p^{\pm}(z)$$

where

$$\varepsilon_p^{\pm}(z) = \int_{-\infty}^{\infty} (\tau - \tfrac{z}{|z|})^{-(n+2)} (|z|^{-1} + |\tau|)^{\frac{n+m(p)}{2}} \delta_p^{\pm}(|z|\tau)\, d\tau.$$

Since $\left|\operatorname{Im}\frac{z}{|z|}\right| \geq \sin\varepsilon$ in the domains under consideration and the functions $\delta_p^{\pm}(|z|\tau)$ tend to zero as $|z| \to \infty$ for all $\tau \neq 0$ then according to the Lebesgue theorem

$$\lim_{|z|\to\infty} \varepsilon_0^{\pm}(z) = 0 \qquad (n \geq 0)$$

$$\lim_{|z|\to\infty} \varepsilon_{\infty}^{\pm}(z) = 0 \qquad (n \geq 1). \tag{32}$$

Therefore the equalities (20′) for $p = 0$ are proved for any $n \geq 0$ from which it follows in particular that in the domains under consideration the Weyl functions $m_0^{\pm}(z)$ of the operators $L_0^{\pm}$ with arbitrary potentials satisfy the equality

$$\lim_{|z|\to\infty} z\left(m_0^{\pm}(z) \mp \frac{i}{\sqrt{z}}\right) = 0$$

and the Weyl functions $m_{\infty}^{\pm}(z) = -(m_0^{\pm}(z))^{-1}$ of the operators $L_{\infty}^{\pm}$ satisfy the equality

$$\lim_{|z|\to\infty} (m_{\infty}^{\pm}(z) \mp i\sqrt{z}) = 0,$$

which proves the equality (20) for $n = 0$. If $n \geq 1$ then from the last equality, formula (31) and the equality (32) it follows that the constants $C(n,h) = C$ are equal to zero and

$$m_{\infty}^{\pm}(z) - m_{\infty}^{\pm}(z,h) = (n+1)|z|^{-\frac{n}{2}}\varepsilon_{\infty}^{\pm}(z).$$

Therefore the equalities (20), (20′) are proved for all $n \geq 0$ and $p = 0, \infty$ which completes the proof of the theorem. □

Acknowledgment

We are thankful to the IHES which made possible this joint work.

References

[1] V.A. Marchenko, On inversion formulas generated by a linear second-order differentiable operator, *Dokl. Akadauber. Nauk SSSR*, **74**, 4, 1950, pp. 657–660.

[2] B.M. Levitan, On the asymptotic behaviour of spectral function, *Izv. Akad. Nauk SSSR*, Ser. Math., **19**, 1955, pp. 33-58.

[3] V.A. Marchenko, *Sturm-Liouville operators and applications*, Birkhäuser Verlag, 1986.

[4] V.A. Marchenko, I.V. Ostrovsky, Approximation of periodic by finite-zone potentials, *Selecta Math. Sovietica*, **6**, 2, 1987, pp. 101-136.

[5] V.A. Marchenko, The Tauberian theorems in the spectral analysis of differential operators, *Izv. Akad. Nauk SSSR*, Ser. Math., **19**, 1955, pp. 381-422.

Anne Boutet de Monvel,
Institut de Mathématiques de Jussieu, CNRS UMR 9994,
Laboratoire de Physique Mathématique et Géométrie,
case 7012, Université Paris 7 Denis Diderot,
2 place Jussieu, F-75251 Paris Cedex 05

Vladimir Marchenko,
Mathematical Division, Institute for Low Temperature Physics,
47 Lenin Ave, 310164, Kharkiv, Ukraine

AMS classification # 34B30, 34L05

Operator Theory
Advances and Applications, Vol. 98

THE GLAZMAN-KREIN-NAIMARK THEOREM FOR ORDINARY DIFFERENTIAL OPERATORS

W.N. Everitt L. Markus

Dedicated to the memory of
Izrail Markovich Glazman

The original form of symmetric boundary conditions for Lagrange symmetric (formally self-adjoint) ordinary linear differential expressions, was obtained by I.M. Glazman in his seminal paper of 1950. This result described all self-adjoint differential operators, in the underlying Hilbert function space, generated by real even-order differential expressions.

The proof was based on earlier work of M.G. Krein, and was re-formulated by M.A. Naimark for the first edition of his book on linear differential operators published in 1952.

Since then the theorem has been extended to both real- and complex-valued symmetric differential expressions of all positive integer orders, and has been named the Glazman-Krein-Naimark (GKN) theorem.

The complete proof of the GKN result has been known for some time but is not readily available in the literature. The proof is now presented in this paper as a contribution to the memorial volume in the name of I.M. Glazman.

1 Introduction

We have dedicated this paper to the memory of I.M. Glazman who, in the years following the Second World War until his untimely death in 1968, contributed with lasting significance to many aspects of the theory of linear operators in Hilbert space. These contributions are to be read in the collection of original papers and books that bear his name as author or co-author.

Two contributions in particular are significant for the purpose of this paper. Firstly, the original and seminal paper [10] that appeared in 1950; this work characterised all self-adjoint extensions of Lagrange symmetric (formally self-adjoint), real-valued (and then essentially even-ordered), quasi-differential expressions, and solved the range problem for the deficiency indexes of such expressions.

Secondly, in collaboration with N.I. Akhiezer, the book [1] on linear operators in Hilbert space; this appeared in three Russian editions and was twice translated into English. This second English translation dated 1980 was made by our late colleague E.R. Dawson[1], and contains an expanded version of the 1950 paper [10]; see [1, Appendix II]. The progress of this second English translation stimulated the work for the third Russian edition.

There is an alternative account of the original Glazman results in the book of M.A. Naimark [14]; this appeared originally in Russian in 1952, with an English translation in 1968.

Earlier work in the general area of both real- and complex-valued quasi-differential expressions had been made by Shin [15], [16] and [17], and by Halperin [11].

The study of classical (*i.e.* with some form of smooth coefficients), symmetric differential expressions of even-order and real-valued coefficients was advanced by the work of Kodaira [13]. The extension of such results for classical expressions with real- and complex-valued coefficients and of even or odd order was made by Kimura and Takahasi [12] (although this work seems never to have been completed), and by Dunford and Schwartz [2].

The results of Shin for arbitrary quasi-differential expressions were re-discovered independently by Zettl in 1965 but not published until the paper Zettl [19]. For this reason Everitt and Markus in [5] proposed to call the collection of such expressions by the name of Shin-Zettl.

The work of [19] was continued by Everitt and Zettl in [8] and then in [9].

All the classical differential expressions appear as special cases of quasi-differential expressions, but this statement requires the support of detailed analysis; there are two accounts dedicated to this end. Firstly the paper by Everitt and Race [7] and, secondly, the later work of Everitt and Markus in the memoir [6].

The theorem to be discussed in this paper is the extension of the original Glazman theorem in [10] on the characterisation of all self-adjoint extensions but now generated by arbitrary symmetric Shin-Zettl quasi-differential extensions of all orders, including the first-order expressions. This structural theorem stems from the early influence of M.G. Krein, the work of I.M. Glazman [10] and the reformulation given by M.A. Naimark in [14, Chapter V]. As Naimark states in [14, Chapter V, end of Section 19.4] "the results in ... are due essentially to I.M. Glazman and M.G. Krein" but here we may release M.A. Naimark from his gentle modesty and record that the final form of the theorem owes much to his labours in his remarkable book, that has had such a seminal influence on the subject of linear ordinary differential operators. It is for these reasons that in writing the paper [9] the authors named the theorem and its multi-interval extensions after Glazman, Krein and Naimark (GKN) and we have continued this 'tradition' here in this paper dedicated to I.M. Glazman.

Although the result that we state here, in Section 3 below, is proved in [9], the proof is hidden under the extensive generalization to a countable number of intervals therein considered. Here we give the most general form of the theorem with a proof restricted to the case of one interval and hope thereby that this contribution forms a standard reference for this structural result.

In Section 2 we give some notations and then in Section 3 state the general properties

[1] Died 1995

of ordinary differential expressions and operators. The general form of the GKN theorem is given in Section 4. The proof of the GKN theorem requires three Lemmas 1,2, and 3 given in Sections 5,6 and 7 respectively. Section 8 brings together the final form of the proof of the GKN theorem.

2 Notations

$\mathbb{R}$ and $\mathbb{C}$ denote the real and complex number fields; $\mathbb{R}_0^+$ denotes the non-negative real numbers; $\mathbb{N}_0$ denotes the non-negative integers and $\mathbb{N}$ the positive integers.

Let (a,b) and $[\alpha,\beta]$ denote open and compact intervals of $\mathbb{R}$.

If $w:(a,b)\to\mathbb{R}_0^+$ and is Lebesgue measurable with w positive almost everywhere on (a,b) then $L^2((a,b):w)$ denotes the Hilbert function space of all functions $f:(a,b)\to\mathbb{C}$ such that

$$\int_a^b w(x)|f(x)|^2\,dx < +\infty \tag{2.1}$$

with inner-product

$$(f,g)_w := \int_a^b w(x)f(x)\overline{g}(x)\,dx. \tag{2.2}$$

The symbol '$(\,x\in K\,)$' is to be read as 'for all the elements x of the set K'.

3 Quasi-differential expressions and operators

The most general ordinary linear differential expressions so far defined, for order $n\in\mathbb{N}$ and $n\geq 2$, are the Shin-Zettl quasi-differential expressions; for details, in order of date of publication, see [9], [4], [7] and [6, Appendix A]. Quasi-differential expressions of order 1 need to be defined separately and the required definition is given below.

Let (a,b) be an arbitrary open interval of the real line $\mathbb{R}$. For $n\geq 2$ let $A\in Z_n(a,b)$ be a Shin-Zettl complex-valued matrix of order n and let M_A be the corresponding quasi-differential expression generated by A. The domain $D(M_A)$ of M_A is defined by

$$D(M_A) := \{\,f:(a,b)\to\mathbb{C} : f_A^{[r-1]}\in AC_{\mathrm{loc}}(a,b) \text{ for } r=1,2,\dots,n\,\} \tag{3.1}$$

with

$$M_A[f] := i^n f_A^{[n]} \quad (\,f\in D(M_A)\,). \tag{3.2}$$

Here $\{\,f_A^{[r]} : r=0,1,2,\dots.n\,\}$ are the quasi-derivatives of f with respect to A. We shall assume, in the notation of [4] and [7], that M_A is Lagrange symmetric (in the older notation this extends the idea of formal self-adjointness of classical differential expressions); that is $A^+=A$ and $M_A^+=M_A$; this being the case the Green's formula for M_A has the form

$$\int_\alpha^\beta \{\overline{g}M_A[f]-f\overline{M_A[g]}\} = [f,g]_A(\beta)-[f,g]_A(\alpha) \quad (\,f,g\in D(M_A)\,) \tag{3.3}$$

for any compact sub-interval $[\alpha, \beta]$ of (a, b). Here the skew-symmetric sesquilinear form $[\cdot,\cdot]_A$ maps $D(M_A)\times D(M_A) \to \mathbb{C}$. The explicit form of $[\cdot,\cdot]_A$ is given by

$$[f,g]_A(x) = i^n \sum_{r=1}^{n} (-1)^{r-1} f_A^{[n-r]}(x)\overline{g_A^{[r-1]}}(x) \quad (x \in (a,b) \text{ and } f,g \in D(M_A)).$$

It is to be noted that in [10] and [14, Chapter V] the order of the quasi-differential expressions are *even* and the coefficients are all *real-valued*; such expressions are special cases of the general Shin-Zettl theory.

In the case when the order $n = 1$ the quasi-differential expressions have to be defined separately; there are no Shin-Zettl matrices of order 1. The general Lagrange symmetric expression on the interval (a, b) is of the form

$$M_1[f] := i(\rho f)' + i\rho f' + qf \quad (f \in D(M_1)) \tag{3.4}$$

where

$$\begin{array}{ll} (i) & \rho, q : (a,b) \to \mathbb{R} \\ (ii) & \rho \in AC_{\text{loc}}(a,b) \quad \text{and} \quad \rho(x) > 0 \ (x \in (a,b)) \\ (iii) & q \in L^1_{\text{loc}}(a,b) \end{array} \tag{3.5}$$

and

$$D(M_1) := \{ f : (a,b) \to \mathbb{C} : f \in AC_{\text{loc}}(a,b) \}. \tag{3.6}$$

For this expression the bilinear form is given by

$$[f,g]_1(x) = 2i\rho(x)f(x)\overline{g}(x) \quad (x \in (a,b) \text{ and } f,g \in D(M_1)) \tag{3.7}$$

and the Green's formula takes the same form as (3.3).

We now collect these notations together and for all $n \in \mathbb{N}$ we denote the set of all quasi-differential expressions on the interval (a, b) by $\{ M_n : n \in \mathbb{N} \}$ with domains $\{D(M_n) : n \in \mathbb{N}\}$ and sesquilinear forms $\{ [\cdot,\cdot]_n : n \in \mathbb{N} \}$. Here for $n = 1$ the definitions are made by (3.4) to (3.7); for $n \geq 2$ the definitions are made by (3.1) to (3.3) with $A \in Z_n(a, b)$, but we suppress the explicit mention of the matrix A.

To define the collection of differential operators generated by this collection $\{M_n : n \in \mathbb{N}\}$ we assume given a non-negative weight

$$w : (a,b) \to \mathbb{R} \text{ with } w \in L^1_{\text{loc}}(a,b) \text{ and } w(x) > 0 \ (\text{almost all } x \in (a,b)). \tag{3.8}$$

This weight defines the space $L^2((a,b) : w)$ as in (2.1) and (2.2).

From the Green's formula (3.3) it follows the limits

$$[f,g]_n(a) := \lim_{x \to a+} [f,g]_n(x) \quad \text{and} \quad [f,g]_n(b) := \lim_{x \to b-} [f,g]_n(x) \tag{3.9}$$

both exist and are finite in $\mathbb{C}$.

The spectral differential equations associated with the pairs $\{ M_n, w \}$ is

$$M_n[y] = \lambda w y \quad \text{on} \quad (a,b). \tag{3.10}$$

where the spectral parameter $\lambda \in \mathbb{C}$. The solutions of this linear equation of order n are considered within the space $L^2((a,b):w)$ and determine the deficiency indices of the symmetric operators considered below.

For some $n \in \mathbb{N}$ consider the pair $\{M_n, w\}$. Following the theory developed in [14, Chapter V] we define two unbounded differential operators $T_{\min}$ and $T_{\max}$ in $L^2((a,b):w)$ generated by this pair; they have the following given definitions and properties, using the definition (3.9) above,

(*i*)

$$D(T_{\max}) := \{ f \in D(M_n) : f, w^{-1}M_n[f] \in L^2((a,b):w)$$
$$T_{\max} f := w^{-1}M_n[f] \quad (f \in D(T_{\max}))$$
$$D(T_{\min}) := \{ f \in D(T_{\max}) : [f,g]_n(b) - [f,g]_n(a) = 0 \; (g \in D(T_{\max}))$$
$$T_{\min} f := w^{-1}M_n[f] \quad (f \in D(T_{\min}))$$

(*ii*) $D(T_{\min})$ is dense in $L^2((a,b):w)$, $D(T_{\min}) \subseteq D(T_{\max})$ and $T_{\min} \subseteq T_{\max}$

(*iii*) $T_{\min}$ is closed and symmetric, and $T_{\max}$ is closed in $L^2((a,b):w)$

(*iv*) $T^*_{\min} = T_{\max}$, $T^*_{\max} = T_{\min}$ where * denotes an adjoint operator in $L^2((a,b):w)$

(*v*) if the deficiency indices of $T_{\min}$ are denoted by (d_n^+, d_n^-) then

$$d_n^{\pm} := \dim\{f \in D(T_{\max}) : M_n[f] = \pm iwf \text{ on } (a,b)\} \tag{3.11}$$

$$d_n^{\pm} \in \mathbb{N}_0 \text{ and } 0 \leq d_n^{\pm} \leq n.$$

For details of the proof of these results see [14, Sections 17 to 19]; the methods used in the case of real-valued, even-ordered quasi-differential expressions extend to the more general case considered here. See also the account in [18, Chapters 3 and 4].

From the general theory of symmetric operators in Hilbert space the minimal operator $T_{\min}$ has self-adjoint extensions S (equivalently the maximal operator $T_{\max}$ has self-adjoint restrictions S) if and only if

$$d_n^+ = d_n^-. \tag{3.12}$$

In the work of [10], [13] and [14, Chapter V] the quasi-differential expressions are real-valued and from this condition on M_n, with n necessarily even, it follows that (3.12) is satisfied. In the more general case of the complex-valued expressions it is possible, notwithstanding the Lagrange symmetry $M_A^+ = M_A$, for $d_n^+ \neq d_n^-$; see the account in the survey paper [3]. Thus, in general, we have to impose (3.12) as a condition to be satisfied by the quasi-differential expression M_A if there are to be self-adjoint extensions of the operator $T_{\min}$.

For a modern survey of the deficiency index conjecture see [6, Sections V.1 and V.2].

4 The GKN theorem

This theorem is one of the major results in the theory of linear ordinary differential operators, particularly in view of the fact that the form of the boundary conditions are both necessary and sufficient for the result to hold.

Theorem 1 *Let $n \in \mathbb{N}$; let M_n be a Lagrange symmetric quasi-differential expression of order n on the open interval (a,b); let the weight w be given on (a,b) with properties (3.8), and so define the space $L^2((a,b):w)$; let the deficiency indices $\{d_n^+, d_n^-\}$ of the generated minimal closed symmetric operator T_{min} in $L^2((a,b):w)$ satisfy the condition*

$$d_n^+ = d_n^- = d \ (say) \tag{4.1}$$

so that $0 \le d \le n$. Then the following results hold:

(1) *Let a linear manifold $D \subset L^2((a,b):w)$ be determined by a set of functions*

$$\{\varphi_s : s = 1,2,\ldots,d\}$$

with the properties

(*i*) $\varphi_s \in D(T_{max}) \quad (s = 1,2,\ldots,d)$

(*ii*) *the set $\{\varphi_s : s = 1,2,\ldots,d\}$ is linearly independent in $D(T_{max})$ modulo $D(T_{min})$*

(*iii*) *the set $\{\varphi_s : s = 1,2,\ldots,d\}$ satisfies the symmetry (formally self-adjoint) conditions*

$$[\varphi_s, \varphi_t]_n(b) - [\varphi_s, \varphi_t]_n(a) = 0 \quad (s,t = 1,2,\ldots,d). \tag{4.2}$$

Define

$$D := \{f \in D(T_{max}) : [f, \varphi_s]_n(b) - [f, \varphi_s]_n(a) = 0 \quad (s = 1,2,\ldots,d)\}. \tag{4.3}$$

Then the operator $S : D(S) \subseteq L^2((a,b):w) \to L^2((a,b):w)$ determined by

$$D(S) := D \quad and \quad Sf := w^{-1}M_n[f] \ (f \in D(S)) \tag{4.4}$$

is a self-adjoint extension of the operator T_{min}.

(2) *Conversely, if S is a self-adjoint extension of T_{min} then there exists a set $\{\varphi_s : s = 1,2,\ldots,d\}$, with the properties (i), (ii) and (iii) given above, such that $D(S)$ and S are determined by (4.3) and (4.4).*

Proof. For the proof of this theorem see the following sections. ■

Remarks 1.We have chosen to model the proof of the GKN theorem upon the proof for the real case given by Naimark in [14, Section 18.1].

2. We call the set $\{\varphi_s : s = 1,2,\ldots,d\}$ a GKN set for the pair of operators $\{T_{\min}, T_{\max}\}$.

3. If the common index $d = 0$ then $T_{\min}^* = T_{\max} = T_{\min}$ and $T_{\min}$ is self-adjoint; in this case $T_{\min}$ is the unique self-adjoint operator generated by the pair $\{M_n, w\}$ in $L^2((a,b):w)$. In all other cases the pair $\{M_n, w\}$ generates a continuum of self-adjoint operators in $L^2((a,b):w)$.

5 Statement and proof of Lemma 1

Lemma 1 *Let all the required conditions of Theorem* 1 *hold. Then a linear manifold D' of $L^2((a,b):w)$ is the domain of a self-adjoint extension of T_{min} if and only if*

(*i*) $D(T_{min}) \subseteq D' \subseteq D(T_{max})$

(*ii*) $[f,g]_n(b) - [f,g]_n(a) = 0 \quad (f,g \in D')$

(*iii*) *if* $f \in D(T_{max})$ *and* $[f,g]_n(b) - [f,g]_n(a) = 0$ $(g \in D')$ *then* $f \in D'$.

Proof. Firstly, suppose that S with domain $D(S)$ is a self-adjoint extension of $T_{\min}$ and let $D' = D(S)$. Then:

(a) we have $D(T_{\min}) \subseteq D'$, that is $S = S^* \subseteq T^*_{\min} = T_{\max}$ and so $D' \subseteq D(T_{\max})$

(b) let $f,g \in D'$; the since S is a restriction of $T_{\max}$, we can apply the Green's formula (3.3) on (a,b) to give

$$[f,g]_n(b) - [f,g]_n(a) = (Sf,g)_w - (f,Sg)_w = 0$$

(c) supposing the conditions of (*iii*) to hold, let $f \in D(T_{\max})$ and again using Green's formula for all $g \in D(S) = D'$, it follows that

$$0 = [f,g]_n(b) - [f,g]_n(a) = (T_{\max}f,g)_w - (f,T_{\max}g)_w = (T_{\max}f,g)_w - (f,Sg)_w;$$

thus $f \in D(S^*) = D(S)$ (since $S^* = S$) and, by definition above, $D' = D(S)$ to give $f \in D'$.

Secondly, suppose that the linear manifold D' satisfies (*i*), (*ii*) and (*iii*) above; then D' is dense in $L^2((a,b):w)$. Define $S : D(S) \to L^2((a,b):w)$ by $D(S) := D'$ and $Sf := T_{\max}f$ $(f \in D(S))$; this is possible since $D' \subseteq D(T_{\max})$. From (*ii*) and the Green's formula it follows that S is hermitian and hence symmetric in $L^2((a,b):w)$; thus $S \subseteq S^*$ and $D(S) \subseteq D(S^*)$. From (*iii*) and the Green's formula the element $f \in D(S^*)$; but then the first part of (*iii*) implies that $f \in D' = D(S)$, that is $D(S^*) \subseteq D(S)$. Thus $D(S^*) = D(S)$ and S is self-adjoint. ■

Remark In the terminology of [6, Section III.2, especially Proposition 1] the quotient space $D(T_{\max})/D(T_{\min})$, which is linearly isomorphic to $N_i \dotplus N_{-i}$ as in the von Neumann formula (6.2) below, can be endowed with the skew-hermitian form $[\cdot,\cdot]_n(b) - [\cdot,\cdot]_n(a)$ from (3.3) and (3.9), to define the complex symplectic space S of dimension $2d$. Then our Lemma 1 above asserts that each self-adjoint extension S of $T_{\min}$ has a domain $D(S) \subseteq D(T_{\max})$ determined by a Lagrangian subspace $L = D(S)/D(T_{\min}) \subseteq \mathsf{S}$, provided that L (or equally well $D(S)$) has the maximizing property (*iii*) above – and the converse holds. We note that the proof of Lemma 1 depends only on basic properties of closed symmetric operators and their adjoints in the complex Hilbert space $L^2((a,b):w)$.

The remaining lemmas, Lemma 2 and Lemma 3, involved in the proof of Theorem 1 show that a Lagrangian space $L \subseteq \mathsf{S}$ has the dimension d if and only if condition (*iii*) above holds, in which case a GKN set $\{\varphi_s : s = 1, 2, \ldots, d\}$ (mod $D(T_{\min})$) specifies a basis for L. The methods given here depend on the von Neumann unitary maps of N_i onto N_{-i} as in (6.1) below; these methods are to be compared with the algebraic treatment in [6, Section III.1, Theorem 1, and Lemma 2 before Theorem 2].

6 Statement and proof of Lemma 2

The deficiency subspaces $N_{\pm i}$ of the closed symmetric operator $T_{\min}$ are defined as, see (3.11)

$$N_{\pm i} := \{f \in D(T_{\max}) : T_{\max} f = \pm i f\};$$

from the definition of $T_{\max}$ it follows that

$$N_{\pm i} = \{y \in D(M_n) \cap L^2((a,b) : w) : M_n[y] = \pm i w y\}.$$

Since, from (4.1), we have $d_n^+ = d_n^- = d$ there exist unitary maps U, see [14, Section 10.5], with the properties

$$U : N_i \overset{\text{onto}}{\rightarrow} N_{-i} \quad \text{and} \quad U^* = U^{-1} : N_{-i} \overset{\text{onto}}{\rightarrow} N_i. \tag{6.1}$$

From the classical von Neumann formula, see [14, Section 14.4, Theorem 4], the domain $D(T_{\max})$ can be represented by the direct sum formula

$$D(T_{\max}) \equiv D(T_{\min}^*) = D(T_{\min}) \dot{+} N_i \dot{+} N_{-i}, \tag{6.2}$$

where the three linear manifolds on the right-hand side are linearly independent in $L^2((a,b) : w)$. From [14, Section 14.8] we have the result that an operator S is a self-adjoint extension of $T_{\min}$ if and only if the domain $D(S)$ is determined by, for some unitary map U with the properties (6.1),

$$D(S) = \{f \in D(T_{\max}) : f = h + \psi + U\psi \quad (h \in D(T_{\min}) \text{ and } \psi \in N_i)\}. \tag{6.3}$$

Let $\{\psi_r : r = 1, 2, \ldots, d\}$ be an orthonormal basis for the finite-dimensional subspace N_i of $L^2((a,b) : w)$; then $\{U\psi_r : 1, 2, \ldots, d\}$ is an orthonormal basis for N_{-i}.

Lemma 2 *Let S be a self-adjoint extension of T_{min} with $D(S)$ determined by* (6.3) *for some unitary map U; let $\{\psi_r\}$ and $\{U\psi_r\}$ be orthonormal bases for N_i and N_{-i} respectively as described above. Then the domain $D(S)$ may also be represented by*

$$D(S) = \{f \in D(T_{max}) : f = h + \sum\nolimits_{r=1}^{d} \alpha_r \tilde{\psi}_r \quad (h \in D(T_{min}) \text{ and } \alpha_r \in \mathbb{C}\ (r = 1, 2, \ldots, d))\} \tag{6.4}$$

where

$$\tilde{\psi}_r := \psi_r + U\psi_r \quad (r = 1, 2, \ldots, d). \tag{6.5}$$

Proof. It is sufficient to show that the two representations (6.3) and (6.4) of $D(S)$ are identical.

If $\psi \in N_i$ then for some set $\{\alpha_r : r = 1, 2, \ldots, d\}$

$$\psi = \sum_{r=1}^{d} \alpha_r \psi_r \quad \text{and} \quad U\psi = \sum_{r=1}^{d} \alpha_r U\psi_r$$

and so

$$\psi + U\psi = \sum_{r=1}^{d} \alpha_r \tilde{\psi}_r$$

and the required result follows. ∎

7 Statement and proof of Lemma 3

Lemma 3 *Let S be a self-adjoint extension of T_{min}; let $D(S)$ be represented in terms of the set $\{\tilde{\psi}_r : r = 1, 2, \ldots, d\}$ in terms of (6.4) and (6.5). Then $D(S)$ may also be represented by*

$$D(S) = \{f \in D(T_{max}) : [f, \tilde{\psi}_r]_n(b) - [f, \tilde{\psi}_r]_n(a) = 0 \quad (r = 1, 2, \ldots, d)\}. \tag{7.1}$$

Corollary 1 *Let U be a unitary map with the properties (6.1) and let the set $\{\tilde{\psi}_r : r = 1, 2, \ldots, d\}$ be determined by (6.5) for some orthonormal set $\{\psi_r : r = 1, 2, \ldots, d\}$ of N_i. Then*

(i) $\tilde{\psi}_r \in D(T_{max}) \quad (r = 1, 2, \ldots, d)$

(ii) the set $\{\tilde{\psi}_r\}$ is linearly independent in $D(T_{max})$ modulo $D(T_{min})$

(iii) $[\tilde{\psi}_r, \tilde{\psi}_s]_n(b) - [\tilde{\psi}_r, \tilde{\psi}_s]_n(a) = 0 \quad (r, s = 1, 2, \ldots, d)$

and $\{\tilde{\psi}_r\}$ is a GKN set for the pair $\{T_{min}, T_{max}\}$.

Remark This Corollary shows that the collection of GKN sets for any pair $\{T_{\min}, T_{\max}\}$ is not empty.

Proof. To prove Lemma 3 let $D(S)$ be the domain of S and define the linear manifold D of $L^2((a, b : w)$ by

$$D := \{f \in D(T_{\max}) : [f, \tilde{\psi}_r]_n(b) - [f, \tilde{\psi}_r]_n(a) = 0 \quad (r = 1, 2, \ldots, d)\}. \tag{7.2}$$

We note from the representation (6.4) of $D(S)$ and the definition (6.5) of the set $\{\tilde{\psi}_r\}$ that $\tilde{\psi}_r \in D(S)$ $(r = 1, 2, \ldots, d)$.

Now let $f \in D(S)$; then from the Green's formula

$$[f, \tilde{\psi}_r]_n(b) - [f, \tilde{\psi}_r]_n(a) = (Sf, \tilde{\psi}_r)_w - (f, S\tilde{\psi}_r)_w = 0 \ (r = 1, 2, \ldots, d)$$

and so $f \in D$. Thus $D(S) \subseteq D$.

To prove the converse result $D \subseteq D(S)$ we start by proving that

$$[f, g]_n(b) - [f, g]_n(a) = 0 \quad (f \in D, \ g \in D(S)). \tag{7.3}$$

From Lemma 2 we can write, since with $g \in D(S)$ it follows that $g = h + \sum_{r=1}^{d} \alpha_r \tilde{\psi}_r$,

$$\begin{aligned}[f,g]_n(b) - [f,g]_n(a) &= [f,h]_n(b) - [f,h]_n(a) \\ &\quad + \left[f, \sum_{r=1}^{d} \alpha_r \tilde{\psi}_r\right]_n (b) - \left[f, \sum_{r=1}^{d} \alpha_r \tilde{\psi}_r\right]_n (a).\end{aligned}$$

Now from the operator properties of Section 3 above we have $[f,h]_n(b) - [f,h]_n(a) = 0$. Also

$$\begin{aligned}\left[f, \sum_{r=1}^{d} \alpha_r \tilde{\psi}_r\right]_n (b) - \left[f, \sum_{r=1}^{d} \alpha_r \tilde{\psi}_r\right]_n (a) &= \sum_{r=1}^{d} \left\{[f, \alpha_r \tilde{\psi}_r]_n(b) - [f, \alpha_r \tilde{\psi}_r]_n(a)\right\} \\ &= \sum_{r=1}^{d} \overline{\alpha}_r \left\{[f, \tilde{\psi}_r]_n(b) - [f, \tilde{\psi}_r]_n(a)\right\} = 0\end{aligned}$$

since $f \in D$ and so (7.3) follows from the definition (7.2). Finally take $f \in D$; then for all $g \in D(S)$ we obtain, using the Green's formula and (7.3),

$$(T_{\max}, g)_w - (f, Sg)_w = [f,g]_n(b) - [f,g]_n(a) = 0;$$

hence $f \in D(S^*) = D(S)$. Thus $D \subseteq D(S)$ and the proof of Lemma 3 is complete.

To prove Corollary 1 we note that (i) and (ii) follow from the definition of the set $\{\tilde{\psi}_r : (r = 1, 2, \ldots, d)\}$.

The required property (iii) follows from $\tilde{\psi}_r \in D(S)$ and the result (7.1) of Lemma 3. ■

8 Proof of the GKN theorem

The proof of (2) of the Theorem follows at once from the previous definitions and results. For if S is a self-adjoint extension of $T_{\min}$ determined by a unitary map $U : N_i \to N_{-i}$ then let the set $\{\varphi_r : r = 1, 2, \ldots, d\}$ be defined, from (6.5), by

$$\varphi_r := \tilde{\psi}_r \quad (r = 1, 2, \ldots, d).$$

Then it follows from Corollary 1 and Lemmas 2 and 3 that the set $\{\varphi_r\}$ is a GKN set for the pair $\{T_{\min}, T_{\max}\}$, and also determines the given self-adjoint operator S in the form given by (4.3) and (4.4).

To prove (1) of the theorem suppose given the set $\{\varphi_r : (r = 1, 2, \ldots, d)\}$ satisfying conditions $(i), (ii)$ and (iii). Let the linear manifold D be defined by

$$D := \{f \in T_{\max} : [f, \varphi_r]_n(b) - [f, \varphi_r]_n(a) = 0 \quad (r = 1, 2, \ldots, d)\}. \tag{8.1}$$

The dimension of D modulo $D(T_{\min})$ is d; for we can suppose, without loss of generality, that $\varphi_r \in N_i \dot{+} N_{-i}$ for $r = 1, 2, \ldots, d$; this direct sum linear manifold $N_i \dot{+} N_{-i}$ is of dimension $2d$; with the conditions (8.1) imposing d linearly independent constraints this dimension is reduced to d.

Now define the linear manifold D' by

$$D' := \{f \in D(T_{\max}) : f = h + \sum_{r=1}^{d} \alpha_r \varphi_r \ (h \in D(T_{\min}) \text{ and } \alpha_r \in \mathbb{C} \ (r = 1, 2, \ldots, d))\};$$

from this definition and the linear independence of the set $\{\varphi_r\}$ it follows the dimension of D' modulo $D(T_{\min})$ is also d .

We show now that

$$D = D'. \tag{8.2}$$

Clearly if $f \in D'$ the $f \in D$ from the operator properties of $T_{\min}$ and $T_{\max}$, and the given properties of the set $\{\varphi_r\}$; thus $D' \subseteq D$. However $\dim D = \dim D'$ and so (8.2) must follow.

Finally we show that D' satisfies conditions $(i), (ii)$ and (iii) of Lemma 1.

Clearly from the definition of D' condition (i) is satisfied.

If $f, g \in D'$ then condition (ii) is satisfied since then $g \in D$ and so

$$\begin{aligned}
[f,g]_n(b) - [f,g]_n(a) &= [f, h + \sum\nolimits_{r=1}^{d} \alpha_r \varphi_r]_n(b) - [f, h + \sum\nolimits_{r=1}^{d} \alpha_r \varphi_r]_n(a) \\
&= \sum\nolimits_{r=1}^{d} \overline{\alpha}_r \{[f,\varphi_r]_n(b) - [f,\varphi_r]_n(a)\} = 0.
\end{aligned}$$

For condition (iii) let $f \in D(T_{\max})$ and suppose that

$$[f,g]_n(b) - [f,g]_n(a) = 0 \ (g \in D'); \tag{8.3}$$

since $\varphi_r \in D = D'$ for $r = 1, 2, \ldots, d$ it follows from (8.3) that

$$[f,\varphi_r]_n(b) - [f,\varphi_r]_n(a) = 0 \quad (r = 1, 2, \ldots, d)$$

and so $f \in D$ and hence $f \in D'$.

Thus all the conditions of Lemma 1 are satisfied and $D = D'$ is the domain of a self-adjoint extension of $T_{\min}$.

This completes the proof of the GKN Theorem. ■

9 Acknowledgement

The authors thank the Professors Yuri Lyubich and Israel Gohberg for the invitation to write this paper for the Glazman memorial volume, and for their advice and help during the preparation of the manuscript.

W.N. Everitt thanks his colleague Anton Zettl, Northern Illinois University, for the many years of collaboration that led to the results given in [9]; these results significantly influenced the contents of this paper.

References

[1] N.I. Akhiezer and I.M. Glazman. *Theory of linear operators in Hilbert space*; **I** and **II**. (Pitman and Scottish Academic Press, London and Edinburgh; 1980: translated, from the material prepared for the third Russian edition of 1978, by E.R. Dawson and edited by W.N. Everitt.)

[2] N. Dunford and J.T. Schwartz. *Linear operators* **II** : *Spectral theory.* (Wiley, New York; 1963.)

[3] W.N. Everitt. 'On the deficiency index problem for ordinary differential operators.' Lecture notes for the *1977 International Conference: Differential Equations*, Uppsala, Sweden; pages 62 to 81. (University of Uppsala, Sweden, 1977; distributed by Almquist and Wiskell International, Stockholm, Sweden.)

[4] W.N. Everitt. 'Linear ordinary quasi-differential expressions'. Lecture notes for the *Fourth International Symposium on Differential Equations and Differential Geometry*, Beijing, Peoples' Republic of China 1983; pages 1 to 28. (Department of Mathematics, University of Beijing, Peoples' Republic of China; 1986.)

[5] W.N. Everitt and L. Markus. 'Controllability of [r]-matrix quasi-differential equations. *J. of Diff. Equations.* **89** (1991), 95-109.

[6] W.N. Everitt and L. Markus. *Boundary value problems and symplectic geometry for ordinary differential and quasi-differential equations.* (In manuscript; to be submitted for publication in 1996.)

[7] W.N. Everitt and D. Race. 'Some remarks on linear ordinary quasi-differential expressions.' *Proc. London Math. Soc.* (3) **54** (1987), 300-320.

[8] W.N. Everitt and A. Zettl. 'Generalized symmetric ordinary differential expressions **I** : the general theory.' *Nieuw Arch. Wisk.* (3) **27** (1979), 363-397.

[9] W.N. Everitt and A. Zettl. 'Differential operators generated by a countable number of quasi-differential expressions on the real line.' *Proc. London Math. Soc.* (3) **64** (1991), 524-544.

[10] I.M. Glazman. 'On the theory of singular differential operators.' *Uspehi Math. Nauk.* **40** (1950), 102-135. (English translation in *Amer. Math. Soc. Translations* (1) **4** (1962), 331-372.)

[11] I. Halperin, 'Closures and adjoints of linear differential operators.' *Ann. of Math.* **38** (1937), 880-919.

[12] T. Kimura and M. Takahasi. 'Sur les operateurs differentiels ordinaires lineaires formellement autoadjoint.' *Funkcial. Ekvac.* **7** (1965), 35-90.

[13] K. Kodaira. 'On ordinary differential equations of any even order, and the corresponding eigenfunction expansions.' *Amer. J. of Math.* **72** (1950), 502-544.

[14] M.A. Naimark. *Linear differential operators*; **II**. (Ungar, New York; 1968: translated from the second Russian edition of 1966 by E.R. Dawson, and edited by W.N. Everitt.)

[15] D. Shin. 'Existence theorems for quasi-differential equations of order *n*.' *Doklad. Akad. Nauk SSSR.* **18** (1938), 515-518.

[16] D. Shin. 'On quasi-differential operators in Hilbert space.' *Doklad. Akad. Nauk. SSSR.* **18** (1938), 523-526.

[17] D. Shin. 'On the solutions of a linear quasi-differential equation of order n.' *Mat. Sb.* **7** (1940), 479-532.

[18] J. Weidmann. *Spectral theory of ordinary differential operators.* (Lecture Notes in Mathematics, **1259**; Springer-Verlag, Berlin and Heidelberg; 1987.)

[19] A. Zettl. 'Formally self-adjoint quasi-differential operators.' *Rocky Mountain J. Math.* **5** (1975), 453-474.

W.N. Everitt
School of Mathematics and Statistics
University of Birmingham
Edgbaston, Birmingham B15 2TT
England, UK.
e-mail: w.n.everitt@bham.ac.uk

L. Markus
School of Mathematics
University of Minnesota
Minneapolis
Minnesota 55455-0488, USA.
e-mail: markus@math.umn.edu

1991 *Mathematics Subject Classification.* Primary: 34B05 34L05. Secondary: 47B25 58F05.

Operator Theory
Advances and Applications, Vol. 98

METRIC CRITICAL POINT THEORY 2. DEFORMATION TECHNIQUES *

A. IOFFE and E. SCHWARTZMAN

The paper contains several results relating to deformation techniques in critical point theory for continuous functions on metric spaces, including two deformation theorems, an extension of the mountain pass theorem and a dense solvability theorem for potential set-valued operators associated with critical points of Lipschitz perturbations of quadratic functionals in Hilbert spaces.

1 Introduction

This is the second in a series of three papers devoted to the critical point theory for continuous functions on metric spaces. Two close versions of the theory were developed recently by Corvellec-Degiovanni-Marzocchi [6, 7] and Katriel [10], on the one hand, and by the authors [9], on the other. As in the classical "smooth" situation, deformation techniques are in the heart of the proofs of many basic results of metric critical point theory. The main difference is that gradient and gradient-like fields are no longer available and trajectories of deformational homotopies are no longer defined by flows associated with differential equations.

Still, suitable deformational techniques have been developed also in general metric spaces setting , sufficient to extend many basic results of the classical critical point theory. Moreover, these new techniques offer additional mechanisms to control corresponding deformations and this allows to construct proofs which, even in the general situation of a continuous function on a complete metric space, are much simpler and more direct than the available pseudo-gradient fields based proofs of parallel facts in the classical smooth setting.

The purpose of the paper is to present extensions of certain basic results of the critical point theory (deformation theorems, a minimax theorem and an existence theorem for critical levels of a Lipschitz perturbation of a quadratic functional), some containing new information even in the smooth situation, and to demonstrate the work and power of the new deformation techniques.

*This research was supported by Technion V.R.P. Fund and by B. and G. Greenberg Research Fund (Ottawa)

2 Deformation techniques - the basic structures

Throughout the paper, unless a more detailed specification is made, (X, ρ) is a complete metric space and f is a continuous function on f (although certain results, as say Theorem 1 and Lemma 1 below, are valid without the assumptions that X is complete and f is continuous).

Definition 1. By a *deformation* in X we mean a continuous mapping $\Phi: [0,1] \times X \mapsto X$ such that $\Phi(0,x) \equiv x$.

In other words, deformation is a continuous homotopy of the identity mapping. We do not assume any other properties as, say, the semigroup property $\Phi(s+t,x) = \Phi(s,\Phi(t,x))$ which is characteristic for deformation flows of differential equations.

We shall next introduce the basic concepts of the critical point theory for nondifferentiable functions.

Definition 2. Let δ be a positive number. We say that f is $\delta - $*regular* at x if there is a neighborhood U of x and a continuous mapping $\varphi(\lambda, x): [0,1] \times U \mapsto X$ such that

(a) $\varphi(\lambda, u) = u$ if and only if $\lambda = 0$;
(b) $f(u) - f(\varphi(\lambda, u)) \geq \delta\rho(u, \varphi(\lambda, u), \quad \forall\lambda \in [0,1],\ u \in U.$

We denote by $|df|(x)$ the upper bound of all such numbers δ; if no positive δ satisfying Definition 2 at x exists, we set $|df|(x) = 0$ and call x a (*lower*) *critical* point of f.

The explanation for the notation and the terminology lies in the fact that for a continuously differentiable function on a Banach space, $|df|(x)$ coincides with the norm of the derivative of f at x.

Given an open set $G \subset X$ and a closed set $C \subset X \backslash G$, we call a *modulus of regularity* of f on G with respect to C any positive nondecreasing function $\delta(t)$ on $(0,\infty)$ such that $|df|(x) \geq \delta(d(x,C))$ for all $x \in G$. The definition easily extends to the case when $C = \emptyset$ if we agree to set $\rho(x, \emptyset) \equiv 1$.

The following central fact was established in [9].

Theorem 1 (Basic Deformation Lemma) . *Let $G \subset X$ be an open set and $C \subset X \backslash G$ a closed set. Let f be a function on X such that there is a modulus of regularity $\delta(t)$ of f on G with respect to C. Then for any $\varepsilon \in (0,1)$ there is a deformation Φ on X with the following three properties:*

(a) $\Phi(\lambda, x) = x$ *if and only if* $\lambda = 0$ *or* $x \notin G$;

(b) $f(x) - f(\Phi(\lambda, x)) \geq \delta(\rho(x,C)/2)\rho(x, \Phi(\lambda, x)),\ \forall\lambda \in (0,1),\ \forall x \in X$;

(c) $\varepsilon\lambda\rho(x, X\backslash G) \geq \rho(x, \Phi(\lambda, x)),\ \forall\lambda \in (0,1),\ \forall x \in X$.

It seems that the property (c) has not appeared earlier in deformation theorems. It offers a convenient instrument to control the "speed" of approaching the boundary of the domain being deformed. We may refer to the proof of Lemma 2 below in which this condition plays an essential role.

The following proposition makes clearer the concept of the modulus of regularity which can be characterized as "Palais-Smale condition minus compactness".

Let $G \subset X$ be an open set. We say that f satisfies the *Palais-Smale condition* (the (PS)-condition) on G if any sequence $\{x_n\} \subset G$ such that the sequence $\{f(x_n)\}$ is bounded and $|df|(x_n) \to 0$ has a converging subsequence.

Consider the set $C(G, f)$ containing limits of all converging sequences $\{x_n\} \subset G$ such that $|df|(x_n) \to 0$.

Proposition 1 *Let f be continuous on the closure of G. The following two conditions are equivalent:*

(1) f satisfies the Palais-Smale condition on G;

(2) for any $\alpha \in R$, the intersection of the set $L(f,\alpha) = \{x : |f(x)| \le \alpha\}$ with C=$C(G,f)$ is compact and f has a modulus of regularity on $\{x \in G : |f(x)| < \alpha\}$ with respect to C.

Proof. (i) $\Rightarrow$ (ii). Compactness of the intersection of C with any $L(f,\alpha)$ is immediate from the (PS)-condition. Take a $t > 0$ and set

$$\delta_\alpha(t) = \inf\{|df|(t) : x \in G, |f(x)| < \alpha\ \rho(x, C) \ge t\}.$$

Clearly, $\delta_\alpha(t)$ does not decrease and $\delta_\alpha(t) > 0$ for otherwise we would have a sequence $\{x_n\} \subset \{x \in G, |f(x)| < \alpha\}$ converging by (PS) to an element outside of C.

(ii) $\Rightarrow$ (i). Let $\delta_\alpha(t)$ be a modulus of regularity of f on $\{x \in G, |f(x)| < \alpha\}$ with respect to C, that is $|df|(x) \ge \delta_\alpha(\rho(x, C))$, let $\{x_n\}$ belong to the set and $|df|(x_n) \to 0$. This means that the distance from x_n to C goes to zero. By compactness, $\{x_n\}$ must have a converging subsequence.

Having Proposition 1 in mind, we immediately get from Theorem 1 the following nonsmooth extension of a deformation theorem recently offered by Shafrir [14]

Proposition 2 *Suppose that f satisfies the (PS)-condition. Let C be the set of critical points of f, and let G be an open set not meeting C. Then there is a deformation Φ on X such that $\Phi(\lambda, x) = x$ for $x \in (X \backslash G)$ and $f(\Phi(\lambda, x)) < f(x)$ for $x \in G$.*

Henceforth we shall abbreviate the basic deformation lemma as BDL. A convenient particular form of BDL is given by the following proposition.

Proposition 3 (Deformation alternative) . *For any open set $G \subset X$ and any $\delta > 0$ the following alternative holds:*

- *either $|df|(x) < \delta$ for some $x \in G$,*

- *or for any $\varepsilon > 0$ there is a deformation Φ in X such that*

(a) $\Phi(\lambda, x) = x$ if and only if $\lambda = 0$ or $x \notin G$;

(b) $f(x) - f(\Phi(\lambda, x)) \ge \delta\rho(x, \Phi(\lambda, x))$, $\forall \lambda \in (0,1)$, $\forall x \in X$;

(c) $\varepsilon\lambda\rho(x, X \backslash G) \ge \rho(x, \Phi(\lambda, x))$, $\forall \lambda \in (0,1)$, $\forall x \in X$.

We shall next describe some further construction based on BDL which allows to obtain new deformations from given ones.

Given a deformation Φ on X, we define a mapping $\Phi_\infty : [0,\infty) \times X \mapsto X$ as follows

$$\Phi_\infty(\lambda, x) = \Phi(\{\lambda\}, \Phi^{[\lambda]}(x)),$$

where $[\lambda]$ is the integer part of λ, $\{\lambda\} = \lambda - [\lambda]$ and $\Phi(x) = \Phi(1, x)$.

Lemma 1 *Assume that f is bounded below on G and has a modulus of regularity on G with respect to a closed set $C \subset X\backslash G$. Let Φ be a deformation on X satisfying conditions (a)-(c) of BDL for some ε. Then*

(a) $\Phi_\infty(\lambda, x) \in G$ for all $\lambda \in [0,\infty)$ and all $x \in G$;

(b) for any $x \in G$ either $\lim_{\lambda\to\infty} \rho(\Phi_\infty(\lambda, x), C) = 0$ or $\lim_{\lambda\to\infty} \Phi_\infty(\lambda, x) = \lim_{m\to\infty} \Phi^m(x) = y(x)$ exists, not belonging to G, and there is a $\gamma > 0$ such that $f(y(x)) \leq f(\Phi^m(x)) - \gamma\rho(\Phi^m(x), y(x))$ for all m.

In particular, if C is empty or contains finitely many points, then the limit $y(x) \notin G$ exists for all $x \in X$.

Proof. The first statement (a) is immediate from the conditions (a) and (c) of BDL. A weaker version of (b), namely that either $\liminf_{m\to\infty} \rho(\Phi^m(x), C) = 0$ or $y(x) = \lim_{m\to\infty} \Phi^m(x)$ exists was proved in [9] (Proposition 3).

We notice that by the condition (c) of BDL,

$$\rho(\Phi_\infty(\lambda, x), C) \leq (1 + \varepsilon\{\lambda\})\rho(\Phi^{[\lambda]}(x), C) \leq (1+\varepsilon)\rho(\Phi^{[\lambda]}(x), C)$$

as $\lambda \to 0$. So to prove (b) we have to verify first that $\liminf_{m\to\infty} \rho(\Phi^m(x), C) = 0$ implies that actually $\lim_{m\to\infty} \rho(\Phi^m(x), C) = 0$.

Take an $\varepsilon > 0$, let $\delta(t) \geq \sigma > 0$ if $t \geq \varepsilon/2$ and choose an m such that

$$f(\Phi^m(x)) - \lim_{i\to\infty} f(\Phi^i(x)) \leq \varepsilon\sigma/2.$$

(The limit above obviously exists as the sequence $\{f(\Phi^i(x))\}$ is bounded below by (a) and does not increase by BDL.) We shall show that $\rho(\Phi^m(x), C) \leq \varepsilon$ for such m. Indeed, if $\rho(\Phi^m(x), C) > \varepsilon/2$ we take the first $n > m$ such that $\rho(\Phi^n(x), C) \leq \varepsilon/2$ (which can be obviously done). Then by BDL(b)

$$\begin{aligned}\rho(\Phi^n(x), \Phi^m(x)) &\leq \sigma^{-1}(f(\Phi^m(x) - f(\Phi^n(x))) \\ &\leq \sigma^{-1}(f(\Phi^m(x)) - \lim_{i\to\infty} f(\Phi^i(x))) \leq \varepsilon/2.\end{aligned}$$

Therefore $\rho(\Phi^m(x), C) \leq \varepsilon$ as claimed which means that $\rho(\Phi^n(x), C) \to 0$ as $n \to \infty$.

It follows from the condition (b) of BDL that in case $\liminf_{m\to\infty} \rho(\Phi^m(x), C) > 0$, we have $\lim_{m\to\infty} \Phi^m(x) = \lim_{\lambda\to\infty} \Phi_\infty(\lambda, x)$. So by Proposition 3 of [9], $\liminf_{m\to\infty} \rho(\Phi^m(x), C) > 0$ implies that the limit $\lim_{\lambda\to\infty} \Phi_\infty(\lambda, x) = \lim_{m\to\infty} \Phi^m(x) = y(x)$ exists. Clearly, $y(x)$ cannot belong to G. Take finally $\sigma > 0$ so small that $\rho(\Phi^m(x), C) > \sigma$ for all m and set $\gamma = \delta(\sigma)$. Then the same property (b) of BDL gives $f(\Phi^m(x)) - f(\Phi^{m+1}(x)) \geq \gamma\rho(\Phi^m(x), \Phi^{m+1}(x))$ and this immediately implies the desired estimate.

Lemma 2 *Let f be a continuous function on X, let $G \subset \{x : f(x) > c\}$ be an open set, and let $C \subset f^{-1}(c)$ be a finite set. Assume that there is a modulus of regularity $\delta(t)$ of f on G with respect to C, and let $\Phi(\lambda, x)$ be a corresponding deformation satisfying* BDL *with $\varepsilon = 1/2$. Set as above, $y(x) = \lim_{\lambda\to\infty} \Phi_\infty(\lambda, x)$, and let $Q = \{x : \ f(y(x)) = c\}$.*

Then for any $x \in Q$ we have

$$\lim_{\substack{\lambda\to\infty \\ u\to x}} \Phi_\infty(\lambda, u) = y(x).$$

In particular, $y(x)$ is continuous on Q.

Proof. First of all we note that the limit $y(x)$ exists for all x by Lemma 1. In view of condition (b) of Theorem 1 and the fact that $f(y(x)) = c$ for $x \in Q$, it is sufficient to verify that

$$\lim_{\substack{m\to\infty \\ u\to x}} \Phi^m(u) = y(x).$$

Note further that for any x we have either $x \in G$ or $x = y(x)$. We consider two cases.

(**a**) $y(x) \in C$. Then we can choose an $r > 0$ such that the ball of radius $2r$ around $y(x)$ does not contain any other point of C. Given a positive $t < r$, we can find a natural $m(t)$ such that for all $m \geq m(t)$

$$\rho(\Phi^m(x), y(x)) < t/2, \quad f(\Phi^m(x)) < c + t\delta(t/2)/4.$$

(In particular, for the case when $y(x) = x$, we have $m(t) = 0$ for all t.) We can then find a neighborhood $U(t)$ of x such that for all $u \in U(t)$

$$\rho(\Phi^{m(t)}(u), y(x)) < t/2, \quad f(\Phi^{m(t)}(u)) < c + t\delta(t/2)/4,.$$

Fix a $u \in U(t)$, and let $n = \sup\{m \geq m(t) : \ \rho(\Phi^m(u), y(x)) \leq t/2\}$. If $n < \infty$, then by condition (c) of Theorem 1

$$\rho(\Phi^{n+1}(x), \Phi^n(x)) \leq (1/2)\rho(\Phi^n(u), X\backslash G_1) \leq (1/2)\rho(\Phi^n(x), y(x)) \leq t/4.$$

On the other hand, if $\bar{m} \geq n+1$ is such that for all m between $n+1$ and $\bar{m}$ we have $t/2 \leq \rho(\Phi^m(u), y(x)) \leq t$, then

$$\rho(\Phi^{n+1}(u), \Phi^{\bar{m}+1}(u)) \leq \delta^{-1}(t/2)(f(\Phi^{n+1}(u)) - f(\Phi^{\bar{m}+1}(u)) \leq t/4,$$

and combining the inequalities, we see that $\rho(\Phi^{\bar{m}+1}(u), y(x)) \leq t$. It follows that $\rho(\Phi^m(u), y(x)) \leq t$ for all $m \geq m(t)$ and all $u \in U(t)$. From this the desired relation follows immediately.

(**b**) $y(x) \notin C$. Find $\sigma > 0$ and $0 < \tau < \rho(y(x), C)$ such that $|df|(v) \geq \sigma$ if $\rho(v, y(x)) \leq \tau$. Now, given an $t \in (0, \tau/2)$, we apply the same argument as in the first case with σ instead of $\delta(t)$.

3 Deformation outside of an isolated critical set

The theorem to be proved in this section is the simplest consequence of the lemmas in the previous section. This is a deformation theorem of a kind which is typical in proofs of theorems of Ljusternik-Schnirelman type (e.g. [1, 11, 12]). We refer to [6] where a similar result (using a slightly different definition of a critical point) for continuous functions on metric spaces appeared for the first time.

Definition 3. We say that c is a *critical level* of f if there is a sequence $\{x_n\}$ such that $f(x_n) \to c$ and $|df|(x_n) \to 0$. It is said that f satisfies the $(PS)_c$*−condition* (the Palais-Smale condition at level c) if any such sequence contains a converging subsequence.

As in Proposition 1, we get the following.

Proposition 4 *Assume that c is an isolated critical level of f, and let $C \subset \{x : f(x) = c\}$ be the collection of all critical points at level c. Assume also that f satisfies the $(PS)_c$-condition. Then C is compact (possibly empty)and there are $\alpha > 0$ and a modulus of regularity $\delta(t)$ of f on*

$$G_\alpha = \{x : |f(x) - c| < \alpha, \; x \notin C\}$$

with respect to C.

Proof. Let $\alpha > 0$ be such that there are no other critical levels in $[c - \alpha, c + \alpha]$. If f does not have a modulus of regularity on G_α then there are a $t > 0$ and a sequence $\{x_n\} \subset G_\alpha$ such that $\rho(x_n, C) \geq t$ and $|df|(x_n) \to 0$. We may assume that $f(x_n)$ converges to a certain $\beta \in [c - \alpha, c + \alpha]$. By $(PS)_c$, $\{x_n\}$ (or a subsequence) converges to a certain x which is at the distance at least t from C and which is a critical point of f. As c is the only critical level in the segment, it follows that $\beta = c$ and we have a critical point at the level c not belonging to C, which contradicts to the definition of C.

Theorem 2 *Let f and C be as in Proposition 4. Then for any $r > 0$ there are $\alpha > 0$ and a deformation $F(\lambda, x)$ in X such that $F(\lambda, x) = x$ if $x \notin G_\alpha$, and $f(F(\lambda, x)) < c$ for some $\lambda > 0$(depending on x)if $|f(x) - c| \leq \alpha/2$ and $\rho(x, C) \geq 2r$.*

Proof. By Proposition 4 there is an $\alpha > 0$ and a modulus of regularity $\delta(t)$ of f on G_α with respect to C.

Let $\gamma > 0$ be such that $\delta(t) \geq \gamma$ if $t \geq r/2$. Taking a smaller α if necessary we can be sure that $2\alpha < \gamma r$. Let now $\Phi(\lambda, x)$ be a deformation satisfying conditions (a) - (c) of BDL with $G = G_\alpha$, and let $\Phi_\infty(\lambda, x)$ and $\Phi(x)$ be defined by $\Phi(\lambda, x)$ as in Lemma 1. By the lemma for any $x \in G_\alpha$ either $\lim_{m\to\infty} \rho(\Phi^m(x), C) = 0$ or $\lim_{m\to\infty} \Phi^m(x) = y(x)$ exists and $f(y(x)) = c - \alpha$. We claim that only the second possibility may hold for $x \in G_\alpha$ with $\rho(x, C) \geq r$.

Indeed, let $\lim_{m\to\infty} \rho(\Phi^m(x), C) = 0$ for such x. Then (as C is compact by Proposition 4) $f(\Phi^m(x)) \geq c$ for all m. On the other hand, let $\bar{m}$ be the first index for which $\rho(\Phi^{\bar{m}}(x), C) \leq r/2$. Then

$$\alpha < \gamma r/2 \leq \gamma\rho(x, \Phi^{\bar{m}}(x)) \leq f(x) - f(\Phi^{\bar{m}}(x)) < \alpha,$$

which is a contradiction.

Let $\varphi(x)$ be a continuous function on X taking values between zero and one, equal to one if $\rho(x, C) \geq 2r$, and equal to zero if $\rho(x, C) \leq r$. Further let $\mu(t)$ be a homeomorphism from $[0, 1)$ onto $[0, \infty)$ with $\mu(0) = 0$. Set

$$F(\lambda, x) = \begin{cases} \Phi_\infty(\mu(\lambda\varphi(x)), x), & \text{if } \lambda\varphi(x) < 1, \\ y(x) & \text{if } \lambda\varphi(x) = 1. \end{cases} \tag{1}$$

As $f(y(x)) = c - \alpha$ for all x for which $\varphi(x) > 0$, Lemma 2 guarantees continuity of F. On the other hand, if $\varphi(x) = 1$, then $\rho(x, C) \geq 2r$ and therefore $\Phi_\infty(\lambda, x) \to y(x)$, so that $f(F(\lambda, x)) < c$ for some λ. This completes the proof.

Remark. Unlike in the classical case of differentiable functions, we cannot prove that our deformations act homeomorphically (as e.g. in [2, 4]) but this property does not seem to be needed in proofs of various minimax theorems.

4 Deformation near a critical point

The following theorem (which is basically a nonsmooth extension of Second Deformation Lemma of [4]) is a slightly strengthened version of a result which was proved by Corvellec [3] in his study of nonsmooth analogues of the Morse theory. We shall give an alternative and much shorter proof based on BDL (and Lemma 2).

Theorem 3 *Let $f(z) = c$, and let there be an open neighborhood V of z such that f has a modulus of regularity $\delta(t)$ on $V \backslash \{z\}$ with respect to z. Then, given an $r > 0$, there is a $r' \in (0, r)$ and a deformation F in X such that*

(a) $F(\lambda, x) = x$ if either $\rho(x, z) \geq r$, or $x = z$, or $\lambda = 0$;

(b) if $\rho(x, z) < r'$, then either $f(F(1, x)) < c$ or $F(1, x) = z$.

Proof. We can assume (taking a smaller r if necessary) that $B(z, 2r) \subset V$. Let $\gamma = \delta(r)$, take an $0 < \alpha < \gamma r/4$ and set

$$G_1 = \{x : \ \rho(x, z) < 2r, \ \ c < f(x) < c + \alpha\}.$$

Then $\delta(t)$ is a modulus of regularity of f on G_1 with respect to $\{z\}$. Let $\Phi : [0, 1] \times X \to X$ be a corresponding deformation on X satisfying conditions of Theorem 1 with $\varepsilon = 1/4$, Let Φ_∞ be defined by Φ as in Lemma 1, and let $y(x) = \lim_{\lambda \to \infty} \Phi_\infty(\lambda, x)$. By Lemma 1 the limit exists for any $x \in X$ and $y(x) \notin G_1$ (which for $x \in G_1$ means that either $f(y(x)) = c$ or $\rho(y(x), z) \geq 2r$).

By Lemma 2 for any $t \leq r$ there is a neighborhood $U(t)$ of z such that $\rho(y(x), z) \leq t$ if $x \in U(t)$. (recall that $m(t) = 0$ in our case as $y(z) = z$). This means that $f(y(x)) = c$ for $x \in U(t)$. Take an $r' < r/2$ such that the ball of radius $2r'$ around z belongs to $U(r)$. Let $\mu(t)$ be a homeomorphism of $[0, 1)$ onto $[0, \infty)$ with $\mu(0) = 0$, and let $\varphi(x)$ be a continuous function $\varphi(x)$ equal to zero if $\rho(x, z) \geq 2r'$, equal to one if $\rho(x, z) \leq r'$ and taking values between zero and one everywhere. Denote by $F_1(\lambda, x)$ the deformation defined by the same formula (1) of the proof of Theorem 2 (but of course with Φ_∞, φ and μ of this proof).

By Lemma 2, F_1 is continuous and by the choice of r', $f(F_1(x)) = c$ if $\rho(x,z) \le r'$, and $F_1(1,x) = x$ if either $x = z$, or $\rho(x,z) \ge 2r'$.

Now let

$$G_2 = \{x : \ |f(x) - c| < \alpha, \ 0 < \rho(x,z) < 2r\}.$$

Again, $\delta(t)$ is a modulus of regularity of f on G_2 w.r.t. $\{z\}$. Let $F_2(\lambda, x)$ be the corresponding deformation satisfying Theorem 1 (say, with $\varepsilon = 1$). Setting

$$F(\lambda, x) = F_2(\lambda, F_1(\lambda, x)),$$

we get the desired deformation.

5 A minimax theorem – existence of critical levels

In the theorem below we consider

- a topological space W and a closed subspace $V \subset W$ having the property that there is a continuous function on W equal to zero on V and positive on $W \backslash V$;
- a continuous mapping $q : \ V \mapsto X$;
- the collection $\mathcal{P}$ of all continuous mappings $p : \ W \mapsto X$ such that $p|_V = q$.

Set

$$c = \inf_{p \in \mathcal{P}} \sup_{w \in W} f(p(w)).$$

Theorem 4 *Suppose that $|c| < \infty$ and for any $p \in \mathcal{P}$ there is a $w \in W \backslash V$ such that $f(p(w)) \ge c$. Then c is a critical level of f.*

Moreover, given a set $A \subset \{x : \ f(x) \ge c\}$ such that $p(W \backslash V) \bigcap A \ne \emptyset$ for any $p \in \mathcal{P}$, then for any $\varepsilon > 0$ and any $p \in \mathcal{P}$ with $\mathbf{f}(p) = \sup_{w \in W} f(p(w)) < c + \varepsilon/4$, there is an $x \in X$ such that

$$|f(x) - c| < \varepsilon, \ \rho(x, p(W)) < 2\sqrt{\varepsilon}, \ \rho(x, A) \le 2\sqrt{\varepsilon} \tag{1}$$

and

$$|df|(x) \le \sqrt{\varepsilon}.$$

Thus, f has a critical point at the level c belonging to A if, in addition, f satisfies the $(PS)_c$-condition.

Proof. We need to prove only the second statement. The method of the proof is similar to that of Theorem 3. Assuming the contrary, we find an $\varepsilon > 0$ and a $p \in \mathcal{P}$ with $\sup_w f(p(w)) < c + \varepsilon/4$ such that $|df|(x) > \sqrt{\varepsilon}$ for all x satisfying (1). Set

$$G_1 = \{x : \ c < f(x) < c + \varepsilon, \ \rho(x, p(W)) < 2\sqrt{\varepsilon}, \ \rho(x, A) < 2\sqrt{\varepsilon}\}.$$

Then $\delta(t) \equiv \sqrt{\varepsilon}$ is a modulus of regularity of f on G_1 with respect to the empty set. If Φ is a corresponding deformation in X satisfying conditions (a)-(c) of Deformation Alternative

(Proposition 3), then by Lemma 1, $y(x) = \lim_{\lambda\to\infty} \Phi_\infty(\lambda, x)$ exists and does not belong to G_1 for all x.

Take a continuous function $\varphi(x)$ on X assuming values between zero and one and such that

$$\begin{array}{ll}\varphi(x)=1, & \text{if } \ |f(x)-c|\le\varepsilon/4,\ \ \rho(x,p(W))\le\sqrt{\varepsilon}/2,\ \ \rho(x,A)\le\sqrt{\varepsilon}/2;\\ \varphi(x)=0, & \text{if } \ \text{either } |f(x)-c|\ge\varepsilon/2,\ \text{ or } \rho(x,p(W))\ge\sqrt{\varepsilon},\ \text{ or } \rho(x,A)\ge\sqrt{\varepsilon},\end{array}$$

and set

$$F_1(x)=\begin{cases}\Phi_\infty(\mu(\varphi(x)),x), & \text{if } \ \varphi(x)<1,\\ \quad y(x) & \text{if } \ \varphi(x)=1,\end{cases}$$

$\mu(t)$ being a homeomorphism of $[0,1)$ onto $[0,\infty)$ with $\mu(0)=0$.

By Proposition 3(b), $\rho(x, F_1(x)) \le \varepsilon^{-1/2}(f(x)-c)$ for all x. Therefore (as $y(x)\notin G_1$), we conclude that $f(y(x)) = c$ if $\varphi(x)>0$. By Lemma 2, it follows that F_1 is continuous. Set $p_1(w) = F_1(p(w))$. Then $\rho(p_1(w), p(w)) \le \sqrt{\varepsilon}/4$ so that

$$f(p_1(w)) = c \qquad \text{if} \qquad \rho(p_1(w), A) \le \sqrt{\varepsilon}/2. \tag{2}$$

Set further

$$G_2 = \{x:\ |f(x)-c| < \varepsilon/8,\ \rho(x,p(W)) < \sqrt{\varepsilon}/2,\ \rho(x,A) < \sqrt{\varepsilon}/2\},$$

and let $\gamma(w)$ be a continuous function on W equal to zero on V and positive outside of V.

As $\delta(t)\equiv\sqrt{\varepsilon}$ is a modulus of regularity of f also on G_2, we can find a corresponding deformation F_2 in X satisfying Deformation Alternative. Set

$$p'(w) = F_2(\gamma(w), F_1(p(w))).$$

Then p' is continuous and for $w\in V$ we have $p'(w) = F_1(p(w)) = p(w)$ (the latter because $f(p(w)) \le c$ for such w by the very definition of c), hence $p'(\cdot)\in\mathcal{P}$.

Now observe that in case $\rho(p_1(w), A) \ge \sqrt{\varepsilon}/2$ we have $p'(w) = p_1(w) \notin A$. If, on the other hand, $\rho(p_1(w), A) < \sqrt{\varepsilon}/2$, and $w \in W\setminus V$, then $f(p'(w) < f(p_1(w) = c$ as $\gamma(w)>0$. Therefore $p'(w)\notin A$ also in this case and $p(W\setminus V)\bigcap A = \emptyset$ in contradiction with the assumption.

Remark. The theorem is another extension of the famous mountain pass theorem of Ambrosetti-Rabinowitz. For earlier (smooth) versions of the result see e.g [2, 8]. For the first nonsmooth extension (for Lipschitz f) see [15]. In all these papers W is assumed to be a compact space. We also mention a manuscript by Rybarska, Tsachev and Karastanov [13] in which a theorem similar to Theorem 4 was proved for l.s.c. functions (again with a compact W).

In some earlier publications an additional assumption that

$$c > \sup_{w\in V} f(p(w))$$

was used. Under this assumption the proof of the theorem can be noticeably simplified. Indeed, in this case we can assume without loss of generality that $f(q(v)) < c - 2\varepsilon$ for $v \in V$. Then a single deformation on the set

$$G = \{: \ |f(x) - c| < \varepsilon, \ \rho(x, p(W)) < 2\sqrt{\varepsilon}, \ \rho(x, A) < 2\sqrt{\varepsilon}\}$$

satisfying Proposition 3 leads to the desired contradiction.

6 Nonsmooth perturbations of indefinite quadratic functionals

If we analyse the proof of Theorem 4, we can easily observe that we do not need all continuous mappings from W into X to spot a critical level. What is actually needed is that $F_2(\gamma(w), F_1(p(w)))$ be an element of $\mathcal{P}$ for any $p \in \mathcal{P}$. Application of this observation requires a more attentive look at the "construction" of the deformation in the proof of BDL. (We use the quotation mark as the proof given in [9] is based on transfinite induction.) We shall briefly describe it here.

Under the assumptions of BDL, for any $x \in G$ there is a neighborhood $U_x \in G$ and mapping $\varphi_x(\lambda, u) : [0,1] \times U_x \mapsto X$ satisfying the conditions (a), (b) of Definition 1. Take a family $\{x_i, \ i \in I\} \subset G$ such that $\{U_{x_i}, \ i \in I\}$ is a locally finite covering of G, and let $\{\mu_{x_i}, \ i \in I\}$ be a partition of unit subordinated to this covering. Fix a certain linear ordering in I. Then, given an $x \in G$ and $\lambda \in [0,1]$, we set $I(x) = \{i \in I : \ x \in U_{x_i}\} = \{i_1, ..., i_n\}$ (with the order of indices induced by the chosen ordering in I), then set recoursively

$$y_0 = x, ..., y_{k+1} = \varphi_{i_k}(\lambda \mu_{i_k}(x), y_k), ...$$

and define $\Phi(\lambda, x)$ as y_n.

It follows in particular that in case when X is in addition a linear space and there is a subspace $E \subset X$ such that $u - \varphi_x(\lambda, u) \in E$ for any x and any $u \in U_x$, then $x - \Phi(\lambda, x) \in E$ for all $\lambda \in [0,1]$ and all $x \in X$.

In this section we study functionals of the form

$$f(x) = (1/2)(Lx|x) + H(x)$$

on a Hilbert space X. Here $(\cdot|\cdot)$ stands for the inner product in X. Throughout the section we assume that

($\mathbf{A}_1$) L is a bounded self-adjoint linear operator in X;

($\mathbf{A}_2$) H is locally Lipschitz and, moreover, the set-valued mapping $x \mapsto \partial H(x)$ has the following compactness property: for any $r > 0$ there is a norm compact set $Q_r \subset X$ such that $\partial H(x) \subset Q_r$ if $\|x\| \leq r$.

Here $\partial H(x)$ stands for the *Clarke's generalized gradient* of H at x. Recall (see [5]) that in our case when H is Lipschitz near every point, it is defined as follows:

$$\partial H(x) = \{y : \ (y|h) \leq H^\circ(x; h), \ \forall \, h\},$$

where

$$H^\circ(x;h) = \limsup_{\substack{u\to x\\ t\to 0}} t^{-1}(H(u+th) - H(u)).$$

It was mentioned in [6, 9] that for a locally Lipschitz function F the condition $0 \notin \partial f(x)$ is sufficient for regularity with $|df|(x) \geq \rho(0, \partial f(x))$. The condition however is not necessary. An example of a regular point with $0 \in \partial f(x)$ was found in [13]. Still, the condition remains the main analytic device to look for critical points in the absence of differentiability.

We have the decomposition $X = X^- \oplus X^+$ where X^- and X^+ are mutually orthogonal L-invariant subspaces of X such that $(Lx|x) < 0$ for every nonzero $x \in X^-$ and $(Lx|x) \geq 0$ for every $x \in X^+$. We shall also consider increasing systems $\mathcal{E} = \{E_n\}$ of finite dimensional L-invariant subspaces of X which is complete in the sense that $\bigcap E_n^\perp = \{0\}$.

Lemma 3 *For any $r > 0$ and any $\delta > 0$ there is a number $n(r,\delta)$ such that whenever $n \geq n(r,\delta)$, $\|x\| \leq r$, $x \in E_n$, $\rho(0, Lx + \partial H(x)) > \delta$, there is an $h \in E_n$, $\|h\| = 1$ with*

$$(Lx|h) + H^\circ(x;h) < -\delta.$$

Proof. Let $x \in E_n$ be such that $\|x\| \leq r$ and $\rho(0, Lx + \partial H(x)) > \delta$. Take an $\alpha > 0$ with $\rho(0, Lx + \partial H(x)) - \alpha > \delta$ and choose an integer $n = n(r,\delta)$ to ensure that $Q_r \subset E_n + B_\alpha$ (here Q_r comes from (A_2)) for $n > n(r,\delta)$ (which is possible as $\mathcal{E}$ is a complete sequence and every E_n is L-invariant). Choose a nonzero $e \in X$ such that

$$f^\circ(x;e) = (Lx|e) + H^\circ(x;e) = \max\{(u|e) :\ u \in Lx + \partial H(x)\} < -(\delta+\alpha)\|e\|.$$

We have $e = h + e'$, where $h \in E_n$ and $e' \in E_n^\perp$. Then (by convexity of $f^\circ(x;\cdot)$)

$$\begin{array}{rcl} f^\circ(x;h) & \leq & f^\circ(x;e) + f^\circ(x;-e') \\ & < & -(\delta+\alpha)\|e\| + \max\{|(u|e')| :\ u \in Lx + Q_r\} \\ & < & -(\delta+\alpha)\|e\| + \alpha\|e'\| \leq -\delta\|e\| \leq -\delta\|h\|. \end{array}$$

It follows that $h \neq 0$, so dividing both parts of the inequality by $\|h\|$ and writing h instead of $h/\|h\|$, we get $\|h\| = 1$ and

$$Lx + H^\circ(x;h) = f^\circ(x;h) < -\delta.$$

This completes the proof.

As an immediate corollary of the lemma (and BDL), we have the following refinement of the deformation alternative for our functional.

Proposition 5 *We assume ($\mathbf{A}_1$) and ($\mathbf{A}_2$). Then for any bounded open set $G \subset X$ and any $\delta > 0$ the following alternative holds:*

(a) either there is an $x \in G$ with $\rho(0, Lx + \partial H(x)) \leq \delta$;

(b) or there is an integer number $n(G,\delta)$ and a deformation Φ in X with the properties that for any $n \geq n(r,\delta)$, $\Phi(\lambda, x) \in E_n$ whenever $x \in G \bigcap E_n$ and the conditions (a) - (c) of Deformation Alternative are satisfied.

Proof. Assume that (a) does not hold. Take an $r > 0$ such that $G \subset rB$, and choose an $n(r, \delta)$ according to Lemma 3. Now for any $x \in G$ we choose a vector $h_x \in X$ $\|h_x\| = 1$, a small neighborhood U_x of x and a positive number λ_x to make sure that

$$f(u + \lambda h_x) - f(u) < -\lambda\delta, \qquad \forall u \in U_x, \ \forall \lambda \leq \lambda_x$$

as follows:

- if $x \in E_n$ and $n = n(r, \delta)$, we take $h_x \in E_n$ using Lemma 3;

- if $x \in E_n \backslash E_{n-1}$ for some $n > n(r, \delta)$, we, as above take $h_x \in E_n$ using Lemma 3 and choose U_x small enough to guarantee that $U_x \bigcap E_{n-1} = \emptyset$;

- if $x \notin \bigcup E_n$ then we choose $h_x \in \bigcup E_n$ to make sure that $(Lx|h_x) + H^\circ(x; h_x) < -\delta$ (which is possible by definition of Clarke's directional derivative as $\bigcup E_n$ is dense in X) and then, if E_n is the minimal subspace to which h_x belongs, we choose U_x as above to guarantee that $U_x \bigcap E_{n-1} = \emptyset$.

Set $\varphi_x(\lambda, u) = u + \lambda\lambda_x h_x$ for $u \in U_x$, $\lambda \in [0, 1]$. Then (U_x, φ_x) satisfies the conditions of Definition 2 for any $x \in G$ and, in addition, for $n \geq n(r, \delta)$, we have $\varphi_x(\lambda, u) - u \in E_n$, provided $U_x \bigcap E_n \neq \emptyset$. Therefore the deformation $\Phi(\lambda, x)$ constructed as described above satisfies $\Phi(\lambda, x) \in E_n$ if $x \in E_n$, $n \geq n(r, \delta)$.

Remark. Necessity to consider deformations as in Proposition 5 comes from the fact that ordinary deformations like those considered in the preceding section are not sufficient in case of an indefinite functional. Deformations acting in finite dimensional invariant subspaces are needed to guarantee the "linking" property needed to establish estimates like (4) below (cf. e.g. [4]).

Theorem 5 *We assume* $(\mathbf{A}_1)$ *and* $(\mathbf{A}_2)$ *as well as*

$(\mathbf{A}_3)$ $\quad \sup\{\lambda < 0 : \ \lambda \in \sigma(L)\} = -\mu < 0$ *($\sigma(L)$ being the spectrum of L);*

$(\mathbf{A}_4)$ $\quad \limsup_{\|x\| \to \infty} \|x\|^{-2} H(x) < \mu$;

$(\mathbf{A}_5)$ $\quad$ *H is bounded below.*

Then there is a $c \in R$ such that for every $\varepsilon > 0$ the inclusion $0 \in Lx + \partial H(x) + \varepsilon B$ has a solution satisfying $|f(x) - c| < \varepsilon$.

The following is a direct corollary of the theorem:

Corollary 1 *Suppose, in addition to the assumptions of the theorem that*

$$\lim_{\|x\| \to \infty} \|x\|^{-1} H(x) = \infty$$

Then the inclusion

$$y \in Lx + \partial H(x)$$

has a solution for every y of a dense subset of X.

To prove the corollary we only need to observe that under the additional assumption of the corollary the function $H(x) - (y|x)$ satisfies $(\mathbf{A}_4)$, $(\mathbf{A}_5)$ if so does H.

Proof of the theorem. Let S_r^- denote the intersection of the r-sphere with X^-. Choose a sufficiently big $r > 0$ to ensure that

$$\sup_{S_r^-} f(x) < \inf_{X^+} f(x) \tag{3}$$

which is possible by $(\mathbf{A}_3)$ and $(\mathbf{A}_4)$(as $\inf_{X^+} f(x)$ by (A_5)). Further let $\mathcal{P}$ be the collection of continuous mappings from $B_r^- = B_r \bigcap X^-$ into X with the following properties:

(p_1) $p(x) = x$ for $x \in S_r^-$;

(p_2) $p(B_r^-)$ is a bounded set for any $p \in \mathcal{P}$;

(p_3) there is a number $n(p)$ (depending on p) such that $p(x) \in E_n$ for all $x \in B_r^- \bigcap E_n$ if $n \geq n(p)$.

Set finally

$$c = \inf_{p \in \mathcal{P}} \sup_{x \in B_r^-} f(p(x)).$$

We notice first that

$$c \geq \inf_{X^+} f(x). \tag{4}$$

Indeed, by (p_3) the restriction of any $p \in \mathcal{P}$ to the corresponding E_n is a mapping from $B_r^- \bigcap E_n$ into E_n which is the identity map on the boundary of $B_r^- \bigcap E_n$. The standard topological argument now implies that $p(B_r^- \bigcap E_n) \bigcap (E_n \bigcap X^+) \neq \emptyset$ which implies (4). We also observe that $\mathcal{P}$ is nonempty as it contains the identity.

The situation we are now in is similar to what we had in Theorem 4 (with $A = \{x : f(x) \geq c\}$, B_r^- as W and S_r^- as V.) We therefore have to show that for any $\varepsilon > 0$ and any $p \in \mathcal{P}$ with $f(p(x)) \leq \varepsilon^2$ for all $x \in B_r^-$ there is an x such that

$$|f(x) - c| < \varepsilon^2, \ \rho(x, p(B_r^-)) < 2\varepsilon$$

and $\rho(0, Lx + \partial H(x)) \leq \varepsilon$. But in the present case

$$c > \sup_{x \in S_r^-} f(x)$$

which allows to simplify the proof as it was explained in the remark concluding the previous section with the only change that to choose the necessary deformation we have to apply Proposition 5 to make sure that the deformation satisfies the properties (p_1) – (p_3).

References

[1] A. Ambrosetti, *Critical Points and Nonlinear Variational Problems*, Societé Math. de France, publ. 49, Suppl. Bull. S.M.F. Tome 120, 1992

[2] H. Brezis and L. Nirenberg, Remarks on finding critical points, Comm. Pure Appl. Math. **44** (1991), 939-963.

[3] J.-N. Corvellec, Morse theory for continuous functionals, J. Math. Ana. Appl. **196** (1995), 1050-1072.

[4] Chang, Kung Ching, *Infinite Dimensional Morse Theory and Multiple Solution Problem*, Birkhäuser 1993.

[5] F.H. Clarke, *Optimization and Nonsmooth Analysis*, Wiley 1983.

[6] J.-N. Corvellec, M. Degiovanni and M. Marzocchi, Deformation properties of continuous functionals and critical point theory, Topological Methods Nonlinear Anal. **1** (1993), 151–171.

[7] M. Degiovanni and M. Marzocchi, A critical point theory for nonsmooth functionals, Ann. Mat. Pura Appl., Ser. IV **167** (1994), 73-100.

[8] N. Ghoussoub and D. Preiss, A general mountain pass principle for locating and classifying critical points, Ann. Inst. H. Poincaré–Analyse Non-linéaire **6**, no. 5 (1989), 321–330.

[9] A. Ioffe and E. Schwartzman, Metric critical point theory 1. Morse regularity and homotopic stability of a minium, J. Math. Pures Appl. **75** (1996), 125-153.

[10] G. Katriel, Mountain pass theorems and global homeomorphism theorems, Ann. Inst. H. Poincaré—Analyse Non-linéaire **11** (1994), 189-209.

[11] R.S. Palais, Critical point theory and minimax principle, in *Global Analysis*, Proc. Symp. Pure Math. Berkley, 1968, vol XV, Amer. Math. Soc. (1970), 185-212.

[12] P.H. Rabinowitz, *Minimax Methods in Critical Point Theory with Applications to Differential Equations*, CBMS Reg. Conf. Ser. Math., vol 65, Amer. Math. Soc. 1986.

[13] N.K. Rybarska, T.V. Tsachev and M.I. Karastanov, Speculation about mountains, preprint 1995.

[14] I. Shafrir, A deformation lemma, C.R. Acad. Sci. Paris **313** (1991), 599–602.

[15] Shi Shuzhong, Ekeland's variational principle and the mountain pass lemma, Acta Mat. Sinica **4** (1985), 348-355.

Department of Mathematics, Technion, Haifa 32000, Israel

1991 Mathematical Subject Classification 58E05

Operator Theory
Advances and Applications, Vol. 98

ERGODIC METHODS FOR THE CONSTRUCTION OF HOLOMORPHIC RETRACTIONS

VICTOR KHATSKEVICH, SIMEON REICH and DAVID SHOIKHET

Let D be a bounded convex domain in a complex Banach space, and let F be a holomorphic self-mapping of D with a nonempty fixed point set. In this paper we study the flow generated by the mapping $I - F$ on D, and use the asymptotic behavior of its Cesàro averages to construct a holomorphic retraction of D onto the fixed point set of F.

Let X be a complex Banach space and let D be a bounded convex domain in X. A mapping $F : D \to X$ is said to be holomorphic in D if the complex Fréchet derivative of F exists at each point of D. If $F : D \to D$ is a self-mapping of D, we denote by $Fix\ F$ the set of fixed points of F in D, i.e.

$$Fix\ F = \{x \in D : x = F(x)\}.$$

Many mathematicians established the existence of holomorphic retractions of D onto $Fix\ F$ in different situations, e.g. in the finite-dimensional case, in Hilbert space and in Banach spaces. See, for example, [R], [H-S], [B], [G-R], [VE1], [VE2], [VJP1], [AA], [K-S], [VJP2], [M-V], [SI] and [D]. Once the existence of a holomorphic retraction is established, it follows that $Fix\ F$ is a complex analytic connected submanifold of D [C].

Some of the most detailed investigations so far seem to be due to J.-P. Vigué [VJP1, VJP2] and to P. Mazet and J.-P. Vigué [M-V]. To construct the retraction, they considered the Cesàro averages

$$C_n = \frac{1}{n}\sum_{k=0}^{n-1} F^k,$$

and in particular proved the following two results.

1. If $X = \mathbf{C}^n$ and $Fix\ F \neq \emptyset$, then there is a subsequence $\{C_{n_k}\}$ which converges to a mapping ϕ such that its iterates $\{\phi^n\}$ converge to a retraction $\rho : D \to Fix\ F$.

2. If X is an arbitrary Banach space, $Fix\ F \neq \emptyset$ and for some point $a \in Fix\ F$ the following condition holds:

$$(*) \qquad Ker(I - F'(a)) \oplus Im(I - F'(a)) = X,$$

then there exists an integer p such that the sequence $\{(C_p)^n\}$ strongly converges to a retraction $\rho : D \to Fix\ F$ as n tends to infinity.

The analogue of assertion 1 was also established for a reflexive space with weak convergence of the Cesàro averages [VJP1], [M-V].

A simple proof of the existence of a retraction was pointed out even for unbounded domains by Do [D], but his arguments are not constructive.

The following nicer situation, when the retraction can be obtained by simple iteration, was described by E. Vesentini [VE1], [VE2] (see also [M-V] and [A]).

Let $\sigma(A)$ denote the spectrum of a linear operator $A : X \to X$, and let Δ be the open unit disk in $\mathbf{C}$.

As above, assume that $Fix F \neq \emptyset$, and suppose that for each a in $Fix\ F$, either

(i) $\sigma(F'(a)) \subset \Delta;$

or

(ii) $\sigma(F'(a)) = \{1\} \cup \{\Delta \cap \sigma(F'(a))\},$

1 is an isolated point of $\sigma(F'(a))$, and

(**) 1 is a pole of the resolvent of $F'(a)$.

Then the sequence of iterates $\{F^n\}$ converges locally uniformly over D (see, for example, [F-V]) to a retraction $\rho : D \to Fix\ F$.

As a matter of fact, it can be shown that in the presence of (ii) condition (**) is equivalent to condition (*).

However, the situation becomes more complicated if $\sigma(F'(a))$ containes a point $\lambda \in \partial\Delta$, $\lambda \neq 1$. For example, the rotation $F(x) = e^{i\phi}$, $0 < \phi < 2\pi$, has a unique fixed point $x = 0$ which is regular, i.e. 1 is not contained in $\sigma(F'(0))$, but the iterates $\{F^n(x)\}$ do not converge to 0 when $x \neq 0$.

In the present paper we construct retractions by using the solutions of Cauchy problems. This method works in more general situations. More precisely, assuming only (*), we construct a continuous semigroup $\{F_t : t \geq 0\}$ of holomorphic self-mappings of D the averages of which converge to a retraction onto $Fix\ F$. Moreover, if F has a regular fixed point $a \in D$, then $\{F_t\}$ converges to a in the topology of local uniform convergence, as t tends to infinity.

We begin with a definition.

Definition. *A mapping $f : D \to X$ is said to be a generator of a flow on D if there exists a strongly continuous one-parameter semigroup $\{F_t\}_{t\geq 0}$ on D with $F_0 = I$, the identity on X, such that for all $x \in D$ the strong limit*

$$\lim_{t\to 0^+} \frac{x - F_t(x)}{t} = f(x). \tag{1}$$

It can be shown (see [K-R-S], [R-S]) that if the generator f is a holomorphic bounded mapping, then $F_t x$ solves the Cauchy problem

$$\begin{aligned} \frac{\partial F_t(x)}{\partial t} &= -f(F_t(x)) \\ F_0(x) &= x, \quad x \in D, \end{aligned} \tag{2}$$

and hence is the unique semigroup satisfying (1). Moreover, the convergence in (1) is locally uniform over D. In this case we will write $F_t(\cdot) = \exp(-tf)(\cdot)$.

Now we formulate our result.

THEOREM. *Let D be a bounded convex domain in a complex Banach space X, and let $F: D \to D$ be a holomorphic self-mapping such that $Fix\ F \neq \emptyset$. Then*

1) The mapping $f = I - F$ is the generator of a flow on D;

2) If for some $a \in Fix\ F$,

(*) $$Ker(I - F'(a)) \oplus Im(I - F'(a)) = X,$$

then the averages

$$\frac{1}{t}\int_0^t \exp(-s(I-F))\, ds$$

converge in the local uniform topology over D to a holomorphic retraction of D onto $Fix\ F$.

3) Moreover, if 1 is a regular point of the resolvent $(\lambda I - F'(a))^{-1}$ for some $a \in Fix\ F$, then a is unique and $\exp(-t(I-F))$ converges to a in the topology of local uniform convergence over D, as $t \to \infty$.

Returning to the simple example of the rotation $F(x) = ix$, $x \in \Delta \subset \mathbf{C}$, we see here that indeed

$$\exp(-t(I-F))(x) = e^{-t(1-i)}x$$

converges to the fixed point of F (the origin) uniformly in D as $t \to \infty$.

To prove our theorem we need some lemmas.

LEMMA 1 [R – S] *Let D be a bounded convex domain in X, and let $f: D \to X$ be a bounded holomorphic mapping which has a null point in D. Then f is the generator of a flow in D if and only if the equation*

$$x + rf(x) = z, \; z \in D, \tag{3}$$

has a unique solution $x_r(z)$ for each $r > 0$ which holomorphically depends on $z \in D$.

LEMMA 2 *Let D be a bounded domain in X and let $F: [0,\infty) \times D \to D$ be continuous. If for $\tau > 0$ and $x \in D$ there exists the strong limit*

$$G(x) = \lim_{n\to\infty} \frac{1}{n\tau}\int_0^{n\tau} F(t)(x)\, dt,$$

then the strong limit

$$\lim_{s\to\infty} \frac{1}{s}\int_0^s F(t)(x)\, dt$$

also exists and equals $G(x)$.

Proof. Consider

$$G_s(x) = \frac{1}{s}\int_0^s F(t)(x)\, dt.$$

For each $\varepsilon > 0$ there exists $m > 0$ such that

$$\|G(x) - G_{n\tau}(x)\| < \frac{\varepsilon}{2} \quad \text{for all} \quad n \geq m.$$

Setting $M = \sup_{t\in\mathbf{R}^+} \|F(t)(x)\|$ and $s_0 = \max\left[\frac{M\cdot 4\tau}{\varepsilon}, m\tau\right]$ we obtain for all $s > s_0$,

$$\left\|\frac{1}{s}\int_0^s F(t)(x)\,dt - \frac{1}{\left[\frac{s}{\tau}\right]\tau}\int_0^{\left[\frac{s}{\tau}\right]\tau} F(t)(x)\,dt\right\| =$$

$$\left\|\frac{1}{s}\int_0^{\left[\frac{s}{\tau}\right]\tau} F(t)(x)\,dt + \frac{1}{s}\int_{\left[\frac{s}{\tau}\right]\tau}^{s} F(t)(x)\,dt - \frac{1}{\left[\frac{s}{\tau}\right]\tau}\int_0^{\left[\frac{s}{\tau}\right]\tau} F(t)(x)\,dt\right\| \leq$$

$$\left\|\left(\frac{1}{s} - \frac{1}{\left[\frac{s}{\tau}\right]\tau}\right)\int_0^{\left[\frac{s}{\tau}\right]\tau} F(t)(x)\,dt\right\| + \left\|\frac{1}{s}\int_{\left[\frac{s}{\tau}\right]\tau}^{s} F(t)(x)\,dt\right\|$$

$$\leq \frac{\left|\left[\frac{s}{\tau}\right]\tau - s\right|}{s}2M < \frac{\varepsilon}{2}$$

because $\left[\frac{s}{\tau}\right] < \frac{s}{\tau} < \left(\left[\frac{s}{\tau}\right] + 1\right)$ and hence $\left|\left[\frac{s}{\tau}\right]\tau - s\right| < \tau.$

Thus for such s we have

$$\|G(x) - G_s(x)\| \leq \|G(x) - G_q(x)\| + \|G_q(x) - G_s(x)\| < \varepsilon$$

where $q = \left[\frac{s}{\tau}\right]\tau$.

LEMMA 3 *Let $F(t,x) = F_t(x)$, $x \in D$, $t \in \mathbf{R}^+$, be a one-parameter continuous semigroup defined on $\mathbf{R}^+ \times D$ such that $F_t(x) \in D$ for all $x \in D$, $t \geq 0$. Suppose that for some $\tau > 0$ and all $x \in D$ the Cesàro averages*

$$(C_\tau)_n(x) = \frac{1}{n}\sum_{k=0}^{n-1} F_{k\tau}(x)$$

are well defined and that the strong limit $C_\tau(x) = \lim_{n\to\infty}(C_\tau)_n(x)$ exists. Then the averages

$$G_s(x) = \frac{1}{s}\int_0^s F_t(x)\,dt$$

strongly converge in D.

Proof. It follows by Lemma 2 that it is sufficient to prove our assertion for $s = n\tau$. For such s $(= n\tau)$ we have

$$G_{n\tau}(x) = \frac{1}{n\tau}\int_0^{n\tau} F(t,x)\,dt = \frac{1}{n}\int_0^n F(\tau u, x)\,du = \frac{1}{n}\sum_{k=0}^{n-1}\int_k^{k+1} F(\tau u, x)\,du =$$

$$\frac{1}{n}\sum_{k=0}^{n-1}\int_0^1 F(\tau(t+k), x)\,dt = \int_0^1 \frac{1}{n}\sum_{k=0}^{n-1} F(\tau k, F(\tau t, x))\,dt = \int_0^1 (C_\tau)_n(x_t)\,dt,$$

where $x_t = F(\tau t, x) \in D$. Thus,

$$G(x) = \lim_{n\to\infty} G_{n\tau}(x) = \int_0^1 C_\tau(x_t)\,dt$$

exists.

LEMMA 4 $[R-S]$ *Let D be a convex bounded domain in X, and let $f : D \to X$ be a bounded holomorphic mapping which generates a flow on D. Suppose that f has a null point $a \in D$ such that the following condition holds:*

$$Ker\ f'(a) \oplus Im\ f'(a) = X.$$

Then
1) the null point set N of the mapping f in D is a connected analytic submanifold;
2) there exists $\delta > 0$ such that for all $t \in (0, \delta)$

$$Fix\ \exp(-tf) = N.$$

Hence this fixed point set does not depend on $t \in (0, \delta)$.

Finally we formulate our last lemma, which is the key to the proof of our Theorem.

LEMMA 5 *Let $A : X \to X$ be a bounded linear operator such that $\sigma(A) \subseteq \Delta$. If $1 \in \sigma(A)$ and $B = I - A$, then*
1) for each $t > 0$,

$$\sigma(\exp(-tB)) = \{1\} \cup \{\sigma(\exp(-tB)) \cap \Delta\};$$

if $1 \notin \sigma(A)$, then

$$\sigma(\exp(-tB)) \subset \Delta.$$

2) If $Ker\ A \oplus Im\ A = X$, then for $\delta > 0$ small enough

$$Ker(I - \exp(-tB)) \oplus Im(I - \exp(-tB)) = X$$

for all $0 < t < \delta$.

Proof. Part 1) follows from a simple direct computation, while 2) is a consequence of Lemma 4.

Proof of the Theorem. 1) Let D and $F : D \to D$ satisfy the hypotheses. Then $f = I - F$ is bounded and has a null point in D. For arbitrary $r > 0$, setting $t = r(r+1)^{-1}$, we have that equation (3), $x + r(x - F(x)) = z$, is exactly the equation $x = (1-t)z + tF(x)$, which has a unique solution $x = x_r(z)$ for each $z \in D$ and $t \in (0, 1)$ by the Earle-Hamilton theorem because D is convex. Since this solution may be obtained by iteration, it is holomorphic in $z \in D$. Thus by Lemma 1, we have that $I - F$ is the generator of the flow $F_t = \exp(-tf) : D \to D$.

2) Now let $a \in Fix\ F$ and $A = F'(a)$. It follows by the chain rule and the Cauchy inequalities that $\|A^n\| \le M < \infty$ for all $n = 1, 2, \ldots$. Thus $\sigma(A) \subseteq \Delta$. In addition, it is easy to see that $[\exp(-t(I - F))]'_x(a) = \exp(-t(I - A))$.

Note also that it follows from the uniqueness of the solution of the Cauchy problem that $Fix\ F \subseteq Fix\ \exp(-t(I - F))$ for each fixed $t > 0$. Therefore Lemmas 4 and 5 imply that for each fixed $t > 0$ which is small enough, the mapping $F_t = \exp(-t(I - F))$ satisfies all the conditions of Lemma 5.2 in [M-V] in some equivalent norm. Thus we obtain that for such $t > 0$ the sequence $\{F_{nt}\} = \{\exp(-tn(I - F))\}_{n=0}^{\infty}$ converges in the topology of local uniform convergence over D to the mapping $\rho_t : D \to D$ which is a retraction onto the set $Fix\ F_t = Fix\ F$. Now Lemma 3 implies the existence of the strong limit of the averages

$$G(x) = \lim_{t\to\infty} \frac{1}{t} \int_0^t \exp(-s(I-F))(x)\, ds.$$

Next we intend to show that the mapping G is actually a retraction of D onto $Fix\ F$. Indeed, it is evident that $G\,|_{Fix\ F} = Id|_{Fix\ F}$. Simple calculations show that for each $t > 0$ we have $G \circ F_t = G$ and hence $G \circ \rho_t = G$ by continuity. But Lemma 4 shows that for $t \in (0, \delta)$, ρ_t is a retraction onto $FixF$. Thus $G(D) = Fix\ F$ and $G^2 = G$. Assertion 2) is proved.

3) Assume now that $1 \notin \sigma(F'(a))$ for all $a \in Fix\ F$. Then $Re\ \lambda \geq \varepsilon > 0$ for all $\lambda \in \sigma((I - F'(a)))$, i.e., the operator $F'(a) - I$ is a strongly dissipative linear operator. Thus by Theorem VII.2.1 in [D-K] the trivial solution $u(t) = a$ of the differential equation

$$\frac{du(t)}{dt} = F(u(t)) - u(t)$$

is uniformly asymptotically stable in some neighborhood U of the point a, i.e., for each $x \in U$ other solutions $u(t,x)$ converge uniformly in $x \in U$ to a. But by the uniqueness of the solution of the Cauchy problem, $u(t,x) = F_t(x) = \exp(-t(I-F))(x)$, and it follows by the Vitali property of holomorphic mappings (see [F-V]) that $\{F_t\}$ converges to a in the topology of local uniform convergence over D. The Theorem is proved.

Remark. As a matter of fact, we have proved a more general assertion, namely the following ergodic type theorem for semigroups with holomorphic generators.

Let D be a convex bounded domain and let $f : D \to X$ be a bounded holomorphic mapping which generates a flow on D. Suppose that the null point set N of f in D is not empty, and for some $a \in N$ the following hypotheses hold:

$$Ker\ f'(a) \oplus Im\ f'(a) = X, \tag{4}$$

and $\sigma(f'(a))$ does not contain any points of the imaginary axis, except perhaps the origin. Then the averages

$$\frac{1}{t}\int_0^t \exp(-sf)\, ds$$

converge in the local uniform topology over D to a holomorphic retraction of f onto N.

Acknowledgement. The second author was partially supported by the Fund for the Promotion of Research at the Technion and by the Technion VPR Fund - M. and M.L. Bank Mathematics Research Fund.

References

[A] Abate, M., Horospheres and iterates of holomorphic maps, Math. Z. **198** (1988), 225-238.

[AA] Abd-Alla, M., L'ensemble des points fixes d'une application holomorphe dans un produit fini de boules-unités d'espaces de Hilbert est une sous-variété banachique complexe, Ann. Mat. Pura Appl. **153** (1988), 63-75.

[B] Bedford, E., On the automorphism group of a Stein manifold, Math. Ann. **266** (1983), 215-227.

[C] Cartan, H., Sur les rétractions d'une variété, C. R. Acad. Sci. Paris **303** (1986), 715-716.

[D-K] Daleckiǐ, Yu.L. and Kreǐn, M.G., Stability of Solutions of Differential Equations in Banach Spaces, AMS, Providence, Rhode Island, 1970.

[D] Do Duc Thai, The fixed points of holomorphic maps on a convex domain, Annales Polonici Mathematici **56** (1992), 143-148.

[E-H] Earle, C.J. and Hamilton, R.S., A fixed-point theorem for holomorpic mappings, Proc. Symp. Pure Math., Vol. **16**, AMS, Providence, RI, (1970), 61-65.

[F-V] Franzoni, T. and Vesentini, E., Holomorphic Maps and Invariant Distances, Math. Studies **40**, North-Holland, Amsterdam, 1980.

[G-R] Goebel, K. and Reich, S., Uniform Convexity, Hyperbolic Geometry, and Nonexpansive Mappings, Marcel Dekker, New York and Basel, 1984.

[H-S] Heath, L.F. and Suffridge, T.J., Holomorphic retracts in complex n-space, Illinois Journal of Math. **25** (1981), 125-135.

[K-R-S] Khatskevich, V., Reich, S. and Shoikhet, D., Ergodic type theorems for nonlinear semigroups with holomorphic generators, Technion Preprint Series No. MT-999, 1994, in *Recent Developments in Evolution Equations,* Pitman Research Notes in Math., Vol. **324**, Longman, Harlow, 1995, 191-200.

[K-S] Kuczumow, T. and Stachura, A., Iterates of holomorphic and k_D-nonexpansive mappings in convex domains in $\mathbf{C}^n$, Adv. Math. **81** (1990), 90-98.

[L] Lempert, L., Holomorphic retracts and intrinsic metrics in convex domains, Anal. Math. **8** (1982), 257-261.

[M-V] Mazet, P. and Vigué, J.-P., Points fixes d'une application holomorphe d'un domaine borné dans lui-même, Acta Math. **166** (1991), 1-26.

[R] Rudin, W., The fixed-point sets of some holomorphic maps, Bull. Malaysian Math. Soc. **1** (1978), 25-28.

[R-S] Reich, S. and Shoikhet, D., Generation theory for semigroups of holomorphic mappings in Banach spaces, Technion Preprint Series No. MT-1007, 1995.

[SI] Shafrir, I., Coaccretive operators and firmly nonexpansive mappings in the Hilbert ball, Nonlinear Analysis **18** (1992), 637-648.

[VE1] Vesentini, E., Iterates of holomorphic mappings, Uspekhi Mat. Nauk **40** (1985), 13-16.

[VE2] Vesentini, E., Su un teorema di Wolff e Denjoy, Rend. Sem. Mat. Fis. Milano **53** (1983), 17-25 (1986).

[VJP1] Vigué, J.-P., Sur les points fixes d'applications holomorphes, C. R. Acad. Sci. Paris Ser. **303** (1986), 927-930.

[VJP2] Vigué, J.-P., Fixed points of holomorphic mappings in a bounded convex domain in $\mathbf{C}^n$, Proc. Symp. Pure Math., Vol. **52** (1991), Part 2, 579-582.

Victor Khatskevich and David Shoikhet
Department of Applied Mathematics
International College of Technology
20101 Karmiel, Israel

Simeon Reich
Department of Mathematics
The Technion - Israel Institute of Technology
32000 Haifa, Israel

AMS Classification Numbers 32H05, 46G20, 47H10, 47H20.

Operator Theory
Advances and Applications, Vol. 98

NEW PROOF OF TRACE FORMULAS IN CASE OF CLASSICAL STURM-LIOUVILLE PROBLEM

B.M. LEVITAN

A generalized trace formula for Sturm–Liouville operators is proved using the method of the wave equation.

1 Introduction.

The usual meaning of the trace as the sum of eigenvalues has no sense for differential operators, because the eigenvalues (and, consequently, their sum) increase infinitely. Therefore, in the case of differential operators, it is appropriate to use various generalizations of the notion of trace.

For ordinary differential operators, the eigenvalues satisfy some sharp asymptotic relations. By subtracting from every eigenvalue several terms of its asymptotic and summing the differences, we can obtain a convergent series. It is naturally to declare the sum of such series as *generalized trace.*

The problem consists in expressing the generalized trace as a functional of coefficients of the corresponding differential operator.

Let us consider a simple example, namely, the classical Sturm-Liouville problem:

$$-y'' + q(x)y = \lambda y, \qquad 0 \leq x \leq \pi \tag{1.1}$$

$$y'(0) = hy(0), \qquad y'(\pi) = -Hy(\pi), \tag{1.2}$$

where $q(x)$ is a real and twice differentiable function, h and H are finite real numbers. Denote by $\lambda_0 < \lambda_1 < \cdots < \lambda_n < \cdots$ the eigenvalues of the problem (1.1)-(1.2). The classical asymptotic formula

$$\lambda_n = n^2 + c + O(\frac{1}{n^2}), \tag{1.3}$$

is valid where

$$c = \frac{1}{\pi}(h + H + \frac{1}{2}\int_0^\pi q(x)dx).$$

It follows from (1.3) that the series.

$$\sum_{n=0}^{\infty}(\lambda_n - c - n^2)$$

converges. Denote its sum by S_λ. In 1953 Gelfand and Levitan proved [1] that if

$$h = H = 0, \qquad \int_0^\pi q(x)dx = 0 \tag{1.4}$$

then

$$S_\lambda = \frac{1}{4}[q(0) + q(\pi)]. \tag{1.5}$$

Remark. It fact, the condition (1.4) is no restriction at all, since we can write (1.1) in the form

$$-y'' + [q(x) - a] = (\lambda - a)y$$

and set

$$a = \frac{1}{\pi}\int_0^\pi q(x)dx.$$

In this article we propose a new proof of formula (1.5). Our method is based on the wave equation method and corresponding Tauberian theorems. We refer the reader for details about this two objects to our papers [2], [3].

The paper consists of four sections and two Appendices. Section I is introductory. In Section II some important auxiliary results are proved. In Section III formula (1.5) is proved. In Section IV we prove the H. Hochstadt trace formula. In Appendix I we describe our Tauberian theorem and some its corollaries which are important for the present paper. Appendix II contains a proof of an elementary theorem about the connection between Riesz's means and usual convergence of series.

2 Auxiliary results.

In accordance with the wave equation method we consider, along with problem (1.1) - (1.2), the mixed problem

$$\frac{\partial^2 u}{\partial t^2} = \frac{\partial^2 u}{\partial x^2} - q(x)u, \qquad 0 \le x \le \pi, \quad 0 \le t < \infty. \tag{2.1}$$

$$u|_{t=0} = f(x), \qquad \frac{\partial u}{\partial t}|_{t=0} = 0, \tag{2.2}$$

$$(\frac{\partial u}{\partial x} - hu)|_{x=0} = 0, \quad (\frac{\partial u}{\partial x} + Hu)|_{x=\pi} = 0, \tag{2.3}$$

where $f(x)$ is a smooth function which satisfies the boundary conditions

$$f'(0) - hf(0) = 0, \quad f'(\pi) + Hf(\pi) = 0.$$

The solution of problem (2.1)–(2.3) can be obtained in two ways:

1) Using the eigenfunction expansion of problem (1.1)–(1.2), the result being

$$u(x,t) = \sum_{n=0}^{\infty} a_n \cos \mu_n t \, \varphi_n(x), \tag{2.4}$$

where $\varphi_n(x)$ are normalized eigenfunctions of the problem (1.1)–(1.2), λ_n are the corresponding eigenvalues, $\mu_n = \sqrt{\lambda_n}$,

$$a_n = \int_0^\pi f(s)\varphi_n(s)\,ds.$$

2) Using the method of characteristiques, the result being

$$u(x,t) = \frac{1}{2}[f(x+t) + f(x-t)] + \frac{1}{2}\int_{x-t}^{x+t} w(x,t,s)f(s)\,ds. \tag{2.5}$$

Formula (2.5) follows immediately from the classical results, if $0 \le x-t \le x+t \le \pi$. In the case $x - t < 0$ or $x + t > \pi$, formula (2.5) can be obtained as follows.

First, we continue the potential $q(x)$ outside the interval $[0,\pi]$ as an even function, and then (in the case $h = H = 0$) we continue the initial function $f(x)$ outside the interval $[0,\pi]$ also as an even function. [1]

Using the uniqueness of the solution of problem (2.1)-(2.3) we obtain the following important identity

$$\sum_{n=0}^{\infty} a_n \cos \mu_n t \, \varphi_n(x) = \frac{1}{2}[f(x+t) + f(x-t)] + \frac{1}{2}\int_{x-t}^{x+t} w(x,t,s)f(s)\,ds \tag{2.6}$$

where $x \in [0,\pi]$, $t \ge 0$ and $q(x)$ and $f(x)$ are continuations of $q(x)$ and $f(x)$ described above.

Further results of this paper are based on formula (2.5) and special Tauberian theorems.

Denote by $g_\epsilon(t)$, $0 \le t \le \epsilon < \frac{1}{2}\pi$ a function with the following properties:

a) $g_\epsilon(t)$ is even, i.e., $g_\epsilon(-t) = g_\epsilon(t)$,

b) $g_\epsilon(t)$ is smooth,

c) $supp\; g_\epsilon(t) \subseteq (-\epsilon,\epsilon)$.

Let $\tilde{g}_\epsilon(\mu)$ be the cos–Fourier transform of $g_\epsilon(t)$:

$$\tilde{g}_\epsilon(\mu) = \int_0^\epsilon g_\epsilon(t)\cos \mu t \;\, dt.$$

According to the inversion formula, we have

$$g_\epsilon(t) = \frac{2}{\pi}\int_0^\infty \tilde{g}_\epsilon(\mu)\cos \mu t \;\, d\mu \tag{2.7}$$

It follows from a) and b) that $\tilde{g}_\epsilon(\mu)$ is fast decreasing as $\mu \to \infty$.

[1] The case $h \neq 0$ and $H \neq 0$ can be treated similarly, but the continuation of the initial function $f(x)$ is more complicated (for details see our papers [2], [3]).

Let us multiply identity (2.6) by $g_\epsilon(t)$ and integrate over $[0,\infty)$. Changing the order of summation and integration on the left-hand side and the order of integrations on the right-hand side (we remind that $\tilde{g}_\epsilon(\mu)$ decreases fast), we obtain

$$\int_0^\pi f(s)\,(\sum_{n=0}^\infty \tilde{g}_\epsilon(\mu_n)\varphi_n(x)\varphi_n(s))\,ds$$

$$=\frac{1}{2}\int_{x-\epsilon}^{x+\epsilon} g_\epsilon(x-s)\,f(s)\,ds+\frac{1}{2}\int_{x-\epsilon}^{x+\epsilon} f(s)[\int_{|x-s|}^{\epsilon} w(x,t,s)g_\epsilon(t)\,dt\,]\,ds.S \tag{2.8}$$

The formula (2.8) can be used immediately, if $0<t\le\epsilon$ and $x\in[\epsilon,\pi-\epsilon]$ (then $x-t\in[0,\pi]$ and $x+t\in[0,\pi]$). Since $f(x)$ is arbitrary, we obtain the first crucial formula:

$$\sum_{n=0}^\infty \tilde{g}_\epsilon(\mu_n)\;\varphi_n(x)\,\varphi_n(s)$$

$$=\begin{cases} \frac{1}{2}g_\epsilon(x-s)+\frac{1}{2}\int_{|x-s|}^{\epsilon} w(x,t,s)\,g_\epsilon(t)\,dt, & |x-s|\le\epsilon,\quad \epsilon\le x\le\pi-\epsilon \\ 0 & |x-s|\ge\epsilon \end{cases} \tag{2.9}$$

The case $s=x$ is of special interest:

$$\sum_{n=0}^\infty \tilde{g}_\epsilon(\mu_n)\,\varphi^2(x)=\frac{1}{2}g_\epsilon(0)+\frac{1}{2}\int_0^\epsilon w(x,t,x)\,g_\epsilon(t)\,dt,\qquad x\in[\epsilon,\pi-\epsilon] \tag{2.10}$$

Suppose now that $0\le x\le\epsilon$. If $x<t$, then $x-t<0$, and we have to take into account the boundary condition at $x=0$. As we have already explained, the boundary condition (2.3) (in case $h=0$) can be replaced by even continuation of the initial function $f(x)$. Since $x\le\epsilon$, we may now suppose that $f(s)=0$ for $s\ge\epsilon$. It follows from (2.8)

$$\int_0^\epsilon f(s)\,[\sum_{n=0}^\infty \tilde{g}_\epsilon(\mu_n)\,\varphi_n(x)\varphi_n(s)]\,ds=$$

$$=\frac{1}{2}\int_{x-\epsilon}^{0} g_\epsilon(x-s)\,f(s)\,ds+\frac{1}{2}\int_0^\epsilon g(x-s)\,f(s)\,ds+$$

$$+\frac{1}{2}\int_{x-s}^{0} f(s)[\int_{|x-s|}^{\epsilon} w(x,t,s)\,g_\epsilon(t)\,dt]\,ds+\frac{1}{2}\int_0^\epsilon f(s)\,[\int_{|x-s|}^{\epsilon} w(x,t,s)\,g_\epsilon(t)\,dt]\,ds=$$

$$=\frac{1}{2}\int_0^{\epsilon-x} g_\epsilon(x+s)f(s)\,ds+\frac{1}{2}\int_0^\epsilon g_\epsilon(x-s)\,f(s)\,ds+$$

$$+\frac{1}{2}\int_0^{\epsilon-x} f(s)[\int_{x+s}^{\epsilon} w(x,t,-s)\,g_\epsilon(t)\,dt]\,ds+\frac{1}{2}\int_0^\epsilon f(s)[\int_{|x-s|}^{\epsilon} w(x,t,s)\,g_\epsilon(t)\,dt]\,ds.$$

Since $f(s)$ is an arbitrary function, we have

$$\sum_{n=0}^{\infty} \tilde{g}_\epsilon(\mu_n)\, \varphi_n(x)\varphi_n(s) =$$

$$= \begin{cases} \frac{1}{2}g_\epsilon(x+s) + \frac{1}{2}g_\epsilon(x-s) + \\ +\frac{1}{2}\int_{x+s}^{\epsilon} w(x,t,-s)\, g_\epsilon(t)\, dt + \frac{1}{2}\int_{|x-s|}^{\epsilon} w(x,t,s)\, g_\epsilon(t)\, dt, & 0 \le s \le \epsilon - x \\ \frac{1}{2}g_\epsilon(x-s) + \frac{1}{2}\int_{|x-s|}^{\epsilon} w(x,t,s)\, g_\epsilon(t)\, dt & \epsilon - x < s \le \epsilon \end{cases}$$

For $s = x$, we obtain

$$\sum_{n=0}^{\infty} \tilde{g}_\epsilon(\mu_n)\, \varphi_n^2(x)$$

$$= \begin{cases} \frac{1}{2}g_\epsilon(0) + \frac{1}{2}\int_0^\epsilon w(x,t,x)\, g_\epsilon(t)\, dt + \\ +\frac{1}{2}g_\epsilon(2x) + \frac{1}{2}\int_{2x}^{\epsilon} w(x,t,-x)\, g_\epsilon(t)\, dt, & 0 \le x \le \frac{1}{2}\epsilon \\ \frac{1}{2}g_\epsilon(0) + \frac{1}{2}\int_0^{\epsilon} w(x,t,x)\, g_\epsilon(t)\, dt, & \frac{1}{2}\epsilon \le x \le \epsilon. \end{cases} \tag{2.11}$$

An analogous formula is valid in the interval $\pi - \epsilon \le x \le \pi$:

$$\sum_{n=0}^{\infty} \tilde{g}_\epsilon(\mu_n)\, \varphi_n^2(x)$$

$$= \begin{cases} \frac{1}{2}g_\epsilon(0) + \frac{1}{2}\int_0^\epsilon w(x,t,x)\, g_\epsilon(t)\, dt, & \pi - \epsilon \le x \le \pi - \frac{1}{2}\epsilon \\ \frac{1}{2}g_\epsilon(0) + \frac{1}{2}\int_0^{\epsilon} w(x,t,x)\, g_\epsilon(t)\, dt + \\ +\frac{1}{2}g_\epsilon(2\pi - 2x) + \frac{1}{2}\int_{2\pi-2x}^{\epsilon} w(x,t,2\pi - x)\, g_\epsilon(t)\, dt, & \pi - \frac{1}{2}\epsilon \le x \le \pi\,. \end{cases} \tag{2.12}$$

We can conclude from (2.10), (2.11) and (2.12) that

$$\sum_{n=0}^{\infty} \tilde{g}_\epsilon(\mu_n)\, \varphi_n^2(x) = \frac{1}{2}g_\epsilon(0) + \frac{1}{2}\int_0^{\epsilon} w(x,t,x)\, g_\epsilon(t)\, dt, \quad 0 \le x \le \pi, \tag{2.13}$$

and in the interval $0 \le x \le \frac{1}{2}\epsilon$ we have to add the sum

$$\frac{1}{2}g_\epsilon(2x) + \frac{1}{2}\int_{2x}^{\epsilon} w(x,t,-x)\, g_\epsilon(t)\, ,dt \tag{2.14}$$

to the right-hand side of (2.13), while in the interval$\pi - \frac{\epsilon}{2} \le x \le \pi$ we have to add the sum

$$\frac{1}{2}g_\epsilon(2\pi - 2x) + \frac{1}{2}\int_{2\pi-x}^{\epsilon} w(x,t,2\pi - x)\, g_\epsilon(t)\, dt. \tag{2.15}$$

Integrating equation (2.13) we obtain with these additions in limits $[0, \pi]$:

$$\sum_0^\infty \tilde{g}_\epsilon(\mu_n) = \frac{\pi}{2} g_\epsilon(0) + \frac{1}{2}\int_0^\epsilon g_\epsilon(t)[\int_0^\pi w(x,t,x)\,dx]\,dt + \frac{1}{2}\int_0^\epsilon g_\epsilon(t)\,dt +$$

$$+\frac{1}{2}\int_0^\epsilon g_\epsilon(t)[\int_0^{\frac{t}{2}} w(x,t,-x)\,dx]\,dt + \frac{1}{2}\int_0^\epsilon g_\epsilon(t)[\int_0^{\frac{t}{2}} w(\pi - x, t, \pi + x)\,dx]\,dt \tag{2.16}$$

3 Proof of formula (1.5).

In order to obtain the trace formula (1.5), let us replace $g_\epsilon(t)$ in (2.16) by $g_\epsilon''(t)$; therefore, instead of $\tilde{g}_\epsilon(\mu)$ we have to take $-\mu^2\tilde{g}_\epsilon(\mu)$:

$$-\sum_0^\infty \mu_n^2\, \tilde{g}_\epsilon(\mu_n) = -\int_0^\infty \mu^2 \tilde{g}_\epsilon(\mu)\,d\mu + \frac{1}{2}\int_0^\epsilon g_\epsilon''(t)[\int_0^{\frac{t}{2}} w(x,t,-x)\,dx]\,dt$$

$$+\frac{1}{2}\int_0^\epsilon g_\epsilon''(t)[\int_0^{\frac{t}{2}} w(\pi - x, t, \pi + x)\,dx]\,dt + \int_0^\epsilon g_\epsilon''(t)[\int_0^\pi w(x,t,x)\,dx]\,dt. \tag{3.1}$$

The following asymptotic formula was proved in the monograph [5] (Ch. 6, p. 276):

$$w(x,t,s) = -\frac{1}{2}\int_{\frac{1}{2}(x+s-t)}^{\frac{1}{2}(x+s+t)} q(\tau)\,d\tau + O(t^3). \tag{3.2}$$

The estimate $O(t^3), t \downarrow 0$, is uniform with respect to x and s in a fixed interval. Substituting (3.2) in (3.1) and integrating twice by parts, we obtain

$$\sum_{n=0}^\infty \mu_n^2 \tilde{g}_\epsilon(\mu_n)$$

$$= \int_0^\infty \mu^2 \tilde{g}_\epsilon(\mu)\,d\mu + \frac{1}{4}[q(0) + q(\pi)]\int_0^\epsilon g_\epsilon(t)\,dt + \frac{1}{4} A\, g_\epsilon(0) + \int_0^\epsilon h(t)\, g_\epsilon(t)\,dt. \tag{3.3}$$

Here $A = \int_0^\pi q(x)\,dx, \;\; h(t) = O(t), \;\; t \downarrow 0$. If we set $\sigma(\mu) = \sum_{\mu_n < \mu} 1$, then

$$\sum_{n=0}^\infty \mu_n^2 \tilde{g}_\epsilon(\mu_n) = \int_0^\infty \tilde{g}_\epsilon(\mu)\,d\sigma(\mu) + \int_{\lambda_0}^0 \tilde{g}_\epsilon(\sqrt{\lambda})\;\;\lambda\,d\sigma(\sqrt{\lambda}).$$

For $s > 2$, it follows from (3.3) and the Tauberian theorem (see Appendix 1)

$$\int_0^\mu (1 - \frac{\nu^2}{\mu^2})^s[\nu^2\,d\sigma(\nu) - \nu^2\,d\nu - \frac{1}{2\pi} A\,d\nu] = \frac{1}{4}[q(0) + q(\pi)] + o(1) \tag{3.4}$$

as $\mu \to \infty$.

Let us consider now the simplest case of $q(x) = 0, \;\; 0 \le x \le \pi, \;\; h = H = 0$. The eigenvalues are equal to $\rho_n = n^2, \;\; n = 0, 1, 2, \ldots$. If we set

$$\tau(\mu) = \sum_{n<\mu} 1,$$

than (3.4) may be represented in the form

$$\int_0^\mu (1-\frac{\nu^2}{\mu^2})^s(\nu^2\,d\tau(\nu)-\nu^2\,d\nu)=o(1). \tag{3.5}$$

Subtracting (3.5) from (3.4), we obtain

$$\int_0^\mu (1-\frac{\nu^2}{\mu^2})^s\{\nu^2\,d[\sigma(\nu)-\tau(\nu)]-\frac{A}{2\pi}\,d\nu\}=\frac{1}{4}[q(0)+q(\pi)]+o(1).$$

Assuming $A=0,\;\; s=3$ and using Theorem A.2.1, we obtain

$$\sum_{n=0}^{\infty}(\lambda_n-n^2)=\frac{1}{4}[q(0)+q(\pi)]. \tag{3.6}$$

Remark 1. If $q(x)$ is of period π, the right-hand side of formula (3.6) reduces to $\frac{1}{2}q(0)$.

Remark 2. In the case of Dirichlet boundary conditions, the continuation of the function $f(x)$ from the initial condition must be odd. Therefore, the right-hand side in the formula (3.6) gains the sign minus.

4 H. Hochstadt's trace formula.

This trace formula is related to Hill's equation, i.e., Sturm-Liouville equation

$$-y''+q(x)\,y=\lambda y, \quad 0\le x\le\pi, \tag{4.1}$$

with smooth periodic potential $q(x)$.

We consider equation (4.1) with boundary conditions

$$(1) \quad y(0)=y(\pi),\;\; y'(0)=y'(\pi),$$

and

$$(2) \quad y(0)=-y(\pi),\;\; y'(0)=-y'(\pi).$$

The first problem is called *periodic*, the second one is *anti-periodic* or *semi-periodic.* Denote the periodic spectrum by $\alpha_0<\alpha_2\le\beta_2<\alpha_4\le\beta_4<\cdots$ and the semi-periodic spectrum by $\alpha_1\le\beta_1<\alpha_3\le\beta_3<\cdots$. It is well known that

$$\alpha_0<\alpha_1\le\beta_1<\alpha_2\le\beta_3<\cdots$$

Since the periodic and semi-periodic problems are problems without boundary, in the case of periodic $q(x)$ and periodic boundary conditions the term

$$\frac{1}{4}[q(0)+q(\pi)]\int_0^\epsilon g_\epsilon(t)\,dt$$

in formula (3.3) disappears. Let us add the formulas (3.3) for periodic and semi-periodic cases and subtract the doubled formula (3.3) for Dirichlet boundary conditions and the same periodic potential. Using afterwards the Tauberian theorem (see Appendix 1), we obtain H. Hochstad's formula

$$\alpha_0 + \sum_{n=1}^{\infty} (\alpha_n + \beta_n - 2\gamma_n) = q(0), \tag{4.2}$$

where γ_n are eigenvalues of the Dirichlet problem. [2]

Remark. If we replace the potential $q(x)$ by $q(x+t)$, $t \in R$, the eigenvalues of periodic and semi-periodic problems do not change. On the contrary, the eigenvalues of Dirichlet problem γ_n are moving in the corresponding gap. Trace formula (4.2) takes on the form

$$\alpha_0 + \sum_{n=1}^{\infty} [\alpha_n + \beta_n - 2\gamma_n(t)] = q(t).$$

Appendix I. Tauberian theorems.

Theorem A.1.1. Let a function $\sigma(\mu)$, $-\infty < \mu < +\infty$ satisfy the following conditions

1) $\sigma(\mu)$ is odd;
2) $Var_a^{a+1}\{\sigma(\mu)\} \leq C(1+|a|)^r$, $C > 0$, $r \geq 0$, $|a| \to \infty$;
3) if $g_\epsilon(t)$ and $\tilde{g}_\epsilon(\mu)$ are as in Section II, then

$$\int_0^{\infty} \tilde{g}_\epsilon(\mu)\, d\sigma(\mu) = 0$$

for every $g_\epsilon(t)$ with the above-stated properties.

Then

$$\int_0^{\mu} (1 - \frac{\nu^2}{\mu^2})^s\, d\sigma(\nu) = O(\mu^{r-s}), \quad \mu \to \infty \qquad (A.1.1.)$$

Theorem A.1.1. is a basic theorem, but it is not always convenient in applications. In particular, we use in this paper the following its generalization.

Theorem A.1.2. Suppose that $\sigma(\mu)$ satisfies conditions 1 and 2 of the previous theorem, and condition 3 is replaced by the condition

3′) the identity

$$\int_0^{\infty} \tilde{g}_\epsilon(\mu)\, d\sigma(\mu) = \int_{\lambda_0}^{0} \tilde{g}_\epsilon(\sqrt{\lambda})\, d\tau(\lambda) + a g_\epsilon(0) + \int_0^{\epsilon} g_\epsilon(t) h(t)\, dt, \qquad (A.1.2)$$

holds where $Var_{\lambda_0}^{0}\{\tau(\lambda)\} < \infty$, a is a constant, and $h(t)$ is a continuous function.

If conditions 1, 2, and 3′ are fulfilled, then

$$\int_0^{N} (1 - \frac{\mu^2}{N^2})^s\, d[\sigma(\mu) - \frac{2a}{\pi}\mu] = \tau(0) - \tau(\lambda_0) + h(0) + O(N^{r-s}),$$

as $N \to \infty$.

[2] At first, we obtain formula (4.2) in the sense of Riesz summation of order 3, but the series in (4.2) converges, and we can apply Theorem (A.2.1).

Proof. Using the inversion formula and the Parseval identity, we obtain

$$\int_{\lambda_0}^{0} \tilde{g}_\epsilon(\sqrt{\lambda})\, d\tau(\lambda) + a g_\lambda(0) + \int_0^\epsilon g_\epsilon(t) h(t)\, dt =$$

$$= \int_0^\epsilon g_\epsilon(t) H(t)\, dt + \frac{2a}{\pi}\int_0^\infty \tilde{g}_\epsilon(\mu)\, d\mu = \frac{2}{\pi}\int_0^\infty \tilde{g}_\epsilon(\mu)[\tilde{H}(\mu) + a]\, d\mu,$$

where

$$H(t) = h(t) + \int_{\lambda_0}^{0} \cos\sqrt{|\lambda|}t\, d\tau(\lambda), \quad \tilde{H}(\mu) = \int_0^\epsilon H(t)\cos\mu t\, dt. \qquad (A.1.3)$$

From (A.1.2) and (A.1.3) we obtain

$$\int_0^\infty \tilde{g}_\epsilon(\mu)\, d_\mu\, [\sigma(\mu) - \frac{2}{\pi}\int_0^\mu \tilde{H}(\nu)\, d\nu - \frac{2a}{\pi}\mu] = 0$$

Now, for $s > r$, it follows from Theorem A.1.1

$$\int_0^N (1 - \frac{\mu^2}{N^2})^s\, d[\sigma(\mu) - \frac{2a}{\pi}\mu] = \frac{2}{\pi}\int_0^N (1 - \frac{\mu^2}{N^2})^s \tilde{H}(\mu)\, d\mu + O(N^{r-s}) =$$

$$= \tau(0) - \tau(\lambda_0) + h(0) + o(1).$$

Let us remark that, as $N \to \infty$, we have

$$\frac{2}{\pi}\int_0^N (1 - \frac{\mu^2}{N^2})^s \tilde{H}(\mu)\, d\mu = \int_0^\epsilon H(t)[\frac{2}{pi}\int_0^N (1 - \frac{\mu^2}{N^2})^s \cos\mu t\, d\mu]\, dt =$$

$$= \frac{N}{\pi}\Gamma(s+1)\Gamma(\frac{1}{2})\int_0^\epsilon \frac{I_{s+\frac{1}{2}}(Nt)}{(\frac{1}{2}Nt)^{s+\frac{1}{2}}} H(t)\, dt =$$

$$= \frac{1}{\pi}\Gamma(s+1)\Gamma(\frac{1}{2})2^{s+\frac{1}{2}}\int_0^{N\epsilon} \frac{I_{s+\frac{1}{2}}(z)}{z^{s+\frac{1}{2}}} H(\frac{z}{N})\, dz \to H(0),$$

since (see, e.g. [6], Ch.3 and Ch.13)

$$\int_0^\infty \frac{I_{s+\frac{1}{2}}(z)}{z^{s+\frac{1}{2}}}\, dz = \frac{\Gamma(\frac{1}{2})}{2^{s+\frac{1}{2}}\Gamma(s+\frac{1}{2})}.$$

Appendix II.

Theorem. Let $\lambda_n = n^2 + O(\frac{1}{n^2})$, $n \to \infty$ and suppose that

$$I_N = \sum_{n=1}^{N}(1 - \frac{\lambda_n}{N^2})^3\lambda_n - \sum_{n=1}^{N}(1 - \frac{n^2}{N^2})^3 n^2 =$$

$$= A + o(1), \qquad N \to \infty,$$

where A is a constant. Then

$$\sum_{n=1}^{\infty}(\lambda_n - n^2) = A.$$

Proof.

$$I_N = \sum_{n=1}^{N}(\lambda_n - n^2) + \frac{3}{N^2}\sum_{n=1}^{N} O(1) + \frac{3}{N^4}\sum_{n=1}^{N} O(n^2) +$$

$$+\frac{1}{N^6}\sum_{n=1}^{N} O(n^4) = \sum_{n=1}^{N}(\lambda_n - n^2) + O(\frac{1}{N}).$$

References

[1] I. M. Gelfand and B. M. Levitan. *On a simple identity for the eigenvalues of a differential operator of second order.* Dokl. Akad. Nauk USSR 88, (1953), 593-596.

[2] B. M. Levitan. *On the asymptotic behavior of the spectral function of a self-adjoint differential equation of second order.* Izv. Akad. Nauk USSR, Serie Math. 16 (1952), 325-352. English translation: Amer. Math. Soc. Transl. (2) 101 (1973), 192-221.

[3] B. M. Levitan. *On the asymptotic behavior of the spectral function of a self-adjoint differential operator of second order.* Izv. Akad. Nauk USSR, Serie Math. 17 (1953), 331-364. English translation:Amer.Math. Soc. Transl. 102 (1973), 191-229.

[4] H. Hochstadt. *On the determination of Hill's equation from its spectrum.* Arch. Rat. Mech. Anal. 19 (1965), 353-362.

[5] B. M. Levitan and I. S. Sargsjan *Introduction to Spectral Theory.* Transl. of Math. Monographs, Amer. Math. Soc. Providence, (1975)

[6] G. N. Watson. *A Treatise on the Theory of Bessel functions.*

[7] H. P. McKean and P. Van Moerbeke. *The spectrum of Hill's equation.* Invent. Math. 30 (1975), 217-274.

School of Mathematics
University of Minneapolis
127 Vincent Hall, 206 Church Str. S.E.
Minneapolis, MN 55455
USA

AMS classification # 47E05

Operator Theory
Advances and Applications, Vol. 98

COMMUTING NONSELFADJOINT OPERATORS AND A UNIFIED THEORY OF WAVES AND CORPUSCLES

M.S. LIVŠIC

To Israel Glazman, my dearest friend from the very childhood.
May his memory live for ever.

In this paper, which is closely connected with our previous study "What is a particle from the standpoint of systems theory?" [3], we consider the case of pairs of commuting operators with one-dimensional imaginary parts and we develop a unified theory of waves and particles for the case of Dirac equations. Particles emerge in Space as localized fields with singularities at their centers and wave equations appear as compatibility conditions for an overdetermined pair of equations of corresponding open systems.

1.Introduction: Open Systems and Operator Colligations

Let A be a linear operator in a finite or infinite dimensional Hilbert space H [1] and let E be a finite dimensional space, equipped with a scalar product. Consider a **system** defined by the following equations:

$$\left(i\frac{d}{d\xi}+A\right)f(\xi)=\psi[u(\xi)], \tag{1}$$

$$f(\xi_0)=f_0,$$

$$v(\xi)=u(\xi)+\varphi[f(\xi)], \tag{2}$$

where ξ is a real variable and

$$\psi:\ E\to H,\quad \varphi:\ H\to E$$

are linear mappings of E into H and of H into E respectively. The input $u(\xi)$ and the output $v(\xi)$ are vector-functions with values belonging to E, and the corresponding inner (intermediate) state $f(\xi)$ is an H-valued function. One can consider the system (1), (2) as a continuous **automaton** and require that the following **continuity conditions** hold:

$$\frac{d}{d\xi}(f,f) = (Ju,u) - (Jv,v), \tag{3}$$

where $J = J^*$ is an appropriate selfadjoint operator in E. Assuming that condition (3) holds for arbitrary $u(\xi)$, $f(\xi)$, and $v(\xi)$ satisfying conditions (1), (2), we obtain

$$\frac{1}{i}((A-A^*)f,f) + (f,i\psi u) + (i\psi u,f) = (\varphi^* J\varphi f,f) + (f,\varphi^* Ju) + (\varphi^* Ju,f)$$

for arbitrary elements $u \in E$ and $f \in H$. Thus,

$$\frac{1}{i}(A-A^*) = \varphi^* J\varphi, \quad \psi = -i\varphi^* J.$$

Putting $\Phi = i\varphi$ we obtain

$$\frac{1}{i}(A-A^*) = \Phi^* J\Phi \tag{4}$$

and hence the conditions (1), (2) can be written in the following **canonical** form:

$$\begin{cases} (i\frac{d}{d\xi} + A)f = \Phi^* J(u), & f(\xi_0) = f_0 \\ v(\xi) = u(\xi) - i\Phi[f(\xi)] \end{cases} \tag{5}$$

Remark: *For an interpretation of the equality (3) it is convenient to assume that scalar products (Ju,u) and (Jv,v) are flows through the input and the output respectively of some imaginary indivisible units, which will be called the monons in the sequel. The continuity conditions may be considered as a conservation law for monons. Conditions (1), (2) of the automaton describe a continuous chain of* **transitions** *$\xi \to \xi + d\xi$, accompanied by the passages of monons through the input and the output, when monons enter and leave the intermediate state.*

The variable ξ is said to be the **transition index** of the open system (1), (2). Integrating (3) we obtain

$$\int_{\xi_1}^{\xi_2}(Ju,u)d\xi - \int_{\xi_1}^{\xi_2}(Jv,v)d\xi = (f,f)_{\xi_2} - (f,f)_{\xi_1}$$

A collection $(A; H, \Phi, E; J(A))$, where $\Phi : H \to E$ and $J(A) = J^*(A)$ is said to be a **colligation** if condition (4) holds. An arbitrary operator A with finite dimensional imaginary part $ImA = \frac{1}{2i}(A-A^*)$ can be embedded in a colligation with $E = G$, $G = (A-A^*)H$, $J(A) = 2(ImA)|_G$ and $\Phi = P_G$, where P_G is the orthogonal projection of H onto G. In the following we consider **pairs** of commuting operators A_1, A_2 with finite dimensional imaginary parts. A collection $(A_1, A_2; H, \Phi, E; J(A_1), J(A_2))$ is said to be a (two-operator) **colligation** if the conditions

$$\frac{1}{i}(A_k - A_k^*) = \Phi^* J(A_k)\Phi \tag{6}$$

hold. A colligation is said to be **strict** if $\Phi H = E$ and $\ker J(A_1) \cap \ker J(A_2) = 0$. A colligation is said to be **commutative** if $A_1 A_2 = A_2 A_1$. An arbitrary given pair of operators with finite dimensional imaginary parts can be embedded in a strict colligation with $E = G$, where $G = G_1 + G_2$ is the sum of "non-Hermitian" subspaces $G_k = (A_k - A_k^*)H$ the operators $J(A_k)$ are, restrictions of $2Im(A_k)|_G$ and $\Phi = P_G$.

If $A_1 A_2 = A_2 A_1$, then the ranges of the selfadjoint operators

$$\begin{cases} \frac{1}{i}(A_1 A_2^* - A_2 A_1^*) = \frac{1}{i}[(A_1 - A_1^*)A_2^* - (A_2 - A_2^*)A_1^*], \\ \frac{1}{i}(A_2^* A_1 - A_1^* A_2) = \frac{1}{i}[(A_1 - A_1^*)A_2 - (A_2 - A_2^*)A_1] \end{cases} \tag{7}$$

belong to the non-Hermitian subspace $G = G_1 + G_2$. In the case of a strict (commutative) colligation, the operators Φ and Φ^* are one-to-one mappings of G onto E and of E onto G,respectively [3,4]. It is easy to check [3] that in this case there exist two selfadjoint operators $J(A_1 A_2^*)$, $J(A_2^* A_1)$, so that we have the following representations:

$$\begin{cases} \frac{1}{i}(A_1 A_2^* - A_2 A_1^*) = \Phi^* J(A_1 A_2^*)\Phi, \\ \frac{1}{i}(A_2^* A_1 - A_1^* A_2) = \Phi^* J(A_2^* A_1)\Phi \end{cases} \tag{8}$$

These relations together with the colligation conditions imply, that

$$J(A_1)\Phi A_2^* - J(A_2)\Phi A_1^* = J(A_1 A_2^*)\Phi \tag{9}$$

$$J(A_1)\Phi A_2 - J(A_2)\Phi A_1 = J(A_2^* A_1)\Phi \tag{10}$$

Subtracting we obtain

$$J(A_2^* A_1) = J(A_1 A_2^*) + i[J(A_1)\Phi\Phi^* J(A_2) - J(A_2)\Phi\Phi^* J(A_1)] \tag{11}$$

If a strict commutative colligation is given, then one may consider the corresponding system

$$\begin{cases} (i\frac{\partial}{\partial \xi} + A_1)f = \Phi^* J(A_1)u & (12) \\ (i\frac{\partial}{\partial \eta} + A_2)f = \Phi^* J(A_2)u & (13) \\ v = u - i\Phi f & (14) \end{cases}$$

where $u(\xi,\eta)$, $v(\xi,\eta)$ are E-valued vector-functions of the real variables (ξ,η) and $f(\xi,\eta)$ is an H valued function. The equations (12), (13) are overdetermined. If $u(\xi,\eta) \equiv 0$, then they are compatible and the common solution is

$$f(\xi,\eta) = e^{i(\xi A_1 + \eta A_2)} f_0$$

If $u(\xi,\eta)$ is an arbitrary given input, then we have the following compatibility conditions:
Compatibility Theorem 1: *If X is a strict commutative colligation, then the two equations (11) and (12) are consistent iff the input $u(\xi,\eta)$ satisfies the following system of PDEs:*

$$[J(A_2)\frac{\partial}{\partial \xi} - J(A_1)\frac{\partial}{\partial \eta} + iJ(A_1 A_2^*)]u = 0. \tag{15}$$

If $u(\xi,\eta)$ satisfies these equations, then the output field $v(\xi,\eta)$ satisfies the PDEs

$$[J(A_2)\frac{\partial}{\partial\xi} - J(A_1)\frac{\partial}{\partial\eta} + iJ(A_2^*A_1)]v = 0. \tag{16}$$

The fields u, f and v, are said to be the input, the inner (intermediate) and the output states, respectively of an **open field**. The continuity conditions in this case are:

$$\begin{cases} \frac{\partial}{\partial\xi}(f,f) = (J(A_1)u,u) - (J(A_1)v,v) \\ \frac{\partial}{\partial\eta}(f,f) = (J(A_2)u,u) - (J(A_2)v,v) \end{cases}$$

Proof of Theorem 1: It is evident that the equations (12) and (13) are equivalent to the equations

$$i\frac{\partial f_1}{\partial\xi} = e^{-i(\xi A_1+\eta A_2)}\Phi^*J(A_1)u$$
$$i\frac{\partial f_1}{\partial\eta} = e^{-i(\xi A_1+\eta A_2)}\Phi^*J(A_2)u,$$

where $f_1 = e^{-i(\xi A_1+\varphi A_2)}f$. These equations are compatible iff

$$\frac{\partial f_1}{\partial\eta}[e^{-i(\xi A_1+\eta A_2)}\Phi^*J(A_1)u] = \frac{\partial f_1}{\partial\xi}[e^{-i(\xi A_1+\eta A_2)}\Phi^*J(A_2)u].$$

Calculating, we obtain

$$\Phi^*(J(A_2)\frac{\partial}{\partial\xi} - J(A_1)\frac{\partial}{\partial\eta})u + i(A_2\Phi^*J(A_1) - A_1\Phi^*J(A_2))u = 0.$$

Using the equality adjoint to (9) we obtain

$$A_2\Phi^*J(A_1) - A_1\Phi^*J(A_2) = \Phi^*J(A_1A_2^*)$$

and therefore

$$\Phi^*\left(J(A_2)\frac{\partial}{\partial\xi} - J(A_1)\frac{\partial}{\partial\eta} + iJ(A_1A_2^*)\right)u = 0.$$

Canceling Φ^* we obtain the PDE (15). To verify the PDE (16) we use the identity $v = u - i\Phi f$, where u is a solution of the PDE (15). Then

$$\left(J(A_2)\frac{\partial}{\partial\xi} - J(A_1)\frac{\partial}{\partial\eta}\right)v = -iJ(A_1A_2^*)u - i\left(J(A_2)\Phi\frac{\partial f}{\partial\xi} - J(A_1)\Phi\frac{\partial f}{\partial\eta}\right).$$

Multiplying equations (12) and (13) by $J(A_2)\Phi$ and $J(A_1)\Phi$, respectively, subtracting and using the equalities (10) and (11), we obtain (16). An open field defined by conditions (12)-(14) is said to be a **diatron** and the variables (ξ,η) are said to be the **transition indices** of this diatron. The open system defined by conditions (5) is said to be a **monotron**. Hence, a diatron is a consistent system of two monotrons.

2. Pairs of commuting operators with one-dimensional imaginary parts

We are going to describe all the possible pairs of commuting operators such that $\dim(A_k - A_k^*)H = 1$. By considering the operators $\pm A_1$, $\pm A_2$ one can assume without loss of generality that

$$\begin{aligned} \tfrac{1}{i}(A_1 - A_1^*)f &= (f, g_1)g_1 \\ \tfrac{1}{i}(A_2 - A_2^*)f &= -(f, g_2)g_2 \end{aligned} \tag{17}$$

Assume that the so called channel vectors g_1 and g_2 are linearly independent. Then the non-Hermitian subspace $G = G_1 + G_2$ is two-dimensional. We choose $E = \mathbb{C}^2$ and define the mapping $\Phi: H \to \mathbb{C}^2$ as follows:

$$\Phi f = (f, g_1)e_1 + (f, g_2)e_2, \tag{18}$$

where $e_1 = \begin{pmatrix} 1 \\ 0 \end{pmatrix}$, $e_2 = \begin{pmatrix} 0 \\ 1 \end{pmatrix}$. It is easy to check that the adjoint mapping $\Phi^* : \mathbb{C}^2 \to H$ has the form

$$\Phi^* u = u_1 g_1 + u_2 g_2. \tag{19}$$

Indeed, if $u = u_1 e_1 + u_2 e_2$, then

$$(\Phi f, u) = (f, g_1)\overline{(u, e_1)} + (f, g_2)\overline{(u, e_2)} = (f, \Phi^* u).$$

The relations (17) can be written in the form of colligation conditions

$$\frac{1}{i}(A_k - A_k^*) = \Phi^* J(A_k)\Phi, \tag{20}$$

where

$$J(A_1) = \begin{pmatrix} 1 & 0 \\ 0 & 0 \end{pmatrix}, \quad J(A_2) = \begin{pmatrix} 0 & 0 \\ 0 & -1 \end{pmatrix}. \tag{21}$$

In this case the continuity conditions are of the form

$$\begin{aligned} \frac{\partial}{\partial \xi}(f, f) &= |u_1|^2 - |v_1|^2 \\ \frac{\partial}{\partial \eta}(f, f) &= |v_2|^2 - |u_2|^2. \end{aligned}$$

If $\mathbf{u}(\xi, \eta) = 0$, then

$$\begin{aligned} \frac{\partial}{\partial \xi}(f, f) &= -|(f, g_1)|^2 \\ \frac{\partial}{\partial \eta}(f, f) &= |(f, g_2)|^2. \end{aligned}$$

Therefore, the corresponding diatron is a consistent system of two "one-channel" systems (monotrons):

$$(i\frac{\partial}{\partial \xi} + A_1)f = \Phi^* J(A_1)\mathbf{u} \tag{22}$$

$$(i\frac{\partial}{\partial\eta} + A_2)f = \Phi^* J(A_2)\mathbf{u} \tag{23}$$

If the index ξ is growing and η is fixed, then the corresponding monotron **absorbs** monons at the **input** and **emits** monons at the **output** and this is a **direct** monotron. If the index η is growing and ξ is fixed, then the corresponding monotron **emits** monons at the **input** and **absorbs** monons at the **output** and this is an **inverse** monotron.

Consider the following relations:

$$\tfrac{1}{i}(A_1A_2^* - A_2A_1^*)f = \tfrac{1}{i}[(A_1 - A_1^*)A_2^* - (A_2 - A_2^*)A_1^*]f =$$

$$= (A_2^*f, g_1)g_1 + (A_1^*f, g_2)g_2 = (f, A_2g_1)g_1 + (f, A_1g_2)g_2$$

and

$$\tfrac{1}{i}(A_1A_2^* - A_2A_1^*)f = \tfrac{1}{i}[A_2(A_1 - A_1^*) - A_1(A_2 - A_2^*)]f =$$

$$= (f, g_1)A_2g_1 + (f, g_2)A_1g_2$$

Comparing these relations we conclude that A_2g_1 and A_1g_2 are linear combinations of g_1 and g_2:

$$\{A_2g_1, A_1g_2\} = \{g_1, g_2\}\begin{pmatrix} \alpha_{11} & \alpha_{12} \\ \alpha_{21} & \alpha_{22} \end{pmatrix} \tag{24}$$

and that

$$\frac{1}{i}(A_1A_2^* - A_2A_1^*)f = \{g_1, g_2\}\begin{pmatrix} \alpha_{11} & \alpha_{12} \\ \alpha_{21} & \alpha_{22} \end{pmatrix}\begin{pmatrix} (f, g_1) \\ (f, g_2) \end{pmatrix}, \tag{25}$$

where the matrix $\alpha = \alpha^*$ is selfadjoint. Let $A_1' = A_1 - \alpha_{22}I$ and $A_2' = A_2 - \alpha_{11}I$. Then either

$$\{A_2'g_1, A_1'g_2\} = \{g_1, g_2\}\begin{pmatrix} 0 & b \\ b^* & 0 \end{pmatrix}, \quad (b = \alpha_{12})$$

or $A_2'g_1 = b^*g_2$, $A_1'g_2 = bg_1$. Therefore,

$$A_1'A_2'g_k = |b|^2 g_k, \quad (k = 1, 2).$$

The case $b \neq 0$. Letting $A_k'' = |b|^{-1}A_k'$, we obtain

$$A_1''A_2''g_k = g_k \quad (k = 1, 2)$$

and

$$A_1''A_2''(A_1^{n_1}A_2^{n_2}g_k) = A_1^{n_1}A_2^{n_2}g_k \quad (n_1, n_2 = 0, 1, 2, \ldots).$$

Hence, on the "principal" subspace

$$\widehat{H} = \operatorname{span}\{A_1^{n_1}A_2^{n_2}g_k\}$$

the product $A_1''A_2'' = I$ and the original operators A_1 and A_2 satisfy the equation of a hyperbola:

$$(A_1 - c_1I)(A_2 - c_2I) = |b|^2 I \quad (c_1, c_2 \text{ are real}).$$

Remark: It is known [4] that the inner space H admits the orthogonal decomposition $H = \widehat{H} \oplus H_0$, where $\widehat{H}$ and H_0 are invariant subspaces of A_1 and A_2, respectively, and A_1 and A_2 are selfadjoint on H_0. In many cases of interest the subspace H_0 can be dropped. In the following we assume that $H = \widehat{H}$. Such a colligation is said to be **irreducible.**

Putting $Q = |b|^{-1}(A_1 - c_1 I)$ we conclude that in the hyperbolic case the operators A_1 and A_2 have the form

$$A_1 = |b|Q + c_1 I \text{ and } A_2 = |b|Q^{-1} + c_2 I.$$

Perfoming the unitary transformation $f \to e^{i(c_1\xi + c_2\eta)}f$, $u \to e^{i(c_1\xi+c_2\eta)}u$ we can reduce the diatron equations to the form (22), (23) with $A_1 = |b|Q$ and $A_2 = |b|Q^{-1}$. Changing the indices $\xi \to |b|\xi$, $\eta \to |b|\eta$ and transforming $f \to |b|^{1/2}f$, $\Phi \to |b|^{-1/2}\Phi$ we obtain the following equations:

$$\begin{cases} (i\frac{\partial}{\partial\xi} + Q)f = \Phi^* J(Q)u & (26) \\ (i\frac{\partial}{\partial\eta} + Q^{-1})f = \Phi^* J(Q^{-1})u & (27) \end{cases}$$

$$v = u - i\Phi f.$$

In this case $|b| = 1$ and we denote $b = \theta$. The relations (24) have the form

$$Qg_2 = \theta g_1, \quad Q^{-1}g_1 = \theta^* g_2. \tag{28}$$

The corresponding colligation is

$$(A_1 = Q, \;\; A_2 = Q^{-1}; \;\; H, \Phi, \mathbb{C}_2; \;\; J(Q_1), J(Q_2)),$$

where

$$J(Q) = \begin{pmatrix} 1 & 0 \\ 0 & 0 \end{pmatrix}, \;\; J(Q^{-1}) = \begin{pmatrix} 0 & 0 \\ 0 & -1 \end{pmatrix} \text{ and } J(Q(Q^{-1})^*) = \begin{pmatrix} 0 & \theta \\ \theta^* & 0 \end{pmatrix}. \tag{29}$$

Using relations (18), (19) and (11) we obtain

$$\Phi\Phi^* = \begin{pmatrix} (g_1, g_1) & (g_2, g_1) \\ (g_1, g_2) & (g_2, g_2) \end{pmatrix}$$

and

$$J((Q^{-1})^*Q) = \begin{pmatrix} 0 & \theta W(0) \\ [\theta W(0)]^* & 0 \end{pmatrix}, \tag{30}$$

where

$$W(\zeta) = 1 - i((Q - \zeta I)^{-1}g_1, g_1) \tag{31}$$

is the characteristic function of the operator Q.

The case $b = 0$. In this case $A_1'g_2 = A_2'g_1 = 0$ and $A_1'A_2'g_k = 0$. Hence, $A_1'A_2'(A_1'^{n_1}A_2'^{n_2}) = 0$ and $A_1'A_2' = 0$ on the principal subspace $\widehat{H}$. Assuming that $H = \widehat{H}$ we conclude that $A_1'A_2' = 0$. Let us consider the subspaces

$$H_1 = span\left\{A_1'^{n_1}g_1\right\} = span\left\{(A_1'^*)^{n_1}g_1\right\}$$

and

$$H_2 = span\left\{A_2'^{n_2} g_2\right\} = span\left\{(A_2'^*)^{n_2} g_2\right\}.$$

The relation $A_1'A_2' = 0$ implies that

$$\left(Q_1'^{n_1} g_1, (Q_2'^*)^{n_2} g_2\right) = 0 \quad (n_1 > 0,\ n_2 > 0).$$

The equality $\frac{1}{i}(A_1' - A_1'^*)g_2 = (g_2, g_1)g_1$ implies that $A_1'^* g_2 = -i(g_2, g_1)g_1$. Therefore,

$$0 = (g_2, A_1' g_2) = (A_1'^* g_2, g_2) = -i|(g_1, g_2)|^2$$

and hence $(g_1, g_2) = 0$ and $A_1'^* g_2 = 0$. Thus the orthogonality relations $(A_1'^{n_1} g_1, (A_2'^*)^{n_2} g_2) = 0$ hold for all $n_1 \geq 0$ and $n_2 \geq 0$, which implies that the subspaces H_1 and H_2 are orthogonal. Hence,

$$H = span\left\{A_1'^{n_1} A_2'^{n_2} g_k\right\} = H_1 \oplus H_2$$

and $A_1'(H_2) = A_2'(H_1) = 0$. Thus, in the case $b = 0$, the operators A_1, A_2 have the form

$$A_1 = \begin{pmatrix} K_1 & 0 \\ 0 & 0 \end{pmatrix} + c_1 I, \quad A_2 = \begin{pmatrix} 0 & 0 \\ 0 & K_2 \end{pmatrix} + c_2 I, \tag{32}$$

where $K_1,\ K_2$ are arbitrary operators in the spaces $H_1,\ H_2$, respectively, such that

$$\frac{1}{i}(K_1 - K_1^*)f_1 = (f_1, g_1)g_1 \tag{33}$$

and

$$\frac{1}{i}(K_2 - K_2^*)f_2 = -(f_2, g_2)g_2.$$

Without loss of generality, we can assume that $c_1 = c_2 = 0$. In this case $A_1 A_2^* = A_2^* A_1 = 0$ and, therefore,

$$J(A_1 A_2^*) = J(A_2^* A_1) = 0$$

and the corresponding diatron disintegrates into two orthogonal monotrons.

Remark: It is easy to check that if g_1 and g_2 are linearly dependent, then $\dim E = 1$ and the operators $A_1,\ A_2$ are connected by a linear relation of the form $A_2 = c_1 A_1 + c_2 I,\ (c_1 < 0)$.

3. The space-time open fields

As we have already shown, a diatron in the hyperbolic case can be reduced to the form (26),(27), where the "generator" Q is an operator with a one-dimensional imaginary part. The pair of operators $A_1 = Q,\ A_2 = Q^{-1}$ is a "point" on the hyperbola $z_1 z_2 = 1$. Let us perform the transformation

$$A = \frac{a}{2}(Q + Q^{-1}), \quad B = \frac{a}{2c}(Q - Q^{-1}), \tag{34}$$

to the symmetry axes of this hyperbola ($a > 0,\quad c > 0$ are given constants). After this transformation the equations of a diatron will have the form:

$$\begin{cases} (i\frac{\partial}{\partial t} + A)f = \Phi^* J(A)u & (35) \\ (i\frac{\partial}{\partial x} + B)f = \Phi^* J(B)u & (36) \\ v = u - i\Phi f, & (37) \end{cases}$$

where the new transition indices t and x are connected with indices ξ and η as follows:

$$t = \frac{\xi + \eta}{a}, \quad x = \frac{c(\xi - \eta)}{a}. \tag{38}$$

It is obvious that

$$J(A) = \frac{a}{2}\begin{pmatrix} 1 & 0 \\ 0 & -1 \end{pmatrix} \text{ and } J(B) = \frac{a}{2c}\begin{pmatrix} 1 & 0 \\ 0 & 1 \end{pmatrix}. \tag{39}$$

To find $J(AB^*)$, consider

$$\tfrac{1}{i}(AB^* - BA^*) = \tfrac{a^2}{4ci}\left[(Q + Q^{-1})(Q^* - (Q^{-1})^*) - (Q - Q^{-1})(Q^* + (Q^{-1})^*)\right]$$

$$= -\tfrac{a^2}{2c}[Q^{-1}Q^* - Q(Q^{-1})^*] = -\tfrac{a^2}{2c}\Phi^* J(Q(Q^{-1})^*)\Phi.$$

Hence,

$$J(AB^*) = -\frac{a^2}{2c}\begin{pmatrix} 0 & \theta \\ \theta^* & 0 \end{pmatrix} \tag{40}$$

Analogously,

$$J(B^*A) = -\frac{a^2}{2c}\begin{pmatrix} 0 & \theta W(0) \\ (\theta W(0))^* & 0 \end{pmatrix}. \tag{41}$$

Therefore, the compatibility PDEs at the input and at the output have the form

$$\left[\begin{pmatrix} 1 & 0 \\ 0 & 1 \end{pmatrix}\frac{\partial}{\partial t} - c\begin{pmatrix} 1 & 0 \\ 0 & -1 \end{pmatrix}\frac{\partial}{\partial x} - ia\begin{pmatrix} 0 & \theta \\ \theta^* & 0 \end{pmatrix}\right]u = 0 \tag{42}$$

and

$$\left[\begin{pmatrix} 1 & 0 \\ 0 & 1 \end{pmatrix}\frac{\partial}{\partial t} - c\begin{pmatrix} 1 & 0 \\ 0 & -1 \end{pmatrix}\frac{\partial}{\partial x} - ia\begin{pmatrix} 0 & \theta W(0) \\ (\theta W(0))^* & 0 \end{pmatrix}\right]v = 0, \tag{43}$$

respectively. The continuity conditions in this case are

$$\begin{aligned} \tfrac{\partial}{\partial t}(f, f) &= \tfrac{a}{2}(|u_1|^2 - |u_2|^2) - \tfrac{a}{2}(|v_1|^2 - |v_2|^2), \\ \tfrac{\partial}{\partial x}(f, f) &= \tfrac{a}{2c}(|u_1|^2 + |u_2|^2) - \tfrac{a}{2c}(|v_1|^2 + |v_2|^2). \end{aligned} \tag{44}$$

The equalities (34) imply that $A^2 - c^2B^2 = a^2 I$. The imaginary parts of A and B are two-dimensional ($\dim G = 2$), the operator B is dissipative: $J(B) > 0$, but the imaginary part of A is indefinite. The diatron (35),(36),(37) is a two-channel open field.

There exists a simple group of transformations of the hyperbola $Q_1Q_2 = I:\ Q_1' = \rho Q,\ Q_2' = \rho^{-1}Q^{-1}\ (\rho > 0)$. In the terms of the transition indices this transformation is equivalent to the transformations $\xi = \rho\xi',\ \eta = \rho^{-1}\eta'$. In matrix form

$$\begin{pmatrix} \xi \\ \eta \end{pmatrix} = \begin{pmatrix} \rho & 0 \\ 0 & \rho^{-1} \end{pmatrix} \begin{pmatrix} \xi' \\ \eta' \end{pmatrix} \tag{45}$$

Performing the transformations (38) we obtain

$$t' = r_1 t - r_2 \frac{x}{c} \text{ and } \frac{x'}{c} = -r_2 t + r_1 \frac{x}{c}, \tag{46}$$

where $r_1 = \frac{1}{2}(\rho + \rho^{-1})$ and $r_2 = \frac{1}{2}(\rho - \rho^{-1})$. The velocity of the point $x' = 0$ is $\mathrm{v} = cr_2/r_1$ and we obtain:

$$\rho = \left[(c+\mathrm{v})(c-\mathrm{v})^{-1}\right]^{\frac{1}{2}}, \quad (|\mathrm{v}| < c).$$

The relation $r_1^2 - r_2^2 = 1$ implies that

$$r_1 = (1 - \mathrm{v}^2/c^2)^{-1/2}, \quad r_2 = \frac{\mathrm{v}}{c}(1 - \mathrm{v}^2/c^2)^{-1/2} \tag{47}$$

and we obtain the Lorenz transformations

$$t' = \frac{t - x\mathrm{v}/c^2}{(1 - \mathrm{v}^2/c^2)^{1/2}}, \quad x' = \frac{x - \mathrm{v}t}{(1 - \mathrm{v}^2/c^2)^{1/2}}. \tag{48}$$

The corresponding transformations of the operators are:

$$A' = r_1 A + r_2 cB, \quad B' = \frac{r_2}{c} A + r_1 B. \tag{49}$$

The diatron equations with respect to the new indices (t', x') are

$$\begin{cases} (i\frac{\partial}{\partial t'} + A')f' = \Phi^* J(A')u' \\ (i\frac{\partial}{\partial x'} + B')f' = \Phi^* J(B')u' \\ v' = u' - i\Phi f', \end{cases}$$

where

$$J(A') = \begin{pmatrix} \rho & 0 \\ 0 & \rho^{-1} \end{pmatrix} J(A), \quad J(B') = \begin{pmatrix} \rho & 0 \\ 0 & \rho^{-1} \end{pmatrix} J(B).$$

It can be checked that

$$J(A'B'^*) = J(AB^*), \quad J(B'^*A') = J(B^*A).$$

The consistency PDEs are

$$\left[T_\rho \frac{\partial}{\partial t'} - cT_\rho \begin{pmatrix} 1 & 0 \\ 0 & -1 \end{pmatrix} \frac{\partial}{\partial x'} - ia \begin{pmatrix} 0 & \theta \\ \theta^* & 0 \end{pmatrix}\right] u' = 0$$

and

$$\left[T_\rho \frac{\partial}{\partial t'} - cT_\rho \begin{pmatrix} 1 & 0 \\ 0 & -1 \end{pmatrix} \frac{\partial}{\partial x'} - ia \begin{pmatrix} 0 & \theta_1 \\ \theta_1 & 0 \end{pmatrix}\right] v' = 0,$$

where

$$T_\rho = \begin{pmatrix} \rho & 0 \\ 0 & \rho^{-1} \end{pmatrix} \text{ and } \theta_1 = \theta W(0).$$

Multiplying these PDEs from the left by $T_\rho^{-\frac{1}{2}}$ and putting $\mathbf{u} = T_\rho^{1/2} u'$, $\quad \mathbf{v} = T_\rho^{1/2} v'$ we conclude that consistency equations are invariant under the Lorenz transformations.

The relations (38) imply that :

$$x = -ct + x_0 \quad (x_0 = \frac{2c\xi}{a}), \qquad x = ct + \bar{x}_0 \quad (\bar{x}_0 = -\frac{2c\eta}{a}).$$

Therefore, if ξ is a fixed given number, then the velocity $\mathrm{v} = \frac{dx}{dt} = -c$. In the case when η is a given number (ξ is changing) the velocity $\mathrm{v} = \frac{dx}{dt} = c$. Hence, one can distinguish the following four cases:

1) **The system** $(i\frac{\partial}{\partial \xi} + Q)f = \Phi^* J(Q)u$: In this case the system **absorbs** monons with $\mathrm{v} = -c$ at the **input** and **emits** monons with $\mathrm{v} = -c$ at the **output**:

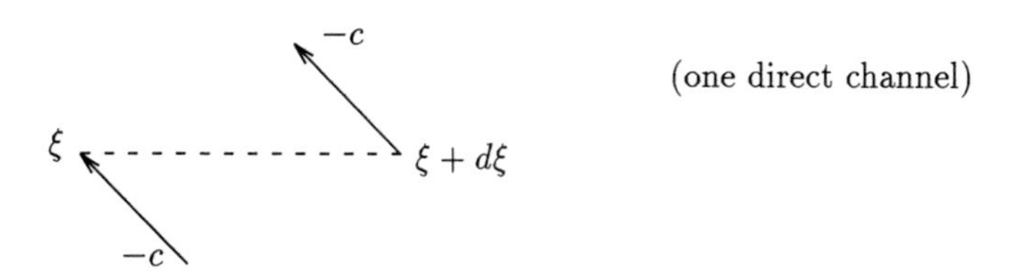

2) **The system** $(i\frac{\partial}{\partial \eta} + Q^{-1})f = \Phi^* J(Q^{-1})u$: In this case the system **emits** monons with $\mathrm{v} = c$ at the **input** and **absorbs** monons with $\mathrm{v} = c$ at the **output**:

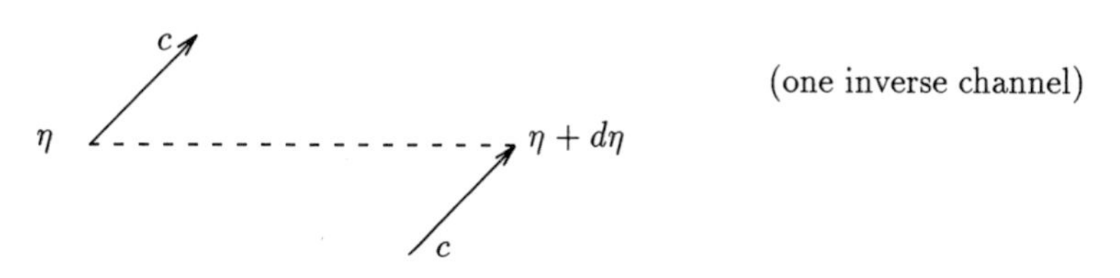

3) **The system** $(i\frac{\partial}{\partial t} + A)f = \Phi^* J(A)u$, where $A = \frac{a}{2}(Q + Q^{-1})$: This system **absorbs** and **emits** monons with $\mathrm{v} = -c$ and $\mathrm{v} = c$ at the input and it **emits** and **absorbs** monons with $\mathrm{v} = -c$ and $\mathrm{v} = c$ at the output, respectively:

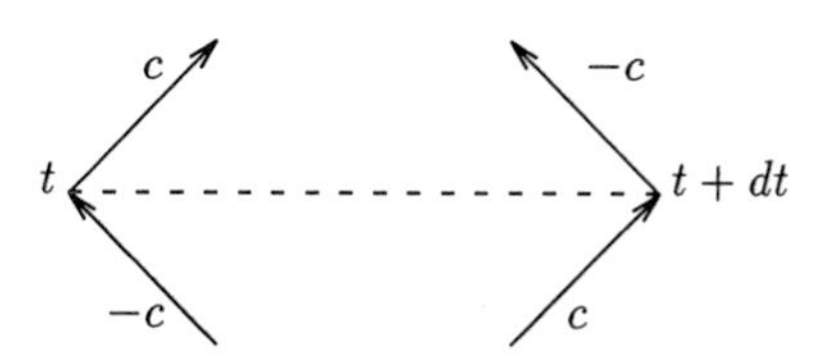

Two channel case (one direct and one inverse channel)

4) **The system** $(i\frac{\partial}{\partial x}+B)f=\Phi^*J(B)u,$ where $B=\frac{a}{2c}(Q-Q^{-1})$: This system absorbs monons with $\mathrm{v}=\pm c$ at the input and emits monons with $\mathrm{v}=\pm c$ at the output:

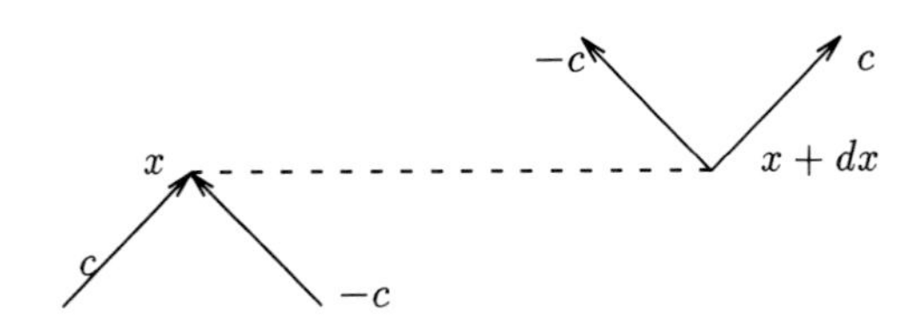

Two direct channels

Remark: In connection with the above considerations I would like to cite Section 47 of Leibniz's "Monalogy" [6]:

"Accordingly, God alone is the primary unity or the original simple substance, of which all the created or derivated monads are produced. They originate, so to speak, through continual fulgurations of the divinity from moment to moment, limited by the receptivity of the created beings, to which it is essential to be limited"

4. Waves and Corpuscles

Let $u(x,t)$ be a solution of the input PDE (42), which admits a Fourier representation of the form

$$u(x,t)=\frac{1}{2\pi}\int\widehat{u}(k,t)e^{ixk}dk. \tag{50}$$

Then

$$\left[\begin{pmatrix}1&0\\0&1\end{pmatrix}\frac{d}{dt}-ick\begin{pmatrix}1&0\\0&-1\end{pmatrix}-ia\begin{pmatrix}0&\theta\\\theta^*&0\end{pmatrix}\right]\widehat{u}=0. \tag{51}$$

Therefore,

$$\widehat{u}=c_1e^{i\lambda_1t}\psi_1+c_2e^{i\lambda_2t}\psi_2,$$

where $\lambda_1,\ \lambda_2$ and $\psi_1,\ \psi_2$ are eigenvalues and eigenvectors, respectively, of the equation

$$[\lambda J(B)-kJ(A)+J(AB^*)]\,\psi=0. \tag{52}$$

The corresponding determinant is

$$\begin{vmatrix}\lambda-ck, & -a\theta\\ -a\theta^*, & \lambda+ck\end{vmatrix}=\lambda^2-c^2k^2-a^2.$$

Hence,

$$\lambda_{1,2}=\pm(c^2k^2+a^2)^{1/2} \tag{53}$$

and c_1, c_2; ψ_1, ψ_2 are functions of k. Then,

$$u(x,t) = \frac{1}{2\pi} \int (\sum_{j=1,2} c_j \psi_j e^{i\lambda_j t}) e^{ixk} dk, \tag{54}$$

$$\begin{aligned} f^{(1)} &= \tfrac{1}{2\pi} \int \left[\sum_{j=1,2} c_j (A - \lambda_j I)^{-1} \Phi^* J(A) \psi_j e^{i\lambda_j t} \right] e^{ixk} dk, \\ f^{(2)} &= \tfrac{1}{2\pi} \int (B - kI)^{-1} \Phi^* J(B) \sum_{j=1,2} c_j \psi_j e^{i\lambda_j t} e^{ixk} dk \end{aligned} \tag{55}$$

are solutions of equations (42), (35) and (36), respectively. To show that $f^{(1)} = f^{(2)}$ we use the following

LEMMA: *If $(A, B,;\ H, \Phi, E;\ J(A), J(B))$ is a commutative strict colligation and ψ is a solution of the linear equation*

$$(\lambda J(B) - \mu J(A) + J(AB^*))\psi = 0, \tag{56}$$

then

$$(A - \lambda I)^{-1} \Phi^* J(A) \psi = (B - \mu I)^{-1} \Phi^* J(B) \psi. \tag{57}$$

Indeed the equality which is adjoint to (9) implies that

$$[B\Phi^* J(A) - A\Phi^* J(B)]\, \psi = \Phi^* J(AB^*) \psi.$$

From (56) it follows that :

$$[\mu \Phi^* J(A) - \lambda \Phi^* J(B)] \psi = \Phi^* J(AB^*) \psi.$$

Subtracting we obtain

$$(B - \mu I) \Phi^* J(A) \psi = (A - \lambda I) \Phi^* J(B) \psi.$$

Assuming that λ and μ are regular points of the corresponding resolvents we obtain (57). Putting $\lambda = \lambda_1$, $\lambda = \lambda_2$ and $\mu = k$ in (56) and using (57) we conclude that $f^{(1)} = f^{(2)} = f$ is a common solution of the diatron equations. Hence, the common intermediate state f has the form

$$f(x,t) = \frac{1}{2\pi} \int (B - kI)^{-1} \Phi^* J(B) \hat{u}(k,t) e^{ixk} dk.$$

Using the convolution theorem, one can write the output $v = u - i\Phi f$ in the form

$$v(x,t) = u(x,t) - \int_{-\infty}^{+\infty} F(x - x')[u(x',t)] dx', \tag{58}$$

where

$$F(x) = \frac{1}{2\pi} \int_{-\infty}^{+\infty} (B - k)^{-1} \Phi^* J(B) e^{ixk} dk. \tag{59}$$

The dissipativity of B ($ImB \geq 0$) implies that the spectrum of B lies in the closed upper halfplane. If $\dim H = N < \infty$ and the colligation is irreducible, then all the eigenvalues μ_k

lie in the upper halfplane: $Im\mu_k > 0 \quad (k = 1, 2, \dots, N)$. Applying the residue formula to the integral (59) we obtain

$$F(x) = \begin{cases} \Phi e^{ixB}\Phi^* J(B), & x > 0 \\ 0, & x < 0. \end{cases}$$

(This result is valid also for a wide class of operators in the infinite dimensional case.) Hence, the integral transformation (58) defines a mapping of the set of solutions of the input PDE onto the set of solutions of the output PDE:

$$\begin{cases} \left[\begin{pmatrix} 1 & 0 \\ 0 & 1 \end{pmatrix}\frac{\partial}{\partial t} - c\begin{pmatrix} 1 & 0 \\ 0 & -1 \end{pmatrix}\frac{\partial}{\partial x} - i\begin{pmatrix} 0 & \theta \\ \theta^* & 0 \end{pmatrix}\right] u = 0 \\ v(x,t) = u(x,t) - \int F(x - x')u(x',t)dx' \\ \left[\begin{pmatrix} 1 & 0 \\ 0 & 1 \end{pmatrix}\frac{\partial}{\partial t} - c\begin{pmatrix} 1 & 0 \\ 0 & -1 \end{pmatrix}\frac{\partial}{\partial x} - i\begin{pmatrix} 0 & \theta_1 \\ \theta_1 & 0 \end{pmatrix}\right] v = 0, \end{cases}$$

where $\theta_1 = \theta W(0)$. If the input at the moment $t = t_0$ is $\mathbf{u}_\delta(x, t_0) = \delta(x - x_0)\mathbf{u}_0$, then **at the same moment $t = t_0$ of time** the output is

$$\mathbf{v}_\delta(x, t_0) = \delta(x - x_0)\mathbf{u}_0 - F(x - x_0)\mathbf{u}_0. \tag{60}$$

Therefore, the output emerges simultaneously at all the points of the real axis at the same moment $t = t_0$ and it appears as a collective entity in Space. The field $F(x - x_0)\mathbf{u}_0$ is said to be the **incarnation** of $\delta(x - x_0)\mathbf{u}_0$. Fields of the form (60) are said to be the **states of a corpuscle**. Hence, waves appear at the input as solutions of the input PDE. These waves provoke "dressed" corpuscles $\mathbf{v}_\delta$ as incarnations of "bare" corpuscles $\mathbf{u}_\delta$ and the output fields is a superposition of "dressed" corpuscles (particles). The incarnation of a corpuscle corresponds to the fixed moment $t = t_0$ of time and therefore the velocity of incarnation $v_{inc} = \infty$. If the input vanishes after the moment $t = t_0$ of the incarnation, then the diatron equations are

$$\begin{cases} \left(i\frac{\partial}{\partial t} + A\right) f = 0 \\ \left(i\frac{\partial}{\partial x} + B\right) f = 0 \\ \mathbf{v} = -i\Phi f \end{cases}$$

The common solution of these equations is

$$\tilde{f}(x,t) = \begin{cases} e^{i[(x-x_0)B + (t-t_0)A]} f_0, & x > x_0, \quad t \geq t_0 \\ 0, & x < x_0, \quad t \geq t_0. \end{cases}$$

The incarnation field $\tilde{v} = -i\Phi\tilde{f}$ satisfies the output PDE. The fields $\tilde{v}(x,t)$ are the states of the corpuscle after its incarnation.

Remark: *Consider the Lorenz transformation*

$$t = \frac{t' - x'\mathrm{v}/c^2}{(1-\mathrm{v}^2/c^2)^{1/2}}, \quad x = \frac{x' - \mathrm{v}t'}{(1-\mathrm{v}^2/c^2)^{1/2}}$$

If $x = x_0$ is fixed, then $\Delta x'/\Delta t' = \mathrm{v} < c$. If $t = t_0$ is fixed, then $\Delta x'/\Delta t' = c^2/\mathrm{v} > c$. Therefore, the velocity of a corpuscle, which is the velocity of its singular point x_0, is less then the velocity of light, and the incarnation velocity is greater then the velocity of light.

Let us recall that the operators A and B have the form

$$A = \frac{a}{2}(Q + Q^{-1}) \text{ and } B = \frac{a}{2c}(Q - Q^{-1}),$$

where the generators Q and Q^{-1} are operators with one-dimensional imaginary parts:

$$2(ImQ)f = (f, g_1)g_1 \text{ and } 2(Im(Q^{-1}))f = -(f, g_2)g_2.$$

It is evident that $ImB \geq 0$ and that all the eigenvalues μ_k $(k = 1, 2, \ldots, N;\ \ \dim H = N < \infty)$ belong to the upper half-plane. Without loss of generality we can assume that $H = span\{Q^n g_1\}$. In this case all the eigenvalues ζ_k are distinct. Then the eigenvalues of A and B are

$$\lambda_k = \frac{a}{2}(\zeta_k + \zeta_k^{-1}) \text{ and } \mu_k = \frac{a}{2c}(\zeta_k - \zeta_k^{-1}).$$

Respectively putting $\zeta_k = |\zeta_k| e^{i\varphi_k}$ $(0 < \varphi_k < \pi)$, we obtain

$$\lambda_k = \frac{a}{2}\left[(|\zeta_k| + |\zeta_k|^{-1})\cos\varphi_k + i(|\zeta_k| - |\zeta_k|^{-1})\sin\varphi_k\right]$$

and

$$\mu_k = \frac{a}{2c}\left[(|\zeta_k| - |\zeta_k|^{-1})\cos\varphi_k + i(|\zeta_k| + |\zeta_k|^{-1})\sin\varphi_k\right].$$

Therefore,

$$\begin{cases} Im\lambda_k > 0 & \text{if} \quad |\zeta_k| > 1 \\ Im\lambda_k < 0 & \text{if} \quad |\zeta_k| < 1 \\ Im\lambda_k = 0 & \text{if} \quad |\zeta_k| = 1. \end{cases}$$

Hence in the case when all the eigenvalues ζ_k are on the unit circle the incarnation field is stable: if $t \to \infty$ all the factors $e^{it\lambda_k}$ are oscillating. In this case, the eigenvalues $\lambda_k = a\cos\varphi_k$ $(0 < \varphi_k < \pi)$ are real and the eigenvalues $\mu_k = i\frac{a}{c}\sin\varphi_k$ of B are purely imaginary. Using the consistency PDEs we conclude that for the plane waves $e^{i(t\lambda + x\mu)}\mathbf{u}_0$ there must be a relation

$$\lambda = \pm(c^2\mu^2 + a^2)^{1/2} \quad (-\infty < \mu < \infty,\ |\lambda| \geq a).$$

It is worth noting that if $|\zeta_k| = 1$, then the eigenvalues λ_k belong to the region $|\lambda_k| < a$.

Remark: *If $\dim H = N < \infty$, then the characteristic function*

$$W(\zeta) = 1 - i((Q - \zeta I)^{-1}g_1, g_1)$$

can be represented as a Blaschke product [5]:

$$W(\zeta) = \prod_{k=1}^{N} \frac{\zeta - \zeta_k^*}{\zeta - \zeta_k}$$

and the triangular model of Q is

$$Q = \begin{pmatrix} \zeta_1, & 0, & \dots, & 0 \\ i\beta_2\beta_1, & \zeta_2, & \dots, & 0 \\ \dots & \dots & \dots, & \cdot \\ i\beta_N\beta_1, & i\beta_N\beta_2 & \dots, & \zeta_N \end{pmatrix},$$

where $\beta_k = (2Im\zeta_k)^{1/2}$. The channel vector g_1 in this case is

$$g_1 = \begin{pmatrix} \beta_1, \\ \vdots \\ \beta_N \end{pmatrix}.$$

Remark: *Our theory of an input wave, that provokes corpuscles at the output of an open field, reminds us at first sight of the concept of a guiding wave, which was introduced by Louis de Broglie [2]. It would be very interesting to investigate the connections between these two different approachs.*

5. Open fields with spin 1/2.

In this case, the equations of the collective states have the form

$$(i\frac{\partial}{\partial t} + A)\begin{pmatrix} f^1 \\ f^2 \end{pmatrix} = \Phi^* J(A)\begin{pmatrix} u^1 \\ u^2 \end{pmatrix}, \tag{61}$$

$$\left[i\left(\sigma_1\frac{\partial}{\partial x_1} + \sigma_2\frac{\partial}{\partial x_2} + \sigma_3\frac{\partial}{\partial x_3}\right) + B\right]\begin{pmatrix} f^1 \\ f^2 \end{pmatrix} = \Phi^* J(B)\begin{pmatrix} u^1 \\ u^2 \end{pmatrix} \tag{62}$$

and

$$\begin{pmatrix} v^1 \\ v^2 \end{pmatrix} = \begin{pmatrix} u^1 \\ u^2 \end{pmatrix} - i\Phi\begin{pmatrix} f^1 \\ f^2 \end{pmatrix}, \tag{63}$$

where the matrices

$$\sigma_1 = \begin{pmatrix} 0 & 1 \\ 1 & 0 \end{pmatrix}, \quad \sigma_2 = \begin{pmatrix} 0 & -i \\ i & 0 \end{pmatrix} \text{ and } \sigma_3 = \begin{pmatrix} 1 & 0 \\ 0 & -1 \end{pmatrix}$$

act on the diatron index $j = 1, 2$. The elements f^ε and u^ε $(\varepsilon = 1, 2)$ belong to the spaces H and $E = \mathbb{C}_2$, respectively. Theorem 1 can be generalized to this case [3]:

THEOREM 2: *The equations (61) and (62) are compatible iff the input* $\mathbf{u} = \{u_j^\varepsilon\}$ *satisfies a PDE of the form*

$$\left[J(B)\frac{\partial}{\partial t} - J(A) \otimes (\vec{\sigma}, \vec{\nabla}) + J(A^*B)\right] \mathbf{u}(\mathbf{x}, t) = 0, \tag{64}$$

where $(\vec{\sigma}, \vec{\nabla}) = \sigma_1 \frac{\partial}{\partial x_1} + \sigma_2 \frac{\partial}{\partial x_2} + \sigma_3 \frac{\partial}{\partial x_3}$. *If* $\mathbf{u}$ *satisfies equation (64), then the output* $\mathbf{v} = \{v_j^\varepsilon\}$ *satisfies the equation*

$$\left[J(B)\frac{\partial}{\partial t} - J(A) \otimes (\vec{\sigma}, \vec{\nabla}) + iJ(B^*A)\right] \mathbf{v}(\mathbf{x}, t) = 0 \tag{65}$$

where

$$J(AB^*) = -\frac{a^2}{2c}\begin{pmatrix} 0 & \theta \\ \theta & 0 \end{pmatrix} \text{ and } J(B^*A) = -\frac{a^2}{2c}\begin{pmatrix} 0 & \theta W(0) \\ \theta W(0) & 0 \end{pmatrix}. \tag{66}$$

The equations (64), (65) have the form of Dirac equations with the constant $a = \frac{mc^2}{h}$. Using the formulas for A and B, we can write equations (61) and (62) in the form

$$[i\hbar\frac{\partial}{\partial t} + \frac{mc^2}{2}(Q + Q^{-1})]\begin{pmatrix} f^1 \\ f^2 \end{pmatrix} = \Phi^* J_1 \begin{pmatrix} u^1 \\ u^2 \end{pmatrix} \tag{67}$$

and

$$[i\hbar(\vec{\sigma}, \vec{\nabla}) + \frac{mc^2}{2}(Q - Q^{-1})]\begin{pmatrix} f^1 \\ f^2 \end{pmatrix} = \Phi^* J_2 \begin{pmatrix} u^1 \\ u^2 \end{pmatrix}, \tag{68}$$

where

$$J_1 = \frac{mc^2}{2}\begin{pmatrix} 1 & 0 \\ 0 & -1 \end{pmatrix} \text{ and } J_2 = \frac{mc}{2}\begin{pmatrix} 1 & 0 \\ 0 & 1 \end{pmatrix}.$$

In this case the continuity conditions have the form [3]:

$$\frac{\partial}{\partial t}[\mathbf{f}, \mathbf{f}] = [J(A)\mathbf{u}, \mathbf{u}] - [J(A)\mathbf{v}, \mathbf{v}] \tag{69}$$

and

$$\mathrm{div}\mathbf{P} = [J(A), \mathbf{u}, \mathbf{u}] - [J(A)\mathbf{v}, \mathbf{v}], \tag{70}$$

where $\mathbf{P}$ is a vector-field with components $\mathbf{P} = [\sigma_k \mathbf{f}, \mathbf{f}]$ and

$$div\mathbf{P} = \sum_{k=1}^{3} \frac{\partial}{\partial x_k}[\sigma_k \mathbf{f}, \mathbf{f}]. \tag{71}$$

The scalar products are:

$$[\mathbf{f}, \mathbf{f}] = (f^1, f^1) + (f^2, f^2) \text{ and } [\mathbf{u}, \mathbf{u}] = (u^1, u^1) + (u^2, u^2).$$

Using Gauss formula, we can write (70) in the form

$$\int\int\int_{(D)}([J(B)\mathbf{u}, \mathbf{u}] - [J(B)\mathbf{v}, \mathbf{v}])d\tau = \int\int_{(S)} P_n dS. \tag{72}$$

Using the formulas for σ and for $J(A)$ and $J(B)$ we can write the following formulas:

$$[J(A)\mathbf{u},\mathbf{u}] = a/2(\sum_{\varepsilon=1,2} |u_1^\varepsilon|^2 - \sum_{\varepsilon=1,2} |u_2^\varepsilon|^2), \tag{73}$$

$$[J(B)\mathbf{u},\mathbf{u}] = \frac{a}{2c} \sum_{\varepsilon,j=1,2} |u_j^\varepsilon|^2, \tag{74}$$

$$P_1 = (f^1, f^2) + \overline{(f^1, f^2)}, \quad P_2 = i[(f^1, f^2) - \overline{(f^1, f^2)}]$$

and

$$P_3 = (f^1, f^1) - (f^2, f^2). \tag{75}$$

The scalar products $[J(B)\mathbf{u},\mathbf{u}]$ and $[J(B)\mathbf{v},\mathbf{v}]$ are the densities of the sources and sinks of the field $\mathbf{P}$.

Remark: In terms of monons, the relation (72) can be formulated as follows: *When the transition indices (x_1, x_2, x_3) are changing in some domain D, then the quantity of monons entering and leaving the intermediate state through the input and the output is equal to the quantity of monons in the intermediate state, which enter and leave the domain D through its boundary.*

As was shown in [3], the following **basic formula** holds (see formula (41) of the paper [2]):

$$\mathbf{v}(x,t) = \mathbf{u}(\mathbf{x},t) - \int F(\mathbf{x}-\mathbf{x})[\mathbf{u}(\mathbf{x}',t)]d\mathbf{x}, \tag{76}$$

where

$$F(\mathbf{x}) = \frac{i}{(2\pi)^3} \int \Phi(B - (\sigma,\mathbf{k}))^{-1} \Phi^* J(B) e^{i(\mathbf{x},\mathbf{k})} d\mathbf{k} \tag{77}$$

and $\mathbf{u}(\mathbf{x},t) = \{u_j^\varepsilon(\mathbf{x},t)\}$ is a solution of the PDE (64). If the input at the moment $t = t_0$ is $\mathbf{u}(x,t_0) = \delta(\mathbf{x}-\mathbf{x}_0)\mathbf{u}_0$, then the output at the same moment is $\delta(\mathbf{x}-\mathbf{x}_0)\mathbf{u}_0 - F(\mathbf{x}-\mathbf{x}_0)\mathbf{u}_0$. Hence the field $F(\mathbf{x}-\mathbf{x}_0)\mathbf{u}_0$ emerges simultaneously at all the points of the three-dimensional space. Just as in the case of the one-dimensional space, which was discussed in §3, the field $F(\mathbf{x}-\mathbf{x}_0)\mathbf{u}_0$ is said fo be the **incarnation** of the field $\delta(\mathbf{x}-\mathbf{x}_0)\mathbf{u}_0$.

Fields of the form

$$\delta(\mathbf{x}-\mathbf{x}_0)\mathbf{u}_0 - F(\mathbf{x}-\mathbf{x}_0)\mathbf{u}_0 \tag{78}$$

are said to be the particle states ("dressed" corpuscles). If the input vanishes identically after the moment $t = t_0$ of the incarnation, then the open field equations have the form

$$\left(i\tfrac{\partial}{\partial t} + A\right)\tilde{\mathbf{f}} = 0$$
$$\left(i(\sigma\,\vec{\nabla}) + B\right)\tilde{\mathbf{f}} = 0 \qquad (t > t_0,\ \mathbf{x} \neq \mathbf{x}_0), \tag{79}$$

and the output $\vec{v} = -F(\mathbf{x}-\mathbf{x}_0,\ t-t_0)\mathbf{u}_0$, where

$$F(\mathbf{x},t) = \frac{i\Phi}{(2\pi)^3} \int (B - (\sigma\mathbf{k}))^{-1} e^{i(\vec{x}\mathbf{k})} d\mathbf{k} e^{itA} \Phi^* J(B) \tag{80}$$

($F(\mathbf{x},t)$ is a 4×4 matrix).

It is easy to see that

$$(B-(\sigma\mathbf{k}))^{-1}=\begin{pmatrix} B+k_3, & k_1-ik_2\\ k_1+ik_2, & B-k_3\end{pmatrix}(B^2-k^2)^{-1}. \tag{81}$$

Introducing the "potential" matrix of the form

$$V(\mathbf{x})=q\frac{i\Phi}{(2\pi)^3}\int(B-(\sigma\mathbf{k}))^{-1}k^{-2}e^{i(\mathbf{xk})}d\mathbf{k}\Phi^*\delta(B)-q\frac{I}{4\pi r},$$

where q is a constant, we obtain

$$\Delta V=\left(\frac{\partial^2}{\partial x_1^2}+\frac{\partial^2}{\partial x_2^2}+\frac{\partial^2}{\partial x_3^2}\right)V=q[\delta(\mathbf{x})I-F(\mathbf{x})]. \tag{82}$$

To calculate $V(\mathbf{x})$ one can use formulas (80) and (81):

$$V(\mathbf{x})=q\left[\begin{pmatrix}1&0\\0&1\end{pmatrix}\otimes\Phi R\Phi^*J(B)-i\begin{pmatrix}\partial_3, & \partial_1-i\partial_2\\ \partial_1+i\partial_2, & -\partial_3\end{pmatrix}\right]\Phi^*B^{-1}R\Phi^*J(B)-\frac{I}{4\pi r},$$

where

$$\begin{aligned}R&=\tfrac{1}{(2\pi)^3}\textstyle\int B(B^2-k^2)^{-1}e^{irk\cos\theta}\sin\theta dkd\theta d\varphi\\ &=\tfrac{1}{(2\pi)^2}\textstyle\int_{-\infty}^{+\infty}(B-k)^{-1}\frac{e^{irk}-e^{-irk}}{2irk}dk.\end{aligned}$$

Using the residue formula we obtain

$$R(\mathbf{x})=\frac{i}{4\pi r}B^{-1}(I-e^{irB}),\quad r=(x_1^2+x_2^2+x_3^2)^{1/2}. \tag{83}$$

Taking the mean value $\overline{\mathbf{V}}=\frac{1}{2}(V^{11}+V^{22})$ with respect to the spin index $\varepsilon=1,2$ we obtain

$$\overline{\mathbf{V}}=q[\Phi R\Phi^*J(B)-\frac{I}{4\pi r}]. \tag{84}$$

Analogously, one can calculate the field $\mathbf{F}(\mathbf{x},t)$ using the corresponding potential $V(\mathbf{x},t)$. In this case we obtain

$$R(\mathbf{x},t)=R(x)e^{itA} \tag{85}$$

and the incarnation field $\tilde{\mathbf{v}}$ can be calculated with the help of the formula

$$qF(\mathbf{x},t)=-\Delta V(\mathbf{x},t), \tag{86}$$

where

$$V(\mathbf{x},t)=q\left[\begin{pmatrix}1&0\\0&1\end{pmatrix}\otimes\Phi R(\mathbf{x},t)\Phi^*J(B)-i\begin{pmatrix}\partial_3, & \partial_1-i\partial_2\\ \partial_1+i\partial_2, & -\partial_3\end{pmatrix}\Phi B^{-1}R(\mathbf{x},t)\Phi^*J(B)\right].$$

Just as in the one-dimensional case, which was discussed in §4, one can conclude that if all the eigenvalues $\zeta_k=e^{i\varphi_k}$ of Q are distinct and are located on the unit circle, then all the eigenvalues $\lambda_k=a\cos\varphi_k$ $(0<\varphi_k<\pi)$ are real and the eigenvalues $\mu_k=i\frac{a}{c}\sin\varphi_k$ are purely imaginary. The incarnation field $F(\mathbf{x},t)\mathbf{u}_0$ is stable: if $t \to \infty$ all the factors $e^{it\lambda_k}$ are oscillating.

Remark: *It is easy to see that if $r \to 0$, then*

$$\overline{\mathbf{V}}(\mathbf{x}) = q\left[-\frac{I}{4\pi r} + \frac{\Phi\Phi^* J(B)}{4\pi} + O(r)\right]$$

and that, if $dimH < \infty$, then the norm $\|e^{irB}\| = O(e^{-lr})$, $(l > 0)$. Hence, the asymptotic expression of $\mathbf{V}(x)$ as $r \to \infty$ is:

$$\mathbf{V}(\mathbf{x}) = q\left[-\frac{1}{4\pi r}S_B(0) + O(e^{lr})\right], \quad l > 0,$$

where

$$S_B(\mu) = I - i\Phi(B - \mu I)^{-1}\Phi^* J(B)$$

is the characteristic operator-function of B ($S_B(0)$ is a 2×2 unitary matrix).

As an example consider the case in which $dimH = N = 2$. Let $\zeta_1 = e^{i\varphi_1}$ and $\zeta_2 = -e^{i\varphi_1}$, $(0 < \varphi_1 < \pi/2)$. Then $\zeta_1 = \alpha_1 + i\frac{\beta_1^2}{2}$ and $zeta_2 = \alpha_2 + i\frac{\beta_2^2}{2}$, where $\alpha_1 = -\alpha_2 = \cos\varphi_1$ and $\beta_1 = \beta_2 = (2\sin\varphi_1)^{1/2}$. The triangular models of Q and Q^{-1} are

$$Q = \begin{pmatrix} e^{i\varphi_1}, & 0 \\ 2i\sin\varphi_1, & -e^{-i\varphi_1} \end{pmatrix} \text{ and } Q^{-1} = \begin{pmatrix} e^{-i\varphi_1}, & 0 \\ 2i\sin\varphi_1, & -e^{i\varphi_1} \end{pmatrix}.$$

and the channel vectors are

$$g_1 = \begin{pmatrix} \beta_1 \\ \beta_2 \end{pmatrix} = (2\sin\varphi_1)^{1/2}\begin{pmatrix} 1 \\ 1 \end{pmatrix} \text{ and } g_2 = Q^{-1}g_1 = e^{-i\varphi_1}(2\sin\varphi_1)^{1/2}\begin{pmatrix} 1 \\ -1 \end{pmatrix}.$$

The operators A and B are given by

$$A = a\begin{pmatrix} \cos\varphi_1, & 0 \\ 2\sin\varphi_1, & -\cos\varphi_1 \end{pmatrix} \text{ and } B = \frac{a}{c}\begin{pmatrix} i\sin\varphi_1, & 0 \\ 0 & i\sin\varphi_1 \end{pmatrix},$$

respectively.

The eigenvalues of A are $\lambda_1 = a\cos\varphi_1$, $\lambda_2 = -a\cos\varphi_1$ and the eigenvalues of B are $\mu_1 = \mu_2 = i\frac{a}{c}\sin\varphi_1$. In this example

$$R(\mathbf{x}) = R_0(r)\begin{pmatrix} 1 & 0 \\ 0 & 1 \end{pmatrix} \text{ and } r = (x_1^2 + x_2^2 + x_3^2)^{1/2},$$

where $R_0(r) = \frac{c}{4\pi r\sin\varphi_1}(1 - e^{-\frac{a}{c}r\sin\varphi_1})\begin{pmatrix} 1 & 0 \\ 0 & 1 \end{pmatrix}$ and

$$\overline{\mathbf{V}}(\mathbf{x}) = \frac{q}{4\pi r}(1 - 2e^{-\frac{a}{c}r\sin\varphi_1})\begin{pmatrix} 1 & 0 \\ 0 & 1 \end{pmatrix}.$$

Then

$$\overline{\mathbf{V}} \simeq \begin{cases} q[-\frac{1}{4\pi r} + \frac{a\sin\varphi_1}{c2\pi}], & r \to 0 \\ \frac{q}{4\pi r}, & r \to \infty \end{cases}$$

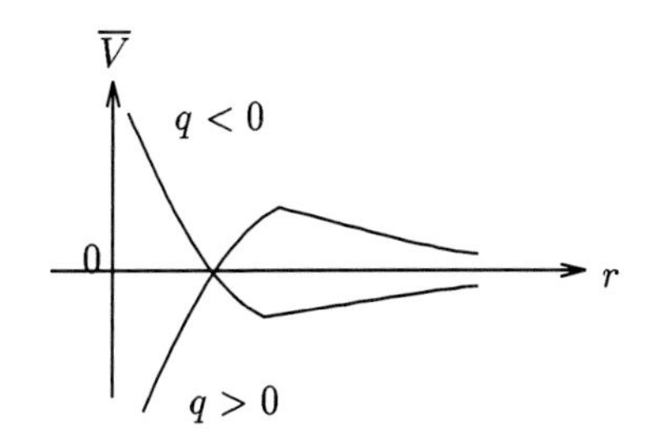

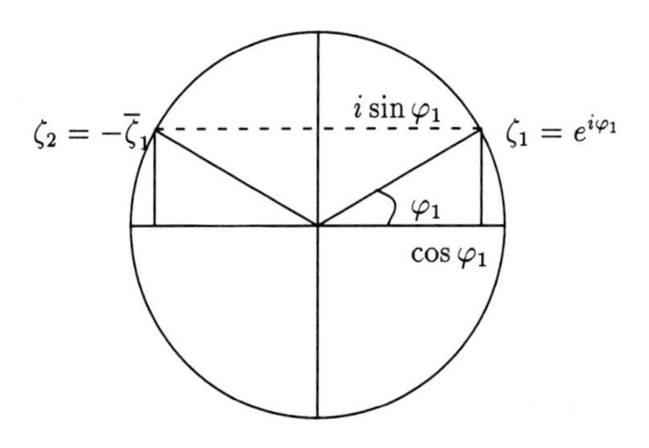

$$\lambda_1 = -\lambda_2 = a \cos \varphi_1$$

$$\mu_1 = \mu_2 = i\frac{a}{c} \sin \varphi_1$$

$$(0 < \varphi_1 < \pi/2)$$

If $\varphi_1 \to \pi/2$, then the limits of the matrices A and B are

$$A = a\begin{pmatrix} 0 & 0 \\ 2 & 0 \end{pmatrix} \text{ and } B = \frac{a}{c}\begin{pmatrix} i & 0 \\ 0 & i \end{pmatrix}.$$

In this case $\lambda_1 = \lambda_2 = 0, \ \ \mu_1 = \mu_2 = ia/c$ and

$$\overline{\mathbf{V}}(\mathbf{x}, t) = \frac{q}{2\pi r}(1 - e^{-\frac{a}{c}r})\begin{pmatrix} 1 & 0 \\ 2at & 1 \end{pmatrix}.$$

Therefore, the element $\mathbf{V}_{1,2}(\mathbf{x}, t)$ tends to infinity, when $t \to \infty$.

To obtain a more general class of examples, let us consider operators of the form

$$A = \begin{pmatrix} cK, & aI \\ aI, & -cK \end{pmatrix} \text{ and } B = \begin{pmatrix} K, & 0 \\ 0, & K \end{pmatrix}, \tag{87}$$

where K is an operator in a Hilbert space $\tilde{H}$ such that

$$\frac{1}{i}(K - K^*)\tilde{f} = (\tilde{f}, \tilde{g})\tilde{g}.$$

The operators A and B are defined on the space $H = \tilde{H} \oplus \tilde{H}$. Introducing $\tilde{F} = \mathbb{C}_1$ and $\tilde{\Phi}\tilde{h} = \tilde{h}\tilde{g}$, we obtain $\tilde{\Phi}^*\tilde{u} = \tilde{u}\tilde{g}$ and

$$\frac{1}{i}(K - K^*) = \tilde{\Phi}^*\tilde{\Phi}.$$

Let us define the operators Q_1, Q_2 as follows:

$$Q_1 = \frac{A + cB}{a} = \begin{pmatrix} 2\frac{c}{a}K, & I \\ I, & 0 \end{pmatrix}, \quad Q_2 = \frac{A - cB}{a} = \begin{pmatrix} 0, & I \\ I, & -2\frac{c}{a}K \end{pmatrix}.$$

It is evident that $Q_1Q_2 = I$ and that

$$\frac{1}{i}(Q_1 - Q_1^*) = 2\frac{c}{a}\frac{1}{i}(K - K^*)\begin{pmatrix} 1 & 0 \\ 0 & 0 \end{pmatrix}, \quad \frac{1}{i}(Q_2 - Q_2^*) = -2\frac{c}{a}\frac{1}{i}(K - K^*)\begin{pmatrix} 0 & 0 \\ 0 & 1 \end{pmatrix}.$$

Hence, the operators A and B have the form (34):

$$A = \frac{a}{2}(Q + Q^{-1}) \text{ and } B = \frac{a}{2c}(Q - Q^{-1}),$$

where $Q = Q_1$ is the generator. From (87) it follows that

$$2ImA = \Phi^*J(A)\Phi, \quad 2ImB = \Phi^*J(B)\Phi \text{ and } 2Im(AB^*) = 2Im(B^*A) = \Phi^*J((AB^*))\Phi,$$

where

$$J(A) = c\begin{pmatrix} 1 & 0 \\ 0 & -1 \end{pmatrix}, \quad J(B) = \begin{pmatrix} 1 & 0 \\ 0 & 1 \end{pmatrix}, \quad J(AB^*) = J(B^*A) = -a\begin{pmatrix} 0 & 1 \\ 1 & 0 \end{pmatrix}$$

and

$$\Phi = \begin{pmatrix} \tilde{\Phi} & 0 \\ 0 & \tilde{\Phi} \end{pmatrix}, \quad H = \tilde{H} \oplus \tilde{H}, \quad E = \tilde{E} \oplus \tilde{E}.$$

In this case the matrix $R(\mathbf{x})$ is

$$R(\mathbf{x}) = \tilde{R}(r)\begin{pmatrix} 1 & 0 \\ 0 & 1 \end{pmatrix},$$

where

$$\tilde{R}(r) = \frac{i}{4\pi r}K^{-1}(I - e^{irK}). \tag{88}$$

Remark: An open field with spin $\frac{1}{2}$ can be considered as a superposition of plane diatrons in all possible directions. Indeed, let us consider the equations (61) and (62) of an open field in the plane-wave case:

$$f^\varepsilon = f(y,t) \otimes (\xi^\pm)^\varepsilon$$

$$u_j^\varepsilon = u_j(y,t) \otimes (\xi^\pm)^\varepsilon,$$

where $y = \alpha_1 x_1 + \alpha_2 x_2 + \alpha_3 x_3$, $(\alpha_1^2 + \alpha_2^2 + \alpha_3^2 = 1)$, α_k are real numbers and the $(\xi^{\pm})^{\epsilon}$ are eigenvectors of $(\vec{\alpha}, \vec{\sigma})$:

$$(\alpha_1\sigma_1 + \alpha_2\sigma_2 + \alpha_3\sigma_3)(\xi^{\pm}) = \pm(\xi^{\pm}).$$

In this case the equations (61) and (62) have the form

$$(i\frac{\partial}{\partial t} + A)f(y,t) = \Phi^* J(A)[u(y,t)]$$

and

$$(\pm i\frac{\partial}{\partial y} + B)f(y,t) = \Phi^* J(B)[u(y,t)],$$

respectively.

Hence, for plane waves, the equations of an open field with spin $\frac{1}{2}$ reduce to the diatron equations.

References

1. N.I.Akhieser and I.M.Glazman: Theory of Linear Operators in Hilbert Space, Pitman Advanced Publishing Program, London, 1980.
2. Louis de Broglie: Non-Linear Wave Mechanics. Elsevier, Amsterdam, 1960.
3. M.S.Livšic: What is a Particle from the Standpoint of Systems Theory?, Integral Equations Operator Theory, **14** (1991), 552–563.
4. M.S.Livšic: Commuting Nonselfadjoint Operators and Collective Motions of Systems, in "Commuting Nonselfadjoint Operators in Hilbert Space", Lecture Notes in Math., **1272**, Springer-Verlag, 1987.
5. M.S.Livšic and A.A.Jancevich: Theory of Operator Colligations in Hilbert Space, J. Wiley, New York, 1979.
6. Nicholas Rescher: G.W.Leibniz's Monadology (an edition for students), London, 1991.

Department of Mathematics and Computer Sciences
Ben-Gurion University of the Negev
Beer-Sheva 84105, Israel

AMS classification # 47A45, 47A48

Operator Theory
Advances and Applications, Vol. 98

ON STABILITY OF NON-NEGATIVE INVARIANT SUBSPACES

V. LOMONOSOV

In the present work we prove theorem on existence of invariant subspaces for a class of operators in a space with indefinite metric.

Let H be an infinite dimensional Hilbert space. Let

$$H = H_1 \oplus H_2 \tag{1}$$

be a direct sum of two orthogonal subspaces. Let P_1 and P_2 be orthogonal projections onto H_1 and H_2 and $J = P_1 - P_2$. Then $J = J^*$, $J^2 = P_1 + P_2 = I$. The operator J determines a space H_J with respect to the indefinite scalar product $< x, y > \stackrel{\text{def}}{=} (Jx, y)$. If $\dim H_2 < \infty$, then H_J is called Pontryagin space, if dim $H_1 = \dim H_2 = \infty$ it is called Krein space.

The vector X is said to be non-negative if $< x, x > \geq 0$. The set β^+ of all non-negative vectors is a cone in H. A subspace $L \subset H_J$ is called non-negative if $L \subset \beta^+$. By Zorn's lemma every non-negative subspace can be extended to a maximal non-negative subspace. The set of these subspaces will be denoted by β^{++}. A linear operator A is called J-non-transitive if it has an invariant subspace $L \in \beta^{++}$. An operator A is called J-selfadjoint if it is selfadjoint with respect to the J-scalar product. That means that $(JAx, y) = (Jx, Ay)$ or $JA = A^*J$. An operator A is called J-dissipative if the operator JA is dissipative, that is $\frac{1}{2i}(JA - A^*J) \geq 0$. Every J-selfadjoint operator is obviously J-dissipative. According to decomposition (1) every operator on H has the form

$$A = \begin{pmatrix} A_{11} & A_{12} \\ A_{21} & A_{22} \end{pmatrix} \tag{2}$$

In 1944 L.S.Pontryagin proved that every J-selfadjoint operator in Pontryagin space is J-non-transitive [1]. In the series of papers [2], [3], [4] (from 1967 to 1964) M.G.Krein proved that the same result is true in Krein space if the operator A_{12} in the decomposition (2) is compact. To prove his result Krein used Tychonov's fixed point theorem for locally convex space. Finally, the combination of Krein's technique with Ky Fan's nice generalization of Tychonov's fixed point theorem [5] proves J-non-transitivity for every J-dissipative operator A with compact A_{12}.

In this paper we will develop a simple approximation procedure that gives all the results mentioned above.

Let $T : H_1 \to H_2$ be a linear contraction, that is $\|T\| \leq 1$. Then a subspace $L \subset H$ of the form $L = \{x + Tx;\ x \in H_1\}$ is maximal and non-negative. It is easy to see that every subspace $L \in \beta^{++}$ is the graph of some contraction $T : H_1 \to H_2$. Let the subspace $L \in \beta^{++}$ be invariant for the operator A. It means that $A(x + Tx) = y + Ty$ holds for every $x \in H_1$. More precisely we have

$$\begin{pmatrix} A_{11} & A_{12} \\ A_{21} & A_{22} \end{pmatrix} \begin{pmatrix} x \\ Tx \end{pmatrix} = \begin{pmatrix} A_{11}x + A_{12}Tx \\ A_{21}x + A_{22}Tx \end{pmatrix} = \begin{pmatrix} y \\ Ty \end{pmatrix}$$

and this gives $T(A_{11}x + A_{12}Tx) = A_{21}x + A_{22}Tx$. Since x is arbitrary, T is a solution of the equation

$$TA_{11} + TA_{12}T - A_{21} - A_{22}T = 0 \tag{3}$$

On the other hand, the graph L of every solution T of the equation (3) is in β^{++} if T is a contraction. Hence to find an invariant subspace $L \in \beta^{++}$ for an operator A and to find a solution T of the equation (3) with $\|T\| \leq 1$ are equivalent problems.

Let $A^{(K)}$ be a sequence of operators in H with decomposition (2)

$$A^{(K)} = \begin{pmatrix} A_{11}^{(K)} & A_{12}^{(K)} \\ & \\ A_{21}^{(K)} & A_{22}^{(K)} \end{pmatrix}$$

We will say that the operator A is a mixed limit of operators $A^{(K)}$ if:

1. $A_{11} = s - \lim\limits_{K\to\infty} A_{11}^{(K)}$
2. $A_{22}^* = s - \lim\limits_{K\to\infty} (A_{22}^{(K)})^*$
3. $A_{21} = w - \lim\limits_{K\to\infty} A_{21}^{(K)}$
4. $\lim\limits_{K\to\infty} \|A_{12} - A_{12}^{(K)}\| = 0.$

Here $s - \lim$ and $w - \lim$ are limits in the strong and weak operator topology. We now prove the following

Theorem. *Let an operator A be the mixed limit of the sequence $\{A^{(K)}\}$ and let A_{12} be compact.*

Let every operator $A^{(K)}$ be J-non-transitive. Then A is J-non-transitive.

Proof. J-non-transitivity of an operator $A^{(K)}$ implies that there is a contraction T_K which satisfies the equation

$$T_K A_{11}^{(K)} + T_K A_{12}^{(K)} T_K - A_{21} - A_{22}^{(K)} T_K = 0. \tag{4}$$

Let T_{K_n} be a subsequence that tends weakly an operator T. Then conditions 1 - 4 give weak convergence of all four terms in the equations (4) to the corresponding terms of (3).

This proves that T is a solution of (3) and hence the operator A is J-non-transitive.

Remark. It is easy to see that if the invariant maximal non-negative subspace of the operator A is unique then we have $T = w - \lim K \to \infty T_K$. Thus the unique invariant subspace $L \in \beta^{++}$ is stable under mixed perturbation.

Corollary. *Let A be J-dissipative and let A_{12} in representation (2) be compact. Then A is J-non-transitive.*

Proof. Let P_n^j be a sequence of n-dimensional projections in the spaces H_i $(i = 1,2)$, and let $Q_n = P_n^1 + P_n^2$ satisfy $s - \lim Q_n = I$. Let $H^{(n)} = Q_n H$, $A_n = Q_n A Q_n$, $J_n = Q_n J Q_n$. A_n is obviously a J_n-dissipative finite dimensional operator in the subspace $H^{(m)}$. Then, according to the corresponding finite dimensional theorem, A_n is J_n-non-transitive. Let $L_n \supset H^{(n)}$ be a non-negative maximal invariant subspace for A_n. Let $M_n = L_n \oplus (1 - P_n^1)H_1$. Then $M_n \in \beta^{++}$ is an invariant subspace for A_n. It is very easy to check that the operators $A_{i,j}^{(n)} = P_n^i A_n P_n^j$ $(i,j = 1,2)$ satisfy the conditions of the theorem.

For example, the equality

$$\lim_{n\to\infty} \|A_{12}^{(n)} - A_{12}\| = 0$$

is true due to the compactness of A_{12}. So the corollary is proved.

Finally it should be mentioned that replacing the uniform limit by the strong limit in the fourth condition would give that every J-dissipative operator is J-non-transitive.

These results were presented in author's talk in the Voronezh Winter Mathematical School in 1986.

REFERENCES

1. Pontryagin, L.S., *Hermitian Operators on a Space with Indefinite Metric*, (Russian) Izv. Akad. Nauk USSR Mat. Vol. 8 (1944) pp. 243-280.

2. Krein, M.G. and Rutman, M.A., *Linear Operators having an Invariant Cone in a Banach Space*, (Russian) Usp. Mat. Nauk Vol. (1948).

3. Krein, M.G., *On an Application of a Fixed Point Theorem to Operator Theory in a Space with Indefinite Metric*, (Russian) Usp. Mat. Nauk Vol. 5 No. 2 (1950) pp. 180-190.

4. Krein, M.G., *On a New Application of a Fixed Point Theorem to Operator Theory in a Space with Indefinite Metric*, (Russian) Dokl, Akad. Nauk USSR, Vol. 154 No. 5 (1964) pp. 1023-1026.

5. Fan, Ky, *Invariant Subspaces of Certain Linear Operators*, Bull. Amer. Math. Soc., 69 (1963) pp. 773-777.

Department of Mathematics & Computer Science,

Kent State University, P.O. Box 5190,

Kent, OH 44242, U.S.A.

AMS Classification 47A15

Operator Theory
Advances and Applications, Vol. 98

THE DUALITY OF SPECTRAL MANIFOLDS AND LOCAL SPECTRAL THEORY

V. LOMONOSOV YU. LYUBICH V. MATSAEV

1 Introduction

Let T be an arbitrary linear bounded operator on a complex Banach space B. As usual, we denote by $\rho(T)$ the resolvent set of T, i.e., $\rho(T)$ is the set of $\lambda \in \mathbf{C}$ such that the resolvent $R_\lambda(T) = (T - \lambda I)^{-1}$ exists in the algebra of all linear bounded operators on B. This set is open and $R_\lambda(T)$ is an analytic operator function on it. The spectrum $\sigma(T) = \mathbf{C}/\rho(T)$ is compact.

In our context the conjugate operator T^* acts in the space B^* consisting of all linear (not antilinear!) continuous functionals on B. By definition, the identity $\varphi(Tx) = (T^*\varphi)(x)$ holds for all $x \in B$, $\varphi \in B^*$. Let us emphasize that $\rho(T^*) = \rho(T)$ and $R_\lambda(T^*) = R_\lambda(T)^*$. Respectively, $\sigma(T^*) = \sigma(T)$.

Any element $\varphi \in B^*$ will be called *covector* as against *vector* $x \in B$ and we write (x, φ) instead of $\varphi(x)$. For any linear subspace $X \subset B$ *its annihilator* $X^\perp \subset B^*$ is defined by condition $x, \varphi) = 0$ for all $x \in B$. Similarly, the *annihilator* $Y_\perp \in B$ is defined for any linear subspace $Y \in B^*$. All of $Y_\perp$ are closed and all of $X^\perp$ are w^*-closed. It is easy to prove that $(X^\perp)_\perp = X$ if X is closed and $(Y_\perp)^\perp = Y$ if Y is w^*-closed.

Let Λ be an arbitrary nonempty subset of the complex plane $\mathbf{C}$. A vector function $f : \Lambda \to B$ is called *analytic* on Λ if there exists an open set $U_f \supset \Lambda$ such that f can be analytically extended to U_f. (If Λ is open, we may take $U_f = \Lambda$). The linear space of all analytic functions $f : \Lambda \to B$ will be denoted by $A(\Lambda, B)$. On the other hand, we consider the Banach space $l_\infty(\Lambda, B)$ of bounded functions $f : \Lambda \to B$ provided with the usual norm $\|f\| = \sup_{\lambda\in\Lambda}\|f(\lambda)\|$. There is the linear subspace $A_\infty(\Lambda, B) = A(\Lambda, B) \bigcap l_\infty(\Lambda, B)$. In general, this is nonclosed, but this is closed if Λ is open.

Every vector $x \in B$ defines the constant function $\hat{x}(\lambda) \equiv x$, so $x \mapsto \hat{x}$ is a natural isometric embedding $B \to l_\infty(\lambda, B)$, $\operatorname{Im} B \subset A_\infty(\Lambda, B)$ and $\operatorname{Im} B$ is closed in $l_\infty(\Lambda, B)$.

For further applications let us consider the linear subspace $\Delta_\Lambda(T) \subset A_\infty(\Lambda, B)$ which consists of the bounded functions on Λ of form $f(\lambda) = (T - \lambda I)g(\lambda)$ with $g \in A(\Lambda, B)$.

At every point $\lambda \in \Lambda \bigcap \rho(T)$ we have $g(\lambda) = R_\lambda(T)f(\lambda)$. Therefore $g(\infty) = 0$ if Λ is unbounded, since $\Lambda \bigcap \rho(T)$ is unbounded in this case, and $\|R_\lambda(T)\| \to 0$ as $\lambda \to \infty$.

The following basic construction is classical and very fruitful in the modern spectral theory.

Definition 1.1. *Let $\Omega \subset \mathbf{C}$, $\Omega \neq \mathbf{C}$ and $\Lambda = \mathbf{C} \setminus \Omega$. The linear subspace*

$$L_\Omega(T) = \{x \mid x \in B, \ \hat{x} \in \Delta_\Lambda(T)\} \tag{1.1}$$

is called prespectral manifold of the operator T corresponding to the set Ω.

Thus, $x \in L_\Omega(T)$ if and only if $x = (T - \lambda I)g(\lambda)$, $\lambda \in \Lambda$, where $g : \Lambda \to B$ is an analytic function, so $g(\lambda) = R_\lambda(T)x$ for $\lambda \in \Lambda \bigcap \rho(T)$. Outside of $\rho(T)$ the function g is, in general, not unique. The uniqueness is provided by SVEP, the *single valued extension property* of T. If T has SVEP and $L_\Omega(T)$ is closed for a closed Ω, then $L_\Omega(T)$ is just a spectral maximal subspace in the Foias sense [3] and $\sigma(T \mid L_\Omega(T)) \subset \Omega \bigcap \sigma(T)$ (see [2], p.23-24). An example of nonclosed $L_\Omega(T)$ for a closed Ω was given in [2, p.25-26].

Note that $L_\Omega(T)$ is, obviously, hyperinvariant.

The dependence $L_\Omega(T)$ of Ω is monotone, i.e., $\Omega_1 \subset \Omega_2 \Rightarrow L_{\Omega_1}(T) \subset L_{\Omega_2}(T)$ because $\Lambda_2 \subset \Lambda_1 \Rightarrow \Delta_{\Lambda_1}(T) \subset \Delta_{\Lambda_2}(T)$.

Following Bishop [1] we consider the *(strong) spectral manifold* $M_\Omega(T)$ as the closure of $L_\Omega(T)$ in B; a wider *(weak) spectral manifold* $N_\Omega(T)$ is defined as the linear subspace in B consisting of the vectors x such that the corresponding constant functions $\hat{x}$ belong to the closure of $\Delta_\Lambda(T)$ in $l_\infty(\Lambda, B)$. Thus, $x \in N_\Omega(T)$ if and only if for every $\varepsilon > 0$ there exists a function $g \in A(\Lambda, B)$ such that

$$\|x - (T - \lambda I)g(\lambda)\| < \varepsilon, \quad \lambda \in \Lambda. \tag{1.2}$$

Obviously, $N_\Omega(T)$ is closed.

Certainly, the previous definitions can be applied to the conjugate operator T^*. Bishop chose this way because he only considered the reflexive case.

In note [6] we modified Bishop's constructions to extend his "duality theory of type 4" to the nonreflexive case. Moreover, we obtained a *local* spectral theorem under a relevant modification of Bishop's *global* condition β.

Definition 1.2. *Let F be a closed subset of $\mathbf{C}$, $F \neq \mathbf{C}$, and $G = \mathbf{C} \setminus F$. We say that T satisfies condition β_F (briefly, $T \in \beta_F$) if for every compact set $K \subset G$ the inequality*

$$\sup_{\lambda \in K} \|g(\lambda)\| \leq C_K \sup_{\lambda \in G} \|(T - \lambda I)g(\lambda)\| \tag{1.3}$$

with some constant C_K holds for all $g \in A(G, B)$.

Bishop's condition β is equivalent to $(\forall F)\beta_F$. Under this condition for T and T^* a theorem on a "weak decomposability" was established in [1]. Now it is known (see [4]) that $(T \in \beta)\,\&\,(T^* \in \beta)$ is equivalent to decomposability in Foias' sense [3].

Obviously, β_F implies the following *local* form of SVEP:

$$g \in A(G, B)\,\&\,(T - \lambda I)g(\lambda) \equiv 0 \Rightarrow g = 0. \tag{1.4}$$

Let us denote property (1.4) by SVEP| G.

Our proofs of the above mentioned results from [6] did still not appear, except for Ph.D. Thesis [5]. Here we present them with a further development. In particular, our technique leads to some exact duality relations (see Section 4 below) instead of inclusions in [1]. These inclusions easily follow from those relations and, moreover, some required conditions are weaker than in [1].

Our approach is based on some explicit descriptions of some analytic functionals and corresponding descriptions of the annihilators $\Delta_F(T)^\perp$ and $\Delta_F(T^*)_\perp$ with closed F (see Section 3). Though Bishop also used some spaces of analytic vector and covector functions, and his method also yields an exact duality relation [1, p.389] (even in the nonreflexive case [4, p.7]), his key point is different (cf. [1, Th.1]).

The main point of Section 5 is the above mentioned local spectral theorem. Now we call it the *spectrum splitting-off theorem* (Theorem 5.13). We obtain this by an "optimal" choice of the set Φ from the following result.

Theorem 5.12 *Let* Φ *be a closed subset of* $\mathbb{C}$ *such that* $\sigma(T) \subset F \bigcup \mathrm{Int}\Phi$. *Suppose that* $T \in \beta_F$ *and* $T^* \in \beta_F$. *Then*

$$\sigma(T) \subset \sigma(T \mid L_F(T)) \bigcup \sigma(T^* \mid L_\Phi(T^*)).$$

Note that the subspaces $L_F(T)$ and $L_\Phi(T^*)$ are closed in this case.

Let us stress that our requirement on $\sigma(T)$ is weaker than usual one, namely, $\sigma(T) \subset \mathrm{Int}F \bigcup \mathrm{Int}\Phi$ (cf. [1]). We are able to do it thanks to the above mentioned improvement of the duality inclusions.

We dedicate this paper to memory of I.M.Glazman, whose powerful activity in Operator Theory inspired us when we began our investigations in this attractive area of mathematics.

2 Spectral manifolds.

Let us start with some simple remarks concerning the prespectral manifolds $L_\Omega(T)$.

Proposition 2.1. $\Delta_{\mathbb{C}}(T) = 0$ *therefore* $L_\emptyset(T) = M_\emptyset(T) = N_\emptyset(T) = 0$.

Proof. All functions from $\Delta_{\mathbb{C}}(T)$ are constant by the Liouville Theorem. If $\hat{x} \in \Delta_{\mathbb{C}}(T)$ for a vector $x \in B$ then $R_\lambda(T)x$ is analytically continued to an entire function. Therefore

$$x = -\frac{1}{2\pi i}\int\limits_{|\lambda|=r} R_\lambda(T)x\,d\lambda$$

with $r > \|T\|$, so $x = 0$. □

On the other hand, we have

Proposition 2.2. *If* $\Omega \supset \sigma(T)$ *then* $L_\Omega(T) = B$.

Proof. Every $f \in A_\infty(\Lambda, T)$ can be represented as $f(\lambda) = (T - \lambda I)g(\lambda)$ with $g(\lambda) = R_\lambda(T)f(\lambda)$, $\lambda \in \Lambda$, because $\Lambda \subset \rho(T)$. Hence, $\Delta_\Lambda(T) = A_\infty(\Lambda, T)$, so

$L_\Omega(T) = B.$ □

Proposition 2.2 suggests to set $L_{\mathbb{C}}(T) = B$ and then $\Delta_\emptyset(T) = B$ in order to rescue Definition 1.1. After that $M_{\mathbb{C}}(T) = N_{\mathbb{C}}(T) = B$ by the standard definitions.

Actually, $L_\Omega(T)$ depends only on the part of Ω lying in the spectrum $\sigma(T)$. In particular, $L_\Omega(T) = \emptyset$ if $\Omega \subset \rho(T)$. Indeed, we have

Proposition 2.3. $L_\Omega(T) = L_{\Omega \bigcap \sigma(T)}(T)$.

Proof. It is sufficient to check that $L_\Omega(T) \subset L_{\Omega \bigcap \sigma(T)}(T)$. Let $x \in B$ and $x = (T - \lambda I)g(\lambda)$, $\lambda \in \Lambda$, $g \in A(\Lambda, B)$. Then the formula

$$\tilde{g}(\lambda) = \begin{cases} g(\lambda), & \lambda \in \Lambda \\ R_\lambda(T)x, & \lambda \in \rho(T) \end{cases}$$

properly defines a function $\tilde{g} \in A(\Lambda \bigcup \rho(T), B)$ such that $x = (T - \lambda I)\tilde{g}(\lambda)$ for all $\lambda \notin \Omega \bigcap \sigma(T)$. □

We now introduce some other *spectral manifolds* $M^*_\Omega(T^*)$ and $N^*_\Omega(T^*)$ instead of $M_\Omega(T^*)$ and $N_\Omega(T^*)$ which are defined in B^* like $M_\Omega(T)$ and $N_\Omega(T)$ in B. Such a modification is relevant in nonreflexive case.

Definition 2.4. $M^*_\Omega(T^*)$ *is the* w^**-closure of* $L_\Omega(T^*)$ *in* B^*.

Thus, $M^*_\Omega(T^*)$ is w^*-closed. Obviously, $M^*_\Omega(T^*) \supset M_\Omega(T^*)$ and they coincide if B is reflexive.

Definition 2.5. $N^*_\Omega(T^*)$ *is the linear subspace in* B^* *consisting of the covectors* $\varphi \in B^*$ *such that the constant functions* $\hat{\varphi}$ *belong to the* w^**-closure of* $\Delta_\Lambda(T^*)$ *in* $l_\infty(\Lambda, B)$.

By these definitions and our agreement above, $M^*_{\mathbb{C}}(T^*) = N^*_{\mathbb{C}}(T^*) = B^*$. By Proposition 2.1, $M^*_\emptyset(T^*) = N^*_\emptyset(T^*) = 0$.

It is easy to check that $N^*_\Omega(T^*) \supset N_\Omega(T^*)$.

Obviously, all the above defined spectral manifolds monotonically depend on Ω.

To explain a formal aspect of Definition 2.5. let us note that the space $l_\infty(\Lambda, B)$ is conjugate to $l_1(\Lambda, B)$, the Banach space consisting of functions $p : \Lambda \to B$ such that

$$\|p\| \equiv \sum_{\lambda \in \Lambda} \|p(\lambda)\| < \infty. \tag{2.1}$$

(in this situation the support $\{\lambda : p(\lambda) \neq 0\}$ must be at most countable).

We have a natural pairing

$$(p, \gamma) = \sum_{\lambda \in \Lambda} (p(\lambda), \gamma(\lambda)) \tag{2.2}$$

with arbitrary $\gamma \in l_\infty(\Lambda, B^*)$. Obviously, $|\,(p, \gamma)\,| \leq \|p\| \cdot \|\gamma\|$. Taking $\gamma(\lambda)$ under the conditions $(p(\lambda), \gamma(\lambda)) = \|p(\lambda)\| \cdot \|\gamma(\lambda)\| = 1$ at every point λ from the support of p and $\gamma(\lambda) = 0$ otherwise, we obtain $(p, \gamma) = \|p\| \cdot \|\gamma\| = 1$. In such a way the space $l_1(\Lambda, B)$ is isometrically embedded into $l_\infty(\Lambda, B^*)$. In fact, $l_\infty(\Lambda, B^*) \equiv l_1(\Lambda, B)^*$, so $l_1(\Lambda, B)$ is canonically embedded into $l_\infty(\Lambda, B^*)$. It is just the same embedding as above.

Proposition 2.6. *w^*-topology on $l_\infty(\Lambda, B^*)$ is stronger than the topology $\tau \equiv \tau_{\Lambda,B^*}$ of the topological power $(B^*_{w^*})^\Lambda$ where $B^*_{w^*}$ is the space B^* provided with w^*-topology. They coincide on every bounded subset $D \subset l_\infty(\Lambda, B^*)$.*

Proof. The first statement follows by the choice $p(\lambda) = x\delta_{\lambda,\lambda_0}$ with a given $x \in B$ and $\lambda_0 \in \Lambda$ (here $\delta_{\lambda,\lambda_0}$ is the Kronecker delta). Then the inequality $|(p,\gamma)| < \varepsilon$ turns into $|(x, \gamma(\lambda_0))| < \varepsilon$.

Conversely, for any $p \in l_1(\Lambda, B)$ and $\varepsilon > 0$ there exists a finite subset $\Lambda_0 \in \Lambda$ such that

$$\sum_{\lambda\in\Lambda\setminus\Lambda_0} \|p(\lambda)\| < \frac{\varepsilon}{2\mu}$$

where $\mu = \sup_{\gamma\in D}\|\gamma\|$. Now if $\gamma \in D$ and

$$\max_{\lambda\in\Lambda_0} |(p(\lambda), \gamma(\lambda))| < \frac{\varepsilon}{2\mathrm{card}\Lambda_0}$$

then $|(p,\gamma)| < \varepsilon$ by virtue of (2.2). □

Proposition 2.7. *$N^*_\Omega(T^*)$ is w^*-closed.*

Proof. Let a covector φ belong to the w^*-closure of $N^*_\Omega(T^*)$. Let $\varepsilon > 0$ and $p \in l_1(\Lambda, B)$ be given. Consider the vector $x = \sum_{\lambda\in\Lambda} p(\lambda)$ and find $\theta \in N^*_\Omega(T^*)$ such that $|(x, \varphi - \theta)| < \varepsilon$. By Definition 2.5 there exists $\alpha \in A(\Lambda, B^*)$ such that $(T^* - \lambda I)\alpha(\lambda)$ is bounded on Λ and

$$\left|\sum_{\lambda\in\Lambda} (p(\lambda), \theta - (T^* - \lambda I)\alpha(\lambda))\right| < \varepsilon,$$

i.e.,

$$\left|(x, \theta) - \sum_{\lambda\in\Lambda} (p(\lambda), (T^* - \lambda I)\alpha(\lambda))\right| < \varepsilon.$$

Then

$$\left|(x, \varphi) - \sum_{\lambda\in\Lambda} (p(\lambda), (T^* - \lambda I)\alpha(\lambda))\right| < 2\varepsilon$$

hence,

$$\left|\sum_{\lambda\in\Lambda} (p(\lambda), \varphi - (T^* - \lambda I)\alpha(\lambda))\right| < 2\varepsilon.$$

This means that $\varphi \in N^*_\Omega(T^*)$. □

Corollary 2.8. $N^*_\Omega(T^*) \supset M^*_\Omega(T^*)$.

Proof. $N^*_\Omega(T^*) \supset L_\Omega(T^*)$ and $N^*_\Omega(T^*)$ is w^*-closed. □

Obviously, $N^*_\Omega(T^*) \supset N_\Omega(T^*)$.

To obtain some stronger conclusions on closure we turn to condition β_F. The following simple property is actually well-known.

Proposition 2.9. *If $T \in \beta_F$ then $\Delta_G(T)$ is closed in $l_\infty(G, B)$.*

Proof. Let $\lim_{n\to\infty}(T-\lambda I)g_n(\lambda) = f(\lambda)$ uniformly on $G, g_n \in A(G,B)$, $f \in l_\infty(G,B)$. It follows from (1.3) that $\{g_n\}$ is a Cauchy sequence with respect to the sup-norm on every compact subset $K \subset G$. Therefore this sequence converges uniformly on every K to a function $g \in A(G,B)$. Obviously, $(T-\lambda I)g(\lambda) = f(\lambda)$ everywhere on G, so $f \in \Delta_G(T)$. □

Corollary 2.10. *If $T \in \beta_F$ then the prespectral manifold $L_F(T)$ is closed and*

$$N_F(T) = M_F(T) = L_F(T). \tag{2.3}$$

Certainly Proposition 2.9 can be applied to T^*. But in this case one can get more.

Theorem 2.11. *If $T^* \in \beta_F$ then the linear subspace $\Delta_G(T^*) \subset l_\infty(G,B)$ is w^*-closed.*

Proof. By a Banach theorem it is sufficient to prove that every w^*-limit point of the unit ball

$$D = \{\varphi \mid \varphi \in \Delta_G(T^*),\ \sup_{\lambda\in G}\|\varphi(\lambda)\| \leq 1\}$$

belongs to the subspace $\Delta_G(T^*)$.

By Proposition 2.6 w^*-topology on D coincides with the topology $\tau \equiv \tau_{G,B^*}$ of pointwise w^*-convergence on G. Therefore, we should prove that D is τ-closed.

Let $\{\varphi_\nu\}$ be a directed set with $\varphi_\nu \in D$ and $\varphi = \lim_\nu \varphi_\nu$ in the topology τ. By definition, $\varphi_\nu(\lambda) = (T^*-\lambda I)\psi_\nu(\lambda)$ $(\lambda \in G)$ where $\psi_\nu \in A(G,B^*)$. It follows from (1.3) that $\sup_{\lambda\in K}\|\psi_\nu(\lambda)\| \leq C_K$ for every compact set $K \subset G$. By a generalization of the classical Vitali Theorem (see Appendix below) there exists a directed subset τ-converging to a function $\psi \in A(G,B^*)$. One can suppose that it is the whole set. Then, for every $x \in B$ and for every $\lambda \in G$,

$$(x,\varphi(\lambda)) = \lim_\nu (x,(T^*-\lambda I)\psi_\nu(\lambda)) = \lim_\nu ((T-\lambda I)x,\psi_\nu(\lambda)) =$$

$$= ((T-\lambda I)x,\psi(\lambda)) = (x,(T^*-\lambda I)\psi(\lambda))$$

hence, $\varphi(\lambda) = (T^*-\lambda I)\psi(\lambda)$, i.e., $\varphi \in \Delta_G(T^*)$. □

Corollary 2.12 *If $T^* \in \beta_F$ then the prespectral manifold $L_F(T^*)$ is w^*-closed and*

$$N_F^*(T^*) = M_F^*(T^*) = L_F(T^*). \tag{2.4}$$

Appendix. First of all we prove a general lemma like the Arzela-Askoli theorem.

Lemma. *Let $\{\theta_\nu\}$ be a directed set of functions $Q \to S$ where Q and S are metric spaces and Q is compact. If this set is equicontinuous and pointwise convergent then it is uniformly convergent.*

Proof. Let us fix some $\delta > 0$ and let $\varepsilon > 0$ be chosen under condition

$$d_Q(q_1, q_2) < \varepsilon \Rightarrow d_S(\theta_\nu(q_1), \theta_\nu(q_2)) < \delta \tag{1}$$

where d_Q, d_S are the distances in Q and S respectively. Let $\theta(q) = \lim_\nu \theta_\nu(q)$ $(q \in Q)$. Passing to limit in (1) we get

$$d_Q(q_1, q_2) < \varepsilon \Rightarrow d_S(\theta(q_1), \theta(q_2)) \leq \delta. \tag{2}$$

Now we choose a finite ε-net $\{ q^j \}_{j=1}^n$ in Q. Let ν_0 be such that

$$\max_j d_S(\theta_\nu(q^j), \theta(q^j)) < \delta$$

for $\nu > \nu_0$. Then for every $q \in Q$ and a suitable q^j

$$d_S(\theta_\nu(q), \theta(q) \leq d_S(\theta_\nu(q), \theta_\nu(q^j)) + d_S(\theta_\nu(q^j), \theta(q^j)) + d_S(\theta(q^j), \theta(q)) < 3\delta. \qquad \square$$

The above used generalization of the Vitali theorem is the following

Theorem. *Let a subset $\Psi \in A(G, B^*)$, $\Psi \neq \emptyset$, be uniformly bounded on every compact set $K \subset G$, i.e.,*

$$\sup_{\psi \in \Psi} \sup_{\lambda \in \Lambda} \|\psi(\lambda)\| < \infty.$$

Then the subset Ψ is relatively compact in $A(G, B^*)_\tau$ provided with the topology $\tau \equiv \tau_{G,B^*}$.

Proof. The subset Ψ is relatively compact in $l_\infty(G, B^*)_\tau$ by combination of Banach-Alaoglu and Tichonov theorems. Let ψ be the τ-limit of a directed set $\{\psi_\nu\}$ with $\psi_\nu \in \Psi$. We should prove that ψ is analytic on G. Without loss of generality, one can assume that G is an open unit disk and

$$m \equiv \sup_\nu \sup_{\|\lambda\|=1} \|\psi_\nu(\lambda)\| < \infty. \tag{3}$$

It is sufficient to prove that ψ is analytic for $|\lambda| < r$ where $0 < r < 1$.

It follows from (3) and the Cauchy formula that

$$\|\psi_\nu'\| \leq C_r \ (|\lambda| \leq r)$$

where $C_r = m(1 - r^2)^{-1}$. Therefore,

$$\|\psi_\nu(\lambda_1) - \psi_\nu(\lambda_2)\| \leq C_r |\lambda_1 - \lambda_2| \quad (|\lambda| \leq r)$$

and

$$|(x, \psi_\nu(\lambda_1)) - (x, \psi_\nu(\lambda_2))| < C_r |\lambda_1 - \lambda_2| \quad (|\lambda| \leq r) \tag{4}$$

for every $x \in B$, $\|x\| = 1$. Fixing x we consider the directed set $\theta_\nu(\lambda) = (x, \psi_\nu(\lambda))$ of functions from $Q = \{\lambda : |\lambda| < r\}$ into $\mathbb{C}$. This is equicontinuous by (4) and this is pointwise convergent by choice of $\{\psi_\nu\}$. Obviously,

$$\lim_\nu \theta_\nu(\lambda) = (x, \psi(\lambda)).$$

By the Lemma, this directed set converges uniformly. Since all θ_ν are analytic, the limit function is analytic as well. This means that the covector function ψ is analytic. $\square$

3 Some analytic functionals

We continue to deal with the pairs of sets F, G in C such that F is closed, $F \neq$ C, and $G =$ C $\setminus F$. Moreover, within this Section we assume that G *is bounded.* Let us establish an explicit form of linear continuous functionals on $A_\infty(F, B)$ (which are just the restrictions to $A_\infty(F, B)$ of linear continuous functionals on $l_\infty(F, B)$).

Lemma 3.1. *For every $\alpha \in A_\infty(F, B)^*$ there exists a unique pair $\{\varphi, \psi\}$ such that $\varphi \in B^*$, $\psi \in A(G, B^*)$ and*

$$(f, \alpha) = (f(\infty), \varphi) + \frac{1}{2\pi i} \int_{\Gamma(f)} (f(\lambda), \psi(\lambda)) d\lambda \tag{3.1}$$

for all $f \in A_\infty(F, B)$.

In (3.1) $\Gamma(f)$ is an oriented rectifiable curve lying in G and such that all singularities of f are inside $\Gamma(f)$. In general, this curve is multiply connected. One can deform it, preserving the above mentioned properties. So $\Gamma(f)$ can be chosen in an arbitrary small neighborhood of the set F. We also note that for every finite set $f_1, \dots, f_m$ there exists a curve Γ playing the role of $\Gamma(f_1), \dots, \Gamma(f_m)$ simultaneously. All these circumstances are the same as for the general Gauchy formula

$$f(\mu) = f(\infty) - \frac{1}{2\pi i} \int_{\Gamma(f)} \frac{f(\lambda)}{\lambda - \mu} d\lambda. \tag{3.2}$$

Proof. If (3.1) takes place, then for an arbitrary $x \in B$ we can choose $f = \hat{x}$ and obtain

$$(x, \varphi) = (\hat{x}, \alpha). \tag{3.3}$$

This shows that the covector φ is unique. Similarly,

$$(x, \psi(\mu)) = (h_{x,\mu}, \alpha) \quad (\mu \in G) \tag{3.4}$$

where

$$h_{x,\mu}(\lambda) = \frac{x}{\lambda - \mu}$$

so $\psi : G \to B^*$ is unique. Note that

$$|(x, \varphi)| \leq \|\alpha\| \cdot \|x\|$$

and

$$|(x, \psi(\mu))| \leq \frac{\|\alpha\| \cdot \|x\|}{\operatorname{dist}(\mu, F)}$$

as follows from (3.3), (3.4).

Conversely, given $\alpha \in A_\infty(F, B)^*$, one can define $\varphi \in B^*$ and $\psi : G \to B^*$ by formulas (3.3) and (3.4). Let us check that ψ is analytic.

For $\mu_0 \in G$ and $|\mu - \mu_0| < \operatorname{dist}(\mu_0, F)$ the expansion

$$h_{x,\mu}(\lambda) = \sum_{k=0}^{\infty} \frac{x}{(\lambda - \mu_0)^{k+1}} (\mu - \mu_0)^k \quad (\lambda \in F)$$

converges uniformly on F, i.e., this series converges in the normed space $A_\infty(F, B)$. By (3.4)

$$(x, \psi(\mu)) = \sum_{k=0}^{\infty} \alpha(\frac{x}{\cdot - \mu_0)^{k+1}})(\mu - \mu_0)^k.$$

We see that ψ is a w^*-analytic covector function on G, so ψ is analytic.

Now let us rewrite formula (3.2) in a more abstract form,

$$f = f(\infty) + \frac{1}{2\pi i} \int_{\Gamma(f)} h_{f(\lambda),\lambda} d\lambda. \tag{3.5}$$

Therefore by (3.3) and (3.4) we get (3.1) since the integral in (3.5) is a limit in $A_\infty(F, B)$ of the Riemannian integral sums. □

Further, we will identify the functionals α on $A_\infty(F, B)$ and the corresponding pairs $\{\varphi, \psi\}$.

We also can explicitly describe the restrictions to $A_\infty(F, B^*)$ of linear functionals on $l_\infty(F, B^*)$ generated by functions $p \in l_1(F, B)$.

Lemma 3.3. *For every $p \in l_1(F, B)$ there exists a unique pair $\{x, g\}$ such that $x \in B$, $g \in A(G, B)$ and*

$$(p, \gamma) = (x, \gamma(\infty)) + \frac{1}{2\pi i} \int_{\Gamma(\gamma)} (g(\lambda), \gamma(\lambda)) d\lambda \tag{3.7}$$

for all $\gamma \in A_\infty(F, B^*)$.

Proof. Let us insert the Cauchy representation

$$\gamma(\mu) = \gamma(\infty) - \frac{1}{2\pi i} \int_{\Gamma(\gamma)} \frac{\gamma(\lambda)}{\lambda - \mu} d\lambda \quad (\mu \in F)$$

into

$$(p, \gamma) = \sum_{\mu \in F} (p(\mu), \gamma(\mu)).$$

This yields (3.7) with

$$x = \sum_{\mu \in F} p(\mu), \quad g(\lambda) = \sum_{\mu \in F} \frac{p(\mu}{\mu - \lambda}. \tag{3.8}$$

By the way, $\|x\| \leq \|p\|$ and

$$\|g(\lambda)\| \leq \frac{\|p\|}{\operatorname{dist}(\lambda, F)}.$$

The only pair $\{x, g\}$ defined by (3.8) yields (3.7) since (3.7) implies

$$(x, \varphi) = (p, \hat{\varphi}) \tag{3.9}$$

for all $\varphi \in B^*$, and

$$(g(\mu), \varphi) = (p, h_{\varphi,\mu}) \quad (\mu \in G). \tag{3.10}$$

□

Lemma 3.3 allows us to identify every $p \in l_1(F, B)$ considered as a functional on $A_\infty(F, B^*)$ with the corresponding pair $\{x, g\}$.

The main consequences of Lemmas 3.1 and 3.3 are the following descriptions of the annihilators $\Delta_F(T)^\perp \subset A_\infty(F, B)^*$ and $\Delta_F(T^*)_\perp \subset l_1(F, B)$.

Proposition 3.4. *The annihilator $\Delta_F(T)^\perp \subset A_\infty(F, B)^*$ consists of all $\alpha \equiv \{\varphi, \psi\}$ with $\varphi \in B^*$, $\psi \in A(G, B^*)$ such that*

$$\varphi = (T^* - \lambda I)\psi(\lambda) \tag{3.11}$$

for all $\lambda \in G$, hence $\hat{\varphi} \in \Delta_G(T^*)$.

Proof. If $\alpha \in \Delta_F(T)^\perp$, then $(f, \alpha) = 0$ for every $f \in A_\infty(F, B)$ such that $f(\lambda) = (T - \lambda I)g(\lambda)$, $g \in A(F, B)$. By Lemma 3.1

$$(f(\infty), \varphi) + \frac{1}{2\pi i}\int_{\Gamma(g)} ((T - \lambda I)g(\lambda), \psi(\lambda))d\lambda = 0, \tag{3.12}$$

since $\Gamma(g)$ is suitable as one of $\Gamma(f)$. Actually,

$$f(\infty) = -\lim_{\lambda\to\infty} \lambda g(\lambda) = -\frac{1}{2\pi i}\int_{\Gamma(g)} g(\lambda)d\lambda. \tag{3.13}$$

It follows from (3.12) and (3.13) that

$$\int_{\Gamma(g)} (g(\lambda), (T^* - \lambda I)\psi(\lambda) - \varphi)d\lambda = 0. \tag{3.14}$$

Setting $g = h_{x,\mu}$ $(x \in B,\ \mu \in G)$ we obtain (3.11). Conversely, (3.11) $\Rightarrow$ (3.14) $\Rightarrow$ (3.12), so $(f, \alpha) = 0$. □

Proposition 3.5. *The annihilator $\Delta_F(T^*)_\perp \subset l_1(F, B)$ consists of all $p \equiv \{x, g\}$ with $x \in B$, $g \in A(G, B)$, such that*

$$x = (T - \lambda I)g(\lambda) \tag{3.15}$$

for all $\lambda \in G$, hence $\hat{x} \in \Delta_G(T)$.

Proof is quite similar. □

Propositions 3.4 and 3.5 are obviously extended to $F = \mathbf{C}$.

4 Duality theorems

In this Section we use the previous descriptions for the annihilators $\Delta_F(T)^\perp$ and $\Delta_F(T^*)_\perp$ to obtain some exact duality relations.

Theorem 4.1. *Let F be closed, $F \neq \mathbb{C}$, and let $G = \mathbb{C} \setminus F$ be bounded. Then*

$$M_F(T)^\perp = N_G^*(T^*) \tag{4.1}$$

and

$$M_F^*(T^*)_\perp = N_G(T). \tag{4.2}$$

Proof. Obviously, $M_F(T)^\perp = L_F(T)^\perp$. By Definition 1.1 and Proposition 3.5,

$$L_F(T) = \{x \mid x \in B, \exists\, g \in A(G,B) : p \equiv \{x,g\} \in \Delta_F(T^*)_\perp\}.$$

By (3.9)

$$L_F(T)^\perp = \{\varphi \mid \varphi \in B^*, \;\; x \in L_F(T) \Rightarrow (p,\hat{\varphi}) = 0\}.$$

Hence,

$$L_F(T)^\perp = \{\varphi \mid \varphi \in B^*, \;\; p \in \Delta_F(T^*)_\perp \Rightarrow (p,\hat{\varphi}) = 0\}.$$

This subspace is just $N_G^*(T^*)$ because of Definition 2.5 and Proposition 2.7.

Now we start with $M_F^*(T^*)_\perp = L_F(T^*)_\perp$. By Definition 1.1 and Proposition 3.4,

$$L_F(T^*) = \{\varphi \mid \varphi \in B^*, \;\; \psi \in A(G,B) : \alpha \equiv \{\varphi,\psi\} \in \Delta_F(T)^\perp\}.$$

By (3.3)

$$L_F(T^*)_\perp = \{x \mid \in B, \;\; \varphi \in L_F(T^*) = 0 \Rightarrow (\hat{x},\alpha) = 0\}.$$

Hence

$$L_F(T^*)_\perp = \{x \mid x \in B, \;\; \alpha \in \Delta_F(T)^\perp \Rightarrow (\hat{x},\alpha) = 0\}.$$

This subspace is just $N_G(T)$. □

Formula (4.2) is closely connected with the Bishop-Lange-Wang formula

$$N^\perp = \mathrm{cl}_{w^*}(M). \tag{4.3}$$

Here the subspaces $N \subset B$ and $M \subset B^*$ are defined like $N_G(T)$ and $L_F(T^*)$, but somewhat differently. Nevertheless, one can show that

$$N = N_G(T), \quad M \subset L_F(T^*). \tag{4.4}$$

Therefore

$$M_F^*(T^*)_\perp = N = \mathrm{cl}_{w^*}(M)_\perp. \tag{4.5}$$

Since $\mathrm{cl}_{w^*}(M) \subset M_F^*(T^*)$ by (4.4) and these subspaces are both w^*-closed, we conclude from (4.5)

$$\mathrm{cl}_{w^*}(M) = M_F^*(T^*). \tag{4.6}$$

As a result, (4.3) takes a more apparent form

$$N_G(T)^{\perp} = M_F^*(T^*). \tag{4.7}$$

Note that though (4.2) is an obvious formal consequence of (4.7) we used (4.2) to get (4.7).

Now we obtain a generalization of Bishop's Theorem 2 concerning the duality theory of type 4. For this purpose we consider a set $H \supset G$ in the complex plane and note that

$$M_F(T)^{\perp} \subset N_H^*(T^*) \tag{4.8}$$

by (4.1) and

$$M_F^*(T^*)_{\perp} \subset N_H(T) \tag{4.9}$$

by (4.2).

Theorem 4.2. *(1) Let F_1 and F_2 be closed subsets of* C *such that $F_1 \bigcup \mathrm{Int}F_2 \supset \sigma(T)$. Then*

$$M_{F_1}(T)^{\perp} \subset N_{F_2}^*(T^*) \tag{4.10}$$

and

$$M_{F_1}^*(T^*)_{\perp} \subset N_{F_2}(T). \tag{4.11}$$

(2) Let F_1 and F_2 be compact subsets of C *such that $F_1 \bigcap F_2 = \emptyset$. Then*

$$M_{F_1}(T)^{\perp} \supset N_{F_2}^*(T^*) \tag{4.12}$$

and

$$M_{F_1}^*(T^*)_{\perp} \supset N_{F_2}(T). \tag{4.13}$$

Proof. (1) One can assume that F_2 is bounded by the monotonicity of the right sides of (4.10) and (4.11). Let us apply (4.8) with $G = \mathrm{Int}F_2, \; F = \mathrm{C} \setminus G, \; H = F_2$.

Since $F \bigcap \sigma(T) \subset F_1$, we have

$$M_F(T) = M_{F \bigcap \sigma(T)}(T) \subset M_{F_1}(T)$$

whence

$$M_{F_1}(T)^{\perp} \subset M_F(T)^{\perp} \subset N_{F_2}^*(T^*),$$

i.e., (4.10) is valid. Similarly, (4.9) implies (4.11).

(2) Taking a bounded open neighborhood G of F_2 such that $G \bigcap F_1 = \emptyset$, we obtain for $F = \mathrm{C} \setminus G \supset F_1$,

$$M_{F_1}(T)^{\perp} \supset M_F(T)^{\perp} = N_G^*(T^*) \supset N_{F_2}^*(T^*)$$

and

$$M_{F_1}^*(T^*)_{\perp} \supset M_F^*(T^*)_{\perp} = N_G(T) \supset N_{F_2}(T).$$

□

Remark 4.3. It follows from (4.10) that

$$M_{F_1}(T) \supset N^*_{F_2}(T^*)_\perp \tag{4.14}$$

if F_1 and F_2 are closed and $F_1 \bigcup \mathrm{Int} F_2 \supset \sigma(T)$. Similarly, (4.12) implies

$$M_{F_1}(T) \subset N^*_{F_2}(T^*)_\perp \tag{4.15}$$

if F_1 and F_2 are compact and $F_1 \bigcap F_2 = \emptyset$.

Remark 4.4. Our requirement $F_1 \bigcup \mathrm{Int} F_2 \supset \sigma(T)$ is weaker that Bishop's $\mathrm{Int} F_1 \bigcup \mathrm{Int} F_2 \supset \sigma(T)$. This leads to an advance in a local spectral theory (see Theorems 5.12 and 5.13).

5 Local spectral analysis

A spectral analysis represented below is local in the sense that it is related to any fixed closed subset $F \subset \mathbf{C}$, $F \neq \mathbf{C}$. (As usual, $G = \mathbf{C} \setminus F$). The following statement is quite general.

Lemma 5.1. *For every point $\mu \in G$ the operator $(T - \mu I) \mid L_F(T)$ is surjective.*

Proof. Let $y \in L_F(T)$, i.e., $y = (T - \lambda I)g(\lambda)$, $\lambda \in G$, $g \in A(G, B)$. Consider the function

$$g_1(\lambda) = \frac{g(\lambda) - g(\mu)}{\lambda - \mu}, \quad \lambda \in G$$

removing the formal singularity at $\lambda = \mu$. Obviously, $g_1 \in A(G, B)$. Since

$$y = (T - \mu I)g(\mu) = (T - \lambda I)g(\lambda)$$

for all $\lambda \in G$, we obtain

$$(T - \lambda I)g_1(\lambda) = g(\mu)$$

which means that $g(\mu) \in L_F(T)$. Thus, $y = (T - \mu I)x$, $x \in L_F(T)$. □

Lemma 5.2. *If T has SVEP$|$ G then for every $\mu \in G$ the operator $(T - \lambda I) \mid L_F(T)$ is injective.*

This means that μ is not an eigenvalue of $T \mid L_F(T)$.

Proof. Let $x \in L_F(T)$ and $Tx = \mu x$. Since $x = (T - \lambda I)g(\lambda)$ where $g \in A(G, B)$, $\lambda \in G$, we obtain

$$(T - \lambda I)(T - \mu I)g(\lambda) = (T - \mu I)x = 0.$$

By SVEP$|$ G, $(T - \mu I)g(\lambda) = 0$ for all $\lambda \in G$. In particular, $(T - \mu I)g(\mu) = 0$, i.e., $x = 0$. □

Corollary 5.3. *If T has SVEP| G then the operator $(T-\mu I) \mid L_F(T)$ is bijective for any $\mu \in G$.*

This does not mean that μ is a regular point since, though the *local resolvent*

$$R_\mu(T,F) = (T-\mu I)^{-1} \mid L_F(T) \tag{5.1}$$

does exist under SVEP| G, we do not know whether this is a bounded operator. Indeed, the linear subspace $L_F(T)$ may be nonclosed, so a standard reference to the Banach inverse operator theorem fails. However, we have

Lemma 5.4. *Let T have SVEP| G. If the local resolvent $R_\mu(T,F)$ is bounded for some $\mu \in G$, then $R_\lambda(F,T)$ is uniformly bounded on a sufficiently small disk $\mid \lambda-\mu \mid < \varepsilon$.*

Proof. The Hilbert identity

$$R_\lambda - R_\mu = (\lambda - \mu) R_\mu R_\lambda \tag{5.2}$$

for the local resolvent yields

$$\|R_\lambda x\| \leq \|R_\mu\| \cdot \|x\| + \varepsilon \|R_\mu\| \cdot \|R_\lambda x\|, \tag{5.3}$$

for $x \in L_F(T)$. Thus,

$$\|R_\lambda x\| \leq \frac{\|R_\mu\|}{1 - \varepsilon\|R_\mu\|} \cdot \|x\|. \qquad \square$$

As a result, we obtain

Theorem 5.5. *Let T have SVEP| G. The local resolvent $R_\mu(T,F)$ is bounded for every $\mu \in G$ if and only if $L_F(T)$ is closed.*

Proof. If $L_F(T)$ is closed then $R_\mu(T,F)$ is bounded by the Banach inverse operator theorem.

Let $R_\mu(T,F)$ be bounded for every $\mu \in G$. Consider a convergent sequence $\{x_k\}_1^\infty \subset L_F(T)$, $x = \lim\limits_{k\to\infty} x_k$. Let $x_k = (T-\lambda I)g_k(\lambda)$ with $g_k \in A(G,B)$, $\lambda \in G$. Then $g_k(\lambda) = R_\lambda(T,F)x_k$. By Lemma 5.4 $\{g_k(\lambda)\}$ locally uniformly converges to a function g. Then $g \in A(G,B)$ and $x = (T-\lambda I)g(\lambda)$, so $x \in L_F(T)$. $\square$

Corollary 5.6. (cf. [2, p.23]) *If T has SVEP| G and $L_F(T)$ is closed then*

$$\sigma(T \mid L_F(T)) \subset F.$$

Moreover,

$$\sigma(T \mid L_F(T)) \subset \sigma(T) \tag{5.4}$$

since $L_F(T)$ is hyperinvariant. Finally,

$$\sigma(T \mid L_F(T)) \subset F \bigcap \sigma(T) \tag{5.5}$$

under the above mentioned conditions.

Remark 5.7. Obviously, if $L_F(T)$ is closed and $\sigma(T \mid L_F(T)) \subset F$ then T has SVEP| G.

Being closed $L_F(T)$, is the greatest subspace among the invariant closed subspaces L such that $\sigma(T \mid L) \subset F$. Indeed, for every $x \in L$ the function $g(\lambda) = R_\lambda(T \mid L)x$ belongs to $A(G, B)$ and $x = (T - \lambda I)g(\lambda)$, so $x \in L_F(T)$.

Thus, under conditions of Corollary 5.6, $L_F(T)$ is a spectral maximal subspace in the Foias sense. In particular, these conditions are fulfilled if $T \in \beta_F$.

A further spectral analysis will be developed assuming $T \in \beta_F$ and $T^* \in \beta_\Phi$ where Φ is complementary to F in some sense. We start with three lemmas of a general character. Two of them are basically algebraic and well-known, so let us omit detailed proofs.

Lemma 5.8. *Let L be an invariant closed subspace and let T/L be the corresponding factor operator in B/L. Then*

$$\sigma(T) \subset \sigma(T \mid L) \bigcup \sigma(T/L). \tag{5.6}$$

Proof. (5.6) is equivalent to the following inclusion for the corresponding resolvent sets

$$\rho(T \mid L) \bigcap \rho(T/L) \subset \rho(T). \tag{5.7}$$

It remains to note that if for a linear operator its restriction to an invariant subspace is invertible jointly with the corresponding factor operator, then the given operator is also invertible. □

Let us consider the annihilator $L^\perp \subset B^*$. This subspace is T^*-invariant and w^*-closed.

Lemma 5.9. *The relation*

$$\sigma(T^* \mid L^\perp) = \sigma(T/L). \tag{5.8}$$

holds.

Proof. There is a natural isomorphism $P : (B/L)^* \to L^\perp$ and $T^* \mid L^\perp = P(T/L)^* P^{-1}$. Thus, $\sigma(T^* \mid L^\perp) = \sigma((T/L)^*) = \sigma(T/L)$. □

Lemma 5.10. *Let L be an invariant closed subspace in B and let L' be a T^*-invariant w^*-closed subspace in B^* such that $L^\perp \subset L'$. If $T^* \mid L'$ is surjective and TL is dense in L then $T^* \mid L^\perp$ is also surjective.*

Proof. Let us show that $T^*(L^\perp)$ is w^*-closed. Let a directed set $\{\varphi_\nu\} \subset L^\perp$ be such that there exists $\psi = w^*\text{-}\lim_\nu T^*\varphi_\nu$. Obviously, $\psi \in L^\perp$. Since $T^* \mid L'$ is surjective, there exists $\varphi \in L'$ such that $\psi = T^*\varphi$. Then $(Tx, \varphi) = (x, \psi) = 0$ for $x \in L$. Hence $\varphi \in L^\perp$ because of density TL in L. This means that $\psi \in T^*L^\perp$.

Now to prove that $T^*(L^\perp) = L^\perp$ it is sufficient to take $x \in (T^*(L)^\perp)_\perp$ and show that $x \in (L^\perp)_\perp$. Note that $(L^\perp)_\perp = L$. If $\varphi \in L^\perp$, so that $T^*\varphi \in T^*(L^\perp)$, then

$(Tx, \varphi) = (x, T^*\varphi) = 0$ thus, $Tx \in L$ hence, $Tx = \lim_{k\to\infty} Tx_k$ where $\{x_k\} \subset L$. If $\psi \in L^\perp$ then $\psi = T^*\varphi$ with $\varphi \in L'$ and $(x, \psi) = \lim_{k\to\infty}(x - x_k, \psi) = \lim_{k\to\infty}(Tx - Tx_k, \varphi) = 0$. □

Corollary 5.11. *Let L and L' be such as in Lemma 5.10. Then*

$$\sigma(T^* \mid L^\perp) \subset \sigma(T \mid L) \bigcup \sigma(T^* \mid L'). \tag{5.9}$$

Proof. Inclusion (5.9) is equivalent to

$$\rho(T \mid L) \bigcap \rho(T^* \mid L') \subset \rho(T^* \mid L^\perp). \tag{5.10}$$

If λ belongs to this intersection then Lemma 5.10 yields the surjectivity of $(T^* - \lambda I) \mid L^\perp$. This operator is also injective because $(T^* - \lambda I) \mid L'$ is so. □

Theorem 5.12. *Let Φ be a closed subset of $\mathbb{C}$ such that $\sigma(T) \subset F \bigcup \mathrm{Int}\Phi$. Suppose that $T \in \beta_F$ and $T^* \in \beta_\Phi$. Then*

$$\sigma(T) \subset \sigma(T \mid L_F(T)) \bigcup \sigma(T^* \mid L_\Phi(T^*)). \tag{5.11}$$

Proof. Let $L = L_F(T)$ and $L' = L_\Phi(T^*)$. The subspace L is closed by Corollary 2.10. The subspace L' is w^*-closed by Corollary 2.12. Moreover, $L = M_F(T)$, $L' = N_\Phi(T^*)$. By Theorem 4.2 $L^\perp \subset L'$. Then Corollary 5.11 yields

$$\sigma(T^* \mid L^\perp) \subset \sigma(T \mid L) \bigcup \sigma(T^* \mid L').$$

By Lemma 5.9 this inclusion takes the form

$$\sigma(T/L) \subset \sigma(T \mid L) \bigcup \sigma(T^* \mid L') \tag{5.12}$$

and (5.11) follows from (5.12) and (5.6). □

Without loss of generality one can assume that $F \subset \sigma(T)$. In this case we choose an "optimal" Φ in Theorem 5.12. For every $Q \subset \sigma(T)$ we denote by $\mathrm{int}Q$ the relative interior of Q with respect to $\sigma(T)$. As a final result we obtain the following "spectrum splitting-off theorem".

Theorem 5.13. *Let $F \subset \sigma(T)$ and $F' = \overline{\sigma(T) \setminus F}$. Let a closed set Φ satisfy the conditions*

$$\Phi \bigcap \sigma(T) = F', \quad \sigma(T) \subset F \bigcup \mathrm{Int}\Phi. \tag{5.13}$$

Suppose that $T \in \beta_F$ and $T^* \in \beta_\Phi$. Then

$$\sigma(T) = \sigma(T \mid L_F(T)) \bigcup \sigma(T^* \mid L_{F'}(T^*)). \tag{5.14}$$

Furthermore,

$$\mathrm{int}\, F \subset \sigma(T \mid L_F(T)) \subset F, \tag{5.15}$$

and

$$\sigma(T^* \mid L_F(T^*)) = F'. \tag{5.16}$$

Proof. By Theorem 5.12

$$\sigma(T) \subset \sigma(T \mid L_F(T)) \bigcup \sigma(T^* \mid L_{F'}(T^*))$$

because $L_{F'}(T^*) = L_\Phi(T^*)$ by Proposition 2.3. On the other hand, since

$$\sigma(T \mid L_F(T)) \subset F, \quad \sigma(T^* \mid L_{F'}(T^*)) \subset F' \tag{5.17}$$

and $F \bigcup F' = \sigma(T)$ we conclude that (5.14) is valid. Since $\mathrm{int} F \bigcap F' = \emptyset$, we get (5.15) from (5.14) and (5.17). Besides, it follows from (5.14) that

$$\sigma(T) \setminus \sigma(T \mid L_F(T)) \subset \sigma(T^* \mid L_{F'}(T)),$$

hence,

$$\sigma(T) \setminus F \subset \sigma(T^* \mid L_{F'}(T^*)) \subset F'.$$

Since $\sigma(T) \setminus F$ is dense in F', we obtain (5.16). □

Inclusions (5.15) mean that $L_F(T)$ is the spectral subspace corresponding to F in our sense [6].

Remark 5.14. A question arises: how does one construct a set Φ serving Theorem 5.13? To answer this let us consider $\mathrm{int} F$ and extend it to an open set H such that $H \bigcap F' = \emptyset$. Then $\Phi = \overline{U \setminus H}$ where U is an arbitrary open set containg $\sigma(T)$.

Remark 5.15. Let $F \subset \sigma(T)$, int $F \neq \emptyset$ and $F \neq \sigma(T)$. Then $L_F(T)$ is a nontrivial hyperinvariant subspace if the conditions of Theorem 5.13 are fulfilled. Moreover, $L_{F'}(T^*)$ is such a subspace for T^* in this situation.

Remark 5.16. Under the same conditions one can show that the set F' is also the spectrum of the factor operator $T/L_F(T)$.

REFERENCES

1. Bishop E.: A duality theorem for an arbitrary operator, Pacif. J. Math., 9:2 (1959), 379-397.

2. Golojoara I. and Foias C,: Theory of Generalized Spectral Operators, Gordon & Breach, N.Y. 1968.

3. Foias C. Spectral maximal spaces and decomposable operators in Banach spaces, Arch. Math. 14 (1963), 341-349.

4. Lange R. and Wang S.: New approaches in spectral decomposition. Contemp. Math., 128, AMS, 1992.

5. Lomonosov V.I.: Some questions of the theory of invariant subspaces, Ph.D. Thesis, Kharkov, 1973.

6. Lomonosov V.I., Lyubich Yu.I. and Matsaev V.I.: Duality of spectral subspaces and conditions for the separation of spectrum of bounded linear operators, Soviet Math. Doklady 15 (1974), 878-881.

V.Lomonosov, Department of Mathematic & Computer Science,
Kent State University, P.O. Box 5190, Kent, OH 44242, U.S.A.

Yu.Lyubich, Department of Mathematics,
Technion - Israel Institute of Technology, 32000 Haifa, Israel.

V.Matsaev, School of Mathematical Sciences, Tel Aviv University,
Ramat Aviv, 69978 Tel Aviv, Israel.

AMS classification # 47A11

Operator Theory
Advances and Applications, Vol. 98

DEGENERATED ELLIPTIC BOUNDARY VALUE PROBLEMS FOR WEAK COUPLED SYSTEMS. SOLVABILITY AND MAXIMUM PRINCIPLE.

B. PANEAH

1. The degenerated elliptic boundary value problems for a *single* equation have been investigated by many mathematicians within the last 30 years and some quite impressive results have been obtained (cf.review [1]). But almost nothing is still known about solvability of such problems for elliptic *systems* of linear differential equations. Even the simplest systems of such a kind are interesting from this point of view because of their applications. For instance, the system

$$L_k u_k + \sum_{j=1}^{N} h_{kj}(u_j - u_k) = F_k(\cdot, u), \quad k = 1, 2, \dots, N \tag{1}$$

(L_k are homogeneous second order elliptic differential operators in a bounded domain $\Omega \subset \mathbb{R}^n$, $n \geq 3$, h_{kj} are positive functions on Ω) appears when describing diffusion and multiplication (or killing) of particles of the first and second types. The lower terms in (1) involved matrix $h = \|h_{jk}\|$ describe transmutation from the first to the second type and vice versa (cf. [2]).

In the scalar case even the (formally) simplest boundary value problem

$$\Delta u = 0 \text{ in } \Omega,$$

$$\partial u / \partial l = \varphi \text{ on } \partial\Omega$$

(the famous oblique derivative problem, where Δ is Laplacian and l is a vector field in a neighborhood of $\partial\Omega$) turns out to be ill-posed if the vector field l is tangent at some points of the boundary $\partial\Omega$. It was shown in ′ 60s to ′ 80s that this and some more general boundary value problems for Δ become well posed after some natural modifications (cf. [1]).

This work deals with two (quite different) kinds of non-elliptic boundary value problems for systems of differential equations. In the next Section we describe one of considered types of boundary value problems and formulate a result concerning the solvability of the modified problem. In Section 3 we deal with a class of boundary value problems including the above mentioned weakly coupled reaction-diffusion equation (1). We prove that under some conditions a strong maximum principle (like Hopf-Giraud maximum principle) for solution of this problem is valid.

1. Let Ω be a bounded domain in R^n, $n \geq 3$, with a smooth boundary $\partial\Omega$ and let $[\Omega] = \Omega \bigcup \partial\Omega$. Let $u = (u_1, u_2, \dots, u_N)$ be a vector-function: $\Omega \to \mathbb{R}^N$. We denote by τ a smooth non-singular vector field on $\partial\Omega$ and by ν a unit vector field of interior normals to $\partial\Omega$. The boundary value problem we deal with at this Section is:

$$\left.\begin{aligned} Lu + Hu &= F \quad \text{in}\,\Omega, \\ \partial u/\partial\tau + A\partial u/\partial\nu + Bu &= f \quad \text{on}\,\partial\Omega. \end{aligned}\right\} \tag{2}$$

Here L is a scalar second order linear elliptic differential operator. In every local coordinate system $z = (z_1, z_2, \dots, z_n)$ in Ω it can be represented as

$$L = \sum_{i,j=1}^{n} a_{ij}(z)\frac{\partial^2}{\partial z_i \partial z_j}.$$

The coefficients $a_{ij}(z)$ are supposed to be smooth functions such that for all vectors $\xi = (\xi_1, \xi_2, \dots, \xi_n) \in \mathbb{R}^n$ the inequality

$$\sum_{i,j=1}^{n} a_{ij}(z)\xi_i\xi_j \geq a\sum_{i=1}^{n} \xi_i^2$$

is valid with a positive constant a for all points $z \in [\Omega]$. The operator H is supposed to be an arbitrary first order differential operator: $C^\infty(\Omega, \mathbb{R}^N) \to C^\infty(\Omega, \mathbb{R}^N)$ and A, B are smooth $N \times N$ matrices on the boundary $\partial\Omega$. Thus, the problem (2) is weak coupled in the domain Ω but not on the boundary $\partial\Omega$ (because of the general character of the matrix A).

It is well-known that the problem (2) is an elliptic one if and only if

$$\det A \neq 0 \quad \text{on}\ \partial\Omega.$$

For this reason we suppose further that the set

$$\mu = \{\, q \in \partial\Omega \mid \det A(q) = 0 \,\}$$

is not empty. Moreover, we assume that *μ is a submanifold in $\partial\Omega$ of codimension 1 and vector field τ is transversal to μ.*

Let $\mathcal{U} \subset \partial\Omega$ be some sufficiently small tubular neighborhood of the submanifold μ. Consider some normal coordinates (t, x) on $\mathcal{U}$ so that $x = (x_1, x_2, \dots, x_n)$ are local coordinates on μ; $(t, x) \in \mu$ iff $t = 0$; t increases along any trajectory of the field τ.

The following two conditions are supposed to be fulfilled on the course of this Section.

Condition 1. *The zero eigensubspace $\mathcal{G}$ of the matrix $A(0, x)$ does not depend on $x \in \mu$ and its dimension is equal to $p \geq 1$.*

It means that in some basis in $\mathbb{R}^N$ the matrix $A(t,x)$ can be represented as

$$\begin{pmatrix} A_{11} & A_{12} \\ A_{21} & A_{22} \end{pmatrix}$$

where A_{11} and A_{22} are square matrices of orders p and $N-p$ respectively and additionally

$$A_{11}(t,x) = t^k \mathcal{A}_{11}(t,x), \quad A_{21}(t,x) = t^k \mathcal{A}_{21}(t,x)$$

for some integer $k \geq 1$ and smooth matrices $\mathcal{A}_{11}$ and $\mathcal{A}_{21}$.

Condition 2. *There exists a smooth function $a(x) \neq 0$ such that*

$$\mathcal{A}_{11}(0,x) = a(x)\Gamma$$

where Γ is a constant matrix without eigenvalues on the imaginary axis.

Theorem 1. *If the number k is even then the problem (2) is Fredholm problem (between suitable space). If k is odd then this problem is not Fredholm.*

The result means in particular that for any odd k the problem (2) has to be modified. It is desirable of course to minimize modifications. We describe below one of possible modifications inspired by corresponding scalar problems (cf. [1]).

Let $\mathcal{G}^+$ and $\mathcal{G}^-$ be spectral subspaces of $\mathbb{R}^N$ corresponding to eigenvalues of the matrix Γ with positive and negative real parts respectively. Denote by Πu the orthogonal projection of a vector $u \in \mathbb{R}^N$ on the subspace $\mathcal{G}$ and by $\Pi_+ u$, $\Pi_- u$ the projections of Πu on $\mathcal{G}^+$ and $\mathcal{G}_-$.

Theorem 2. *If the number k is odd then the modified problem*

$$\left.\begin{aligned} Lu + Hu &= F && \text{in } \Omega, \\ \partial u/\partial\tau + A\partial u/\partial\nu + Bu &= f && \text{on } \partial\Omega \setminus \mu, \\ \Pi_- u &= g && \text{on } \mu \end{aligned}\right\} \tag{3}$$

is Fredholm one (between suitable spaces). Moreover, if all functions F, f, g are sufficiently smooth then the boundary value v of the solution u on $\partial\Omega$ belongs to the Sobolev space $H^1(\partial\Omega \setminus \mu)$ and the jump $[v]_\mu$ of v on μ coincides with the jump $[\Pi_+ v]_\mu$. In other words the solution u satisfies on $\partial\Omega$ the condition

$$\frac{\partial u}{\partial\tau} + A\frac{\partial u}{\partial\nu} + Bu = f + [v]_\mu \otimes \delta(\mu).$$

The full proofs of these two theorems are too long and they will be published elsewhere.

2. This part of the paper is devoted to the strong maximum principle for the boundary value problem of the following type:

$$\left.\begin{aligned} L_k u_k + \sum_{j=1}^{n} h_{kj} u_j &= F_k(\cdot, u) && \text{in } \Omega, \\ \partial u_k/\partial \tau_k + a_k \partial u_k/\partial \nu + \sum_{j=1}^{n} b_{kj} u_j &= f_k(\cdot, u) && \text{on } \partial\Omega,\ k = 1, 2, \dots N \end{aligned}\right\} \tag{4}$$

Here L_k are the operators

$$L_k = \sum_{i,j=1}^{n} a_{ij}^{(k)}(z) \frac{\partial^2}{\partial z_i \partial z_j} + \sum_{j=1}^{n} a_j^{(k)}(z) \frac{\partial}{\partial z_j}$$

whose coefficients smoothly depend on local coordinates z in Ω and the inequalities

$$\sum_{i,j=1}^{n} a_{ij}^{(k)}(z) \xi_i \xi_j \geq a \sum_{i=1}^{n} \xi_i^2$$

with a positive constant a are valid for all points $z \in [\Omega]$ and unit vectors $\xi = (\xi_1, \xi_2, \dots, \xi_n)$. All τ_k are some non-singular vector fields on $\partial\Omega$ and a_k, b_{kj}, h_{kj} are smooth functions. No assumption concerning the smoothness of the functions F_k and f_k are required. This boundary value problem is an elliptic one if and only if none of functions a_k vanishes at any point $q \in \partial\Omega$. Introduce the sets

$$\gamma_k = \{ q \in \partial\Omega \mid a_k(q) = 0 \}, \quad k = 1, 2, \dots N.$$

We assume for brevity that all sets γ_k are not empty. The only condition we impose on sets γ_k, $k = 1, 2, \dots, N$, is:

the set γ_k does not contain any maximal trajectory of the vector field τ_k (5)

It means that any curve

$$z(t) = (z_1(t), z_2(t), \dots, z_n(t)) : \mathbb{R}^1 \to \partial\Omega$$

such that

$$\dot{z}(t) = (\tau_k \circ z)(t), \quad z(0) = p \in \gamma_k$$

leaves the set γ_k as $t \to \pm\infty$ (although later this curve can enter in γ_k once again).

In order to describe the class of matrices $\|h_{ij}\|$ and $\|b_{ij}\|$ which are involved in the problem (4) we introduce two definitions. We say that (arbitrary) *matrix* $c = \|c_{ij}(p)\|$ *satisfies the (rd)-condition* if for all indices i, j at any point p the inequalities

$$(rd) \qquad c_{ii} c_{ij}(p) \leq 0 \quad \text{and} \quad c_{ii} \sum_{j=1}^{n} c_{ij}(p) \geq 0$$

hold. We say that a *matrix* $c(p)$ *satisfies the (zs)-condition* if for all $i, j = 1, 2, \dots, N$

$$(zs) \qquad c_{ij}(p) = 0 \ \text{ implies } c_{ji}(p) = 0.$$

The classical strong maximum principle in the scalar case (due to E.Hopf and G.Giraud) can be formulated as follows:
if a smooth function u satisfies inequalities $Lu + au \geq 0$ in Ω, $\frac{\partial}{\partial \nu}u + bu \geq 0$ on $\partial\Omega$ with non-positive coefficients a, b then u is strictly negative on $[\Omega]$ or $u \equiv$ const.

Of course, the case $a \equiv 0$ and $b \equiv 0$ is excluded.

The following theorem extends this result to matrix-valued boundary value problem (4).

Theorem 3. *Let $u = (u_1, u_2, \dots, u_N)$ be a solution of the problem (4). Assume that*

1° any function a_k does not change its sign on $\partial\Omega$;
2° $a_k b_{kk} \leq 0$ for all $k = 1, 2, \dots, N$;
3° $h_{kk} \leq 0$ for all k;
4° the matrices h and b satisfy both of (rd)- and (zs)-conditions.

If $F_k(\cdot, u) \geq 0$ in Ω and $a_k(\cdot) f_k(\cdot, u) \geq 0$ on $\partial\Omega$ for all k then each component u_k either strictly negative in $[\Omega]$ or equal to a constant.

Proof. First note that without loss of generality one can only consider non-negative functions a_k. Otherwise, we merely change vector-field τ_k by $-\tau_k$. According to 2° $b_{kk} \leq 0$ for all indices $k = 1, 2, \dots, N$. Further, note that if some functions u_k, say $u_{m+1}, u_{m+2}, \dots, u_N$ are strictly negative on $[\Omega]$ then remaining functions $u_1, u_2, \dots, u_m$ are solutions of "reduced" boundary value problem

$$\left.\begin{aligned} L_k u_k + \sum_{j=1}^{m} h_{kj} u_j &= \widetilde{F}_k && \text{in } \Omega, \\ \partial u_k/\partial \tau_k + a_k \partial u_k/\partial \nu + \sum_{j=1}^{m} b_{kj} u_j &= \widetilde{f}_k && \text{on } \partial\Omega, \quad k = 1, 2, \dots, m \end{aligned}\right\} \tag{6}$$

where $\widetilde{F}_k = F_k - \sum_{j=m+1}^{N} h_{kj} u_j$, $\widetilde{f}_k = f_k - \sum_{j=m+1}^{N} b_{kj} u_j$. Due to (rd)-conditions the inequalities $\widetilde{F}_k \geq 0$ and $\widetilde{f}_k \geq 0$ are valied for all $k = 1, 2, \dots, m$. Therefore, all conditions of Theorem 3 are fulfilled for problem (6). Since its solutions $u_1, u_2, \dots, u_m$ are supposed to be non-negative functions at some points $p_1, p_2, \dots, p_m$ of $[\Omega]$, it remains to prove that all those functions are (non-negative) constants. Let $M_k = u_k(p_k) = \max_{[\Omega]} u_k$, $k = 1, 2, \dots, m$. For definiteness we assume that $M_1 = \max\{M_k\}_1^m$. Then the function $v_1 = u_1 - M_1$ satisfies equation

$$L_1 v_1 + h_{11} v_1 = \widetilde{F}_1 + \sum_{k=2}^{m} h_{1k}(M_1 - u_k) - M_1 \sum_{k=1}^{m} h_{1k}$$

and $\max v_1 = v_1(p_1) = 0$. The right-hand side of this equation is non-negative by virtue of (rd)-condition and choice of M_1. Since $h_{11} \leq 0$ we may conclude (cf. Hopf's maximum principle) that $p_1 \in \partial\Omega$ (or $u_1 \equiv M_1$). There are two possibilities: $p_1 \in \gamma_1$ or $p_1 \notin \gamma_1$. If $p_1 \notin \gamma_1$ and $v_1 \not\equiv 0$ then (according to Giraud's theorem) $a_1(p_1)(\partial/\partial\nu)v_1(p_1) < 0$. On the other hand, $(\partial/\partial\tau_1)v_1(p_1) = 0$ which means that at the point p_1

$$0 > (\partial/\partial\tau_1)v_1 + a_1(\partial/\partial\nu)v_1 = \widetilde{f}_1 + \sum_{j=1}^{m} b_{1j}(M_1 - u_j) - M_1\sum_{j=1}^{m} b_{1j} \geq 0.$$

We have used the *(rd)*-condition and inequality $b_{11} \leq 0$. The contradiction proves that the point p_1 does not belong to $\partial\Omega \setminus \gamma_1$. Assume that $p_1 \in \gamma_1$ (and $v_1 \not\equiv 0$). Consider the trajectory $z = z(t)$ of the vector field τ_1 which passes through the point $p_1 \in \gamma_1$ at the moment $t = 0$. Along this trajectory the function v_1 satisfies equation

$$\frac{d}{dt}V_1(t) = f_1 - a_1\frac{\partial v_1}{\partial\nu} + \sum_{k=2}^{m} b_{1k}(M_1 - u_k) - M_1\sum_{k=1}^{m} b_{1k}$$

and initial condition $V_1(0) = 0$. Here $V_1(t) = v_1(z(t))$. According to condition (5) there exists a value $t_+ \geq 0$ such that $a_1(t) = 0$ for all t, $0 \leq t \leq t_+$ but $a_1(t_k) > 0$ for some sequence $t_k \to t_+$, $t_k > t_+$, and therefore

$$\frac{d}{dt}V_1(t_k) > 0.$$

If $t_+ = 0$ then $(d/dt)V_1(t) \geq 0$ in a neighborhood of the point $t = 0$ and $(d/dt)V_1(t_k) > 0$ for some sequence $t_k \to 0$, $t_k > 0$. But this is impossible since $\max V_1(t) = V_1(0)$.

If $t_+ > 0$ then $V_1(t) = 0$ for all values t between 0 and t_+, as $V_1(0) = 0$ is the maximal value of V_1 and the derivative $(d/dt)V_1(t)$ is non-negative on the interval $(0, t_+)$. As in the case $t_+ = 0$ for some sequence $t_k \to t_+$, $t_k > t_+$, we have $a_1(t_k) > 0$ and therefore $(d/dt)V_1(t_k) > 0$. But the latter is impossible because $t = t_+$ is the point of local maximum of the function $V_1(t)$. Thus we proved that $u_1 \equiv M_1$ in Ω. It is obvious that the same is true for all components $u_2, u_3, \dots, u_r$ of the solution u of the problem (6) whose maximal value coincides with M_1. Consider remaining components $u_{r+1}, u_{r+2}, \dots, u_m$. They are solutions of the boundary value problem

$$\left.\begin{aligned} L_k u_k + \sum_{j=r+1}^{m} h_{kj}u_j &= \Phi_k \qquad \text{in } \Omega,\ k = r+1, r+2, \dots, m, \\ \partial u_k/\partial\tau_k + a_k\partial u_k/\partial\nu + \sum_{j=r+1}^{m} b_{kj}u_j &= \varphi_k \qquad \text{on } \partial\Omega \end{aligned}\right\} \tag{7}$$

where $\Phi_k = \widetilde{F}_k - M_1\sum_{j=1}^{r} h_{kj}$, $\varphi_k = \widetilde{f}_k - M_1\sum_{j=1}^{r} b_{kj}$. Since the matrices $\|h_{kj}\|$ and $\|b_{kj}\|$, $r+1 \leq k, j \leq m$, satisfy conditions 2°, 3°, 4° of Theorem 3 it is sufficient to show that $\Phi_k \geq 0$ and $\varphi_k \geq 0$ for $k \geq r+1$. Let us use equations (6). This leads to systems of equations

$$\left.\begin{aligned} M_1\sum_{j=1}^{r} h_{kj} + \sum_{j=r+1}^{m} h_{kj}u_j &= \widetilde{F}_k \qquad \text{in } \Omega,\ k = r+1, r+2, \dots, m, \\ M_1\sum_{j=1}^{r} b_{kj} + \sum_{j=r+1}^{m} b_{kj}u_j &= \widetilde{f}_k \qquad \text{on } \partial\Omega,\ k = r+1, r+2, \dots, m \end{aligned}\right\} \tag{8}$$

Rewrite the first system in the form

$$M_1 \sum_{j=1}^{m} h_{kj} + \sum_{j=r+1}^{m} h_{kj}(u_j - M_1) = \widetilde{F}_k, \quad k = 1, 2, \ldots, r$$

and note that $u_j < M_1$ for $j \geq r+1$ by choice of M_1; all $\widetilde{F}_k$ are non-negative functions and $\sum_{j=1}^{m} h_{kj} \leq 0$ by the (rd)-condition. This immediately gives us that $h_{kj} \equiv 0$ for $k \leq r,\ j \geq r+1$. Then $h_{kj} \equiv 0$ for $k \geq r+1,\ j \leq r$ according to the (zs)-condition and therefore $\Phi_k = F_k \geq 0$ for all $k = r+1, r+2, \ldots, m$. The same arguments when applied to the second equation in (8) lead us to the inequality $\widetilde{\varphi} \geq 0$. This completes the proof of Theorem 3.

Concluding remarks.

1. *Sharpness of the (zs)-condition.* Here we will show that a violation of the *(zs)*-condition can fail the strong maximum principle. Indeed, consider the following elliptic system of differential equations

$$\left.\begin{array}{l} \Delta u = 0 \\ \Delta v + u - v = 0 \end{array}\right\}$$

in the circle $x^2 + y^2 < 1$. This system satisfies all conditions 1° - 4° of Theorem 3 except for *(zs)*-condition. The functions $u = 2,\ v = 2 - \text{ch}y$ are solutions of this system but v is positive in the unit circle.

2. *Coefficients a_k may change signs.* It is not difficult to extend the result of Theorem 3 to the case when some of functions a_k may be both positive and negative on the boundary $\partial\Omega$. The maximum principle in this situation turns out to be valid for solutions of modified boundary value problem of the type (3).

3. *Parabolic systems.* The strong maximum principle can be extended on the initial boundary value problem (4) with operators

$$-(\partial/\partial t) + L_k$$

instead of L_k, unknown functions $U_k(t,x)$ instead of $u_k(x)$ and with additional initial condition $u_k(0,x) = \varphi_k(x),\ k = 1, 2, \ldots, N$.

4. *Weakly coupled Reaction-Diffusion Equations.* (cf. M. Freidlin [2]). This is the class of systems of the form

$$(-\frac{\partial}{\partial t} + L_k)u_k + \sum_{j=1}^{N} d_{kj}(u_j - u_k) = F_k(\cdot, u) \quad \text{in}\ \Omega \times \mathbb{R}^1_+$$

$k = 1, 2, \ldots, N$, with *positive* coefficients d_{kj}. It is clear that the corresponding stationary system satisfies both of (rd)- and (zs)-conditions. Let u be a solution of the boundary value problem (4) which corresponds to this system and let $F_k(\cdot, u) \geq 0$ for all indices $k = 1, 2, \ldots, m$. By Theorem 3 if some component u_j of u is non-negative at least at one point $p_j \in [\Omega]$ then u_j is equal to a constant $M \geq 0$. This constant M does not depend on $j = 1, 2, \ldots, m$.

5. *Other results.* The strong maximum principle for elliptic systems of partial differential equations with Neumann-boundary condition $(\partial/\partial\nu)u = f$ on $\partial\Omega$ under *(rd)*-condition was considered in the book of M. Protter and H. Wainberger [3]. Our Theorem 3 is inspired by their result.

REFERENCES

1. B. Paneah, Degenerate Elliptic Boundary Value Problems, in Encyclopedia of Mathematical Sciences, **63**, Springer-Verlag: Berlin-Heidelberg-New-York, 1993 (162-201).

2. M. Freidlin, Coupled reaction-diffusion equations, The Annals of Probability, 1991, 19 (29-57).

3. M. Protter & H. Weinberger, The Maximum Principles in differential equations, Prentice Hall, 1967.

Department of Mathematics,
Technion,
Haifa, 32000,
Israel

AMS Classification Primary 35J70; Secondary 35S15

Operator Theory
Advances and Applications, Vol. 98

CHARACTERIZATION OF THE PERIODIC AND ANTI-PERIODIC SPECTRA OF NONSELFADJOINT HILL'S OPERATORS

J.-J.SANSUC, V.TKACHENKO*

The necessary and sufficient conditions are given for a sequence of complex numbers to be the periodic (or anti-periodic) spectrum of Hill's operator.

It is well known [1] that the spectrum $\sigma(H)$ of Hill's operator

$$H = -\frac{d^2}{dx^2} + q(x), \qquad x \in \mathbf{R} \tag{1}$$

in the space $L^2(\mathbf{R})$ with a real π-periodic potential $q(x) \in \mathbf{R}$ is real, bounded below and has a band structure

$$\sigma(H) = [\nu_0, \nu_1^-] \bigcup_{k=1}^{\infty} [\nu_k^+, \nu_{k+1}^-]$$

with

$$\nu_0 < \nu_1^- \leq \nu_1^+ < \ldots < \nu_k^- \leq \nu_k^+ < \nu_{k+1}^- \leq \cdots \tag{2}$$

The numbers $\nu_0, \nu_2^-, \nu_2^+, \ldots, \nu_{2k}^-, \nu_{2k}^+ \ldots$ form the spectrum of periodic boundary problem

$$-y''(x) + q(x)y(x) = \mu y(x), \qquad 0 < x < \pi, \tag{3}$$

$$y(0) = y(\pi), \qquad y'(0) = y'(\pi)$$

in the space $L^2[0,\pi]$, while the numbers $\nu_1^-, \nu_1^+, \ldots, \nu_{2k+1}^-, \nu_{2k+1}^+ \ldots$ form the spectrum of anti-periodic boundary problem

$$-y''(x) + q(x)y(x) = \mu y(x), \qquad 0 < x < \pi, \tag{4}$$

$$y(0) = -y(\pi), \qquad y'(0) = -y'(\pi).$$

Both spectra are uniquely determined by the Hill discriminant

$$\Delta(\lambda) = \frac{c(\pi,\lambda) + s'(\pi,\lambda)}{2},$$

*Gvastela Fellow, partially supported by Israel Science Foundation of the Israel Academy of Science and Humanities and by Israel Ministry of Science

where $c(\pi, \lambda)$ and $s(\pi, \lambda)$ are solutions to the equation

$$-y''(x) + q(x)y(x) = \lambda^2 y(x), \qquad 0 < x < \pi,$$

with initial data $c(0,\lambda) = s'(0,\lambda) = 1, c'(0,\lambda) = s(0,\lambda) = 0$. The Hill discriminant of every π-periodic potential $q(x) \in L^2[0,\pi]$ is an even entire function of exponential type π, and points of its periodic and anti-periodic spectra are zeros of $\Delta(\sqrt{\mu}) - 1$ and $\Delta(\sqrt{\mu}) + 1$, respectively. It implies, in particular, that one of these spectra completely determines another.

The exact asymptotic behavior of periodic/anti-periodic spectra (2) of real potentials was described by Marchenko [2] in the following form:

If $q(x) \in L^2[0,\pi]$ and $\Im q(x) = 0$, then eigenvalues $\nu_{2k}^{\pm}$ of periodic problem (3) *and eigenvalues $\nu_{2k+1}^{\pm}$ of anti-periodic problem* (4) *satisfy the asymptotic relations*

$$\sqrt{\nu_k^{\pm}} = k + \frac{Q}{2k} \pm \frac{f_k}{k} + \frac{\omega_k^{\pm}}{k^2}, \qquad k \to +\infty$$

where

$$Q = \frac{1}{\pi} \int_0^{\pi} q(x)dx$$

and

$$\sum^{\infty} |f_k|^2 < \infty, \qquad \sum^{\infty} |\omega_k^{\pm}|^2 < \infty. \tag{5}$$

An additional analysis of the proof of this statement from [2] shows that the above description remains valid for spectra of operators (1) with complex valued potentials $q(x)$.

All real sequences (2) which are periodic and anti-periodic spectra of real π-periodic potentials $q(x) \in L^2[0,\pi]$ are described by Marchenko and Ostrovskii [3] using special conformal mappings. This description shows that no prescribed asymptotic behavior of given sequence $\{\mu_k^{\pm}\}$ of real numbers, however close it is to the corresponding spectrum of trivial operator $H = -d^2/dx^2$ can guarantee that it is the periodic or antiperiodic spectrum of some Hill operator with *real* potential $q(x) \in L^2[0,\pi]$.

The aim of the present paper is to give an intrinsic description of sequences which are periodic or anti-periodic spectra of Hill's operators (1) with complex valued potentials.

Theorem 1. *For a sequence*

$$\mu_0, \mu_1^-, \mu_1^+, \dots, \mu_k^-, \mu_k^+, \dots \tag{6}$$

to be the periodic spectrum of some Hill operator (1) *with a complex valued potential $q(x) \in L^2[0,\pi]$ it is necessary and sufficient that it has the form*

$$\mu_0 = \lambda_0^2, \qquad \mu_k^{\pm} = (\lambda_k^{\pm})^2, \qquad k = 1, 2, \dots$$

where, for all sufficiently large n, each disk $\{\lambda : |\lambda| \leq 2n+1\}$ contains exactly $2n+1$ points $\lambda_0, \lambda_k^{\pm}, k = 1, 2, \dots, n$, and the following conditions are fulfilled:

i) *asymptotic relations*

$$\lambda_k^{\pm} = 2k + \frac{Q}{4k} \pm \frac{f_k}{k} + \frac{\omega_k^{\pm}}{k^2}, \qquad k \to +\infty \tag{7}$$

are satisfied with some constant $Q \in \mathbf{C}$, and (5) *is valid;*

ii) *the identity holds*

$$\mu_0 - Q + \sum_{k=1}^{\infty}\{\mu_k^+ + \mu_k^- - 8k^2 - 2Q\} = 0; \tag{8}$$

iii) *the sequence*

$$\omega_k = \begin{cases} \lambda_k^+ + \lambda_k^- - 4k - \dfrac{Q}{2k} & k > 0 \\ -\omega_{-k} & k < 0 \end{cases}$$

is such that

$$\sum_{n=-\infty}^{\infty}\left|\sum_{k=-\infty}^{\infty}\frac{\omega_k k^2}{2k-2n+1}\right| < \infty. \tag{9}$$

Theorem 2. *For a sequence*

$$\mu_1^-, \mu_1^+, \ldots, \mu_k^-, \mu_k^+, \ldots$$

to be the anti-periodic spectrum of some Hill operator (1) *with a complex valued potential* $q(x) \in L^2[0,\pi]$ *it is necessary and sufficient that it has the form*

$$\mu_k^\pm = (\lambda_k^\pm)^2, \qquad k = 1, 2, \ldots$$

where, for all sufficiently large n, *each disk* $\{\lambda : |\lambda| \leq 2n+2\}$ *contains exactly* $2n$ *points* $\lambda_k^\pm, k = 1, 2, \ldots, n$, *and the following conditions are fulfilled:*

i) *asymptotic relations*

$$\lambda_k^\pm = 2k + 1 + \frac{Q}{2(2k+1)} \pm \frac{f_k}{k} + \frac{\omega_k^\pm}{k^2}, \qquad k \to +\infty$$

are satisfied with some constant $Q \in \mathbf{C}$, and (5) *is valid;*

ii) *the identity holds*

$$\sum_{k=1}^{\infty}\{\mu_k^+ + \mu_k^- - 2(2k+1)^2 - 2Q\} = 0;$$

iii) *the sequence*

$$\omega_k = \begin{cases} \lambda_k^+ + \lambda_k^- - 2(2k+1) - \dfrac{Q}{2k+1} & k > 0 \\ -\omega_{-k} & k < 0 \end{cases}$$

is such that (9) *is satisfied.*

We note that identities ii) from both theorems are the trace formulas recently stated by P.Lax [4] for real potentials.

The proof of both theorems is carried out in the same lines, and here we present only a proof of Theorem 1.

Our starting point is the following description of Hill's discriminants given in [5].

Theorem 3. *For a function $\Delta(\lambda)$ to be the Hill discriminant of some Hill operator (1) with π-periodic complex valued potential $q(x) \in L^2[0,\pi]$ it is necessary and sufficient that $\Delta(\lambda)$ is an even entire function of exponential type π which is representable in the form*

$$\Delta(\lambda) = \cos\lambda\pi + \frac{\pi Q}{2\lambda}\sin\lambda\pi - \frac{\pi^2 Q^2}{8\lambda^2}\cos\lambda\pi + \frac{f(\lambda)}{\lambda^2}, \tag{10}$$

where $f(\lambda)$ is an even entire function of exponential type not bigger than π and such that

$$\int_{-\infty}^{\infty} |f(\lambda)|^2 d\lambda < \infty, \qquad \sum_{n=-\infty}^{\infty} |f(n)| < \infty. \tag{11}$$

Proof of Theorem 1. Sufficiency. Assume that (7) is valid with $Q = 0$. We set $\lambda_0^\mp = \pm\lambda_0$ and $\lambda_{-k}^\pm = -\lambda_k^\mp$ for each $k > 0$. Then for all sufficiently large n each disk $\{\lambda : |\lambda| \le 2n+1\}$ contains exactly $4n+2$ points $\lambda_k^\mp, k = 0, \pm 1, \ldots$ and asymptotic relations are fulfilled

$$\lambda_k^\mp = 2k \pm \frac{f_k}{k} + \frac{\omega_k^\pm}{k^2}, \qquad |k| \to \infty, \tag{12}$$

and

$$\sum_{k\in\mathbf{Z}} \{(\lambda_k^+)^2 + (\lambda_k^-)^2 - 8k^2\} = 0. \tag{13}$$

We set

$$\omega_k = \lambda_k^+ + \lambda_k^- - 4k, \qquad k = 0,1,2,\ldots \tag{14}$$

and

$$\delta_k = (\lambda_k^+ - 2k)(\lambda_k^- - 2k), \quad k = 0,1,2,\ldots \tag{15}$$

According to the definition, $\omega_{-k} = -\omega_k, \delta_{-k} = \delta_k$, and

$$\{\omega_k k^2\} \in l^2, \qquad \{\delta_k k^2\} \in l^1. \tag{16}$$

Let us introduce now the function

$$\Phi(\lambda) = -\frac{\pi^2}{2}(\lambda - \lambda_0^+)(\lambda - \lambda_0^-)\prod_{k\neq 0}\frac{(\lambda - \lambda_k^+)(\lambda - \lambda_k^-)}{4k^2} \tag{17}$$

and prove that the entire function

$$f(\lambda) = \lambda^2(\Phi(\lambda) + 2\sin^2\frac{\pi\lambda}{2}) \tag{18}$$

satisfies (11). According to Theorem 3, it implies that the function

$$\Delta(\lambda) = \Phi(\lambda) + 1 = \cos\lambda\pi + \frac{f(\lambda)}{\lambda^2} \tag{19}$$

is Hill's discriminant of some operator (1) with the periodic spectrum (6).

If $\lambda_k^{\pm} = 2k, k = 0, 1, \ldots$, then (17) takes on the form

$$\Phi(\lambda) = -2\sin^2\frac{\lambda\pi}{2} = -\frac{\pi^2}{2}\lambda^2\prod_{k\neq 0}\frac{(\lambda-2k)^2}{4k^2}, \tag{20}$$

and both functions (17) and (20), for all sufficiently large n, have $4n+2$ zeros in each disk $\{\lambda : |\lambda \leq 2n+1\}$ with account taken of multiplicities. Hence

$$f(\lambda) = 2\lambda^2\sin^2\frac{\lambda\pi}{2}(1-\exp\Omega(\lambda)) \tag{21}$$

where

$$\Omega(\lambda) = \sum_{k=-\infty}^{\infty}\ln\left(1-\frac{\omega_k}{\lambda-2k}+\frac{\delta_k}{(\lambda-2k)^2}\right). \tag{22}$$

Let $\lambda = x+iM, M = 4(1+\max_{k\in\mathbf{N}}(|\omega_k|+|\delta_k|))$. Then

$$\left|\frac{\omega_k}{\lambda-2k}\right| + \left|\frac{\delta_k}{(\lambda-2k)^2}\right| \leq \frac{1}{2},$$

and to estimate $\Omega(\lambda)$ we apply the elementary inequality

$$|\ln(1+z) - z + \frac{z^2}{2} - \frac{z^3}{3}| \leq \frac{|z|^4}{2}, \qquad |z| \leq \frac{1}{2}.$$

It yields the representation

$$\lambda^2\Omega(\lambda) = -\lambda\sum_{k\in\mathbf{Z}}\omega_k + \sum_{k\in\mathbf{Z}}\{-2k\omega_k+\delta_k-\frac{1}{2}\omega_k^2\} + \sum_{k\in\mathbf{Z}}\frac{\gamma_k}{\lambda-2k} + \gamma(\lambda) \tag{23}$$

where

$$\gamma_k = -4k^2\omega_k + 4k\delta_k - 2k\omega_k^2 + \omega_k\delta_k - \frac{1}{3}\omega_k^3 \tag{24}$$

and

$$|\gamma(\lambda)| \leq C\sum_{k\in\mathbf{Z}}\left\{\frac{k^2|\delta_k|+|k\omega_k|^2+|k^2\omega_k|+|\omega_k^3k^2|}{|\lambda-2k|^2} + \frac{|\lambda|^2(|\delta_k|+|\omega_k|)^2}{|\lambda-2k|^4}\right\} \tag{25}$$

with C independent of λ.

The first term in (23) vanishes, since $\{\omega_k\}$ is an odd sequence. According to (14) and (15) we have $\omega_k^2+4k\omega_k-2\delta_k = \mu_{|k|}^+ + \mu_{|k|}^- - 8k^2$, and the second term in (23) vanishes by virtue of condition (13). According to (16) we have $\{\gamma_k\} \in l^2$, and if $\lambda = x+iM, 2n-1 \leq x \leq 2n+1$, then

$$\left|\sum_{k\in\mathbf{Z}}\frac{\gamma_k}{\lambda-2k}\right| \leq |\gamma_n| + C\sum_{k\neq n}\frac{|\gamma_k|}{|n-k|^2} + \left|\sum_{k\neq n}\frac{\gamma_k}{2n-2k}\right|. \tag{26}$$

For the second term on the right-hand side of this inequality we obtain, using Schwartz inequality,

$$\sum_{n\in\mathbf{Z}}\left|\sum_{k\neq n}\frac{|\gamma_k|}{|n-k|^2}\right|^2 \leq \sum_{n\in\mathbf{Z}}\left(\sum_{k\neq n}\frac{|\gamma_k|^2}{|n-k|^2}\sum_{k\neq n}\frac{1}{|n-k|^2}\right) < \infty.$$

The last sum in (26) is the n-th coordinate of the Hilbert transform of the sequence $\{\gamma_k\}$, and since this transform is bounded in l^2, we find

$$\int_{\mathbf{R}+iM} \left| \sum_{k\in\mathbf{Z}} \frac{\gamma_k}{\lambda - 2k} \right|^2 d\lambda < \infty.$$

Integrating now (23) over the line $\{\lambda : \Im\lambda = M\}$ we find that the function (18) satisfies the first inequality in (11).

To prove the second inequality in (11) we note that since the function $\Phi(\lambda)$ is bounded in the strip $\{\lambda : |\Im\lambda| \leq 1\}$, then for all sufficiently large values of $|n|$ we have

$$|f(2n)| \leq Cn^2|\lambda_n^+ - 2n||\lambda_n^- - 2n| \max_{|\lambda-2n|=1} \left| \frac{\Phi(\lambda)}{(\lambda_n^+ - \lambda)(\lambda_n^- - \lambda)} \right| \leq Cn^2|\delta_n|$$

with some constant C not depending on n. It proves that $\{f(2n)\} \in l^1$.

To estimate $f(2n+1)$ we use representations (23) and (24). Now we obtain

$$(2n+1)^2\Omega(2n+1) = -\sum_{k\in\mathbf{Z}} \frac{4k^2\omega_k}{2n-2k+1} + \sum_{k\in\mathbf{Z}} \frac{\sigma_k}{2n-2k+1} + \gamma(2n+1) \tag{27}$$

where

$$\sigma_k = 4k\delta_k - 2k\omega_k^2 + \omega_k\delta_k - \frac{1}{3}\omega_k^3. \tag{28}$$

The first term on the right-hand side of (27) belongs to l^1 by virtue of (9). For the second term we have, taking into account that $\sigma_k = -\sigma_{-k}$,

$$\left| \sum_{k\in\mathbf{Z}} \frac{\sigma_k}{2n-2k+1} \right| \leq \left| \frac{1}{2n+1} \sum_{k\in\mathbf{Z}} \frac{2k\sigma_k}{2n-2k+1} \right|$$

$$\leq |\sigma_n| + \left| \frac{1}{2n+1} \sum_{n\neq k} \frac{k\sigma_k}{n-k} \right| + \left| \frac{1}{2n+1} \sum_{n\neq k} \frac{|k\sigma_k|}{|n-k|^2} \right|$$

and since $\{k\sigma_k\} \in l^1$, we obtain

$$\sum_{k\in\mathbf{Z}} \frac{\sigma_k}{2n-2k+1} \in l^1. \tag{29}$$

Besides, estimate (25) implies $\{\gamma(2n+1)\} \in l^1$ which completes the proof of the second inequality in (11).

Let now (7) and (8) be valid with some $Q \neq 0$. We set $\tilde{\mu}_k^\pm = \mu_k^\pm - Q$ and obtain the sequence $\tilde{\mu}_k^\pm$ satisfying (8) with $Q = 0$. If $\tilde{\lambda}_k^\pm, k = 0, \pm1, \pm2, \ldots,$ is defined by $\tilde{\mu}_k^\pm$ in the same way as $\lambda_k^\pm, k = 0, \pm1, \pm2, \ldots,$ is defined by $\mu_k^\pm$, i.e., $\tilde{\mu}_k^\pm = (\tilde{\lambda}_k^\pm)^2, k = 1, 2, \ldots,$ then

$$\tilde{\lambda}_k^\pm = 2k \pm \frac{f_k}{k} + \frac{\omega_k^\pm}{k^2} + \frac{\tilde{Q}}{k^3} + \frac{\tilde{\omega}_k^\pm}{k^4}, \qquad |k| \to \infty,$$

where $\tilde{Q}$ is a constant, f_k and $\omega_k^{\pm}$ are as described by conditions *ii)* and *iii)* of Theorem 1, and $\{\tilde{\omega}_k^{\pm}\}$ is square integrable sequences. If now $\tilde{\omega}_k = \tilde{\lambda}_k^+ + \tilde{\lambda}_k^- - 4k$, then $\tilde{\omega}_{-k} = -\tilde{\omega}_k$, and

$$\tilde{\omega}_k = \omega_k + \frac{2\tilde{Q}}{k^3} + \frac{\tilde{\gamma}_k}{k^4}, \qquad |k| \to \infty$$

where ω_k is defined by (14) and satisfies (16), and $\{\tilde{\gamma}_k\} \in l^2$. For all $n \in \mathbf{R}$ we have

$$\sum_{k\neq 0} \frac{1}{2k} \frac{1}{2n-2k+1} = \lim_{m\to\infty} \frac{1}{2n+1} \sum_{|k|=1}^{m} \left(\frac{1}{2k} + \frac{1}{2n-2k+1}\right) = 0,$$

and since $\{\tilde{\gamma}_k\}$ is an odd sequence,

$$\sum_{k\neq 0} \frac{\tilde{\gamma}_k}{k^2(2n-2k+1)} = \frac{1}{2n} \sum_{k\neq 0} \frac{(2k-1)\tilde{\gamma}_k}{k^2(2n-2k+1)}.$$

The last sum defines a square integrable sequence and hence $\tilde{\omega}_k$ satisfies (9). We conclude that the function

$$\tilde{\Delta}(\lambda) = \tilde{\Phi}(\lambda) + 1,$$

where $\tilde{\Phi}(\lambda)$, is given by equation (17), corresponding to the sequence $\{\tilde{\lambda}_k^{\pm}\}$ is representable in the form (19). Hence the function $\Delta(\lambda) = \tilde{\Delta}(\sqrt{\lambda^2 - Q})$ has the form (10) and, according to Theorem 3, is the Hill discriminant of some Hill operator (1) with $q(x) \in L^2[0,\pi]$ whose periodic spectrum is $\mu_0, \mu_1^-, \mu_1^+, \ldots$.

Proof of Theorem 1. Necessity. Let $\Delta(\lambda)$ be the Hill discriminant of operator (1) with $q(x) \in L^2[0,\pi]$ such that $Q = 0$. Then according to Theorem 3

$$\Delta(\lambda) = \cos \lambda\pi + \frac{f(\lambda)}{\lambda^2}$$

and conditions (11) are satisfied. Using (10) let us prove that for every such a function solutions $\{\lambda_k^{\pm}\}$ to the equation $\Delta(\lambda) - 1 = 0$ satisfy asymptotic relations (12).

The first inequality in (11) implies that $f(\lambda)$ satisfies the conditions

$$|f(\lambda)| \le C \exp \pi |\Im \lambda|, \quad \int_{\mathbf{R}} (|f'(\lambda)|^2 + |f''(\lambda)|^2) d\lambda < \infty,$$

and the sequence $\{m_k\}$ defined by the relations

$$m_k = \max_{|\lambda - 2k| \le 1/4} \{|f(\lambda)| + |f'(\lambda)| + |f''(\lambda)|\}$$

is square integrable. By virtue of the Rouche theorem, for all sufficiently large n, every disk $\{\lambda : |\lambda| \le 2n+1\}$ contains exactly $4n+2$ zeros $\lambda_k^{\pm}$ of $\Delta(\lambda) - 1$ as the function $\cos \pi\lambda$ does with account taken of multiplicities. If $\lambda_k^{\pm} = 2k + 2\delta_k^{\pm}$ then

$$\sin^2 \pi\delta_k^{\pm} - \frac{f(2k + 2\delta_k^{\pm})}{8(k+\delta_k^{\pm})^2} = 0, \tag{30}$$

which implies an estimate $|\delta_k^\pm| \le C(|k|+1)^{-1}$ with some constant C. Moreover, since $|f(2k+2\delta_k^\pm) - f(2k)| \le 2m_k|\delta_k^\pm|$ we conclude, using the second inequality from (11), that

$$\sum_{k=-\infty}^{\infty} |f(2k+2\delta_k^\pm)| < \infty,$$

and hence

$$\sum_{k=-\infty}^{\infty} |\delta_k^\pm|^2 k^2 \le C \sum_{k=-\infty}^{\infty} |f(2k+2\delta_k^\pm) < \infty.$$

Let us now represent (30) in the form $(\pi\delta_k^\pm)^2 - 2\pi\delta_k^\pm F_k - G_k = 0$ where

$$G_k = \frac{f(2k)}{8k^2}$$

and using the Taylor formula obtain an estimate

$$|F_k| = |\sin^2 \pi\delta_k^\pm - (\pi\delta_k^\pm)^2 - \frac{f(2k+2\delta_k^\pm)}{8(k+\delta_k^\pm)^2} + \frac{f(2k)}{8k^2}||2\pi\delta_k^\pm)|^{-1}$$

$$\le C(\frac{m_k}{|k|^2} + |\delta_k^\pm|^3)$$

with some constant C not depending on k. It follows now

$$\pi\delta_k^\pm = \pm\sqrt{\frac{f(2k)}{8k^2}} + F_k \pm (\sqrt{F_k^2 + G_k} - \sqrt{G_k}).$$

Since

$$\sum_{k=-\infty}^{\infty} (k^2(|F_k| + |\sqrt{F_k^2+G_k} - \sqrt{G_k}|))^2 \le C \sum_{k=-\infty}^{\infty} (|F_k|k^2)^2 < \infty,$$

we arrive at the representation (12) with (5) being fulfilled. Thus (7) is proved with $Q = 0$. Represent now the function $\Phi(\lambda) = \Delta(\lambda) - 1$ in the form (17). If ω_k and δ_k are defined by Eqs.(14) and (15), and $\Omega(\lambda)$ has the form (22), we have, as follows from the representation (23),

$$\lambda^2\Omega(\lambda) = \sum_{k\in\mathbf{Z}} \{-2k\omega_k + \delta_k - \frac{1}{2}\omega_k^2\} + \sum_{k\in\mathbf{Z}} \frac{\gamma_k}{\lambda - 2k} + \gamma(\lambda)$$

with (24) and (25) being valid. According to Theorem 3, the function $f(\lambda)$ is square integrable on the real line. Hence the first sum on the right-hand side of the previous equation vanishes which proves condition *ii*) of Theorem 1 for $Q = 0$.

To prove the necessity of *iii*) we note that by virtue of Theorem 3 the left-hand side of (27), with σ_k defined by (28), belongs to the space l^1. Since (13) holds and $\gamma(\lambda)$ satisfies inequality (25), we conclude that (9) is fulfilled.

If $Q \ne 0$ then it is sufficient to introduce the potential $q(x) = \tilde{q}(x) - Q$ and apply the previous arguments to the function $\Delta(\sqrt{\lambda^2 + Q})$.

References

[1] E. C. Titchmarsh: Eigenfunction expansions associated with second-order differential equations, Oxford, Claredon Press, 1958.

[2] V. A. Marchenko: Sturm-Liouville operators and applications, Birkhäuser, Basel, 1986.

[3] V. A. Marchenko and I. V. Ostrovskii: Characterization of the spectrum of Hill's operator, Mathem.Sborn., 1975, **97**, 4, 540–606; English transl. in Math. USSR-Sb. 26 (175).

[4] P. Lax: Trace formulas for the Schroedinger operator, CPAM, 1994, **47**, 4, 503–512.

[5] V. A. Tkachenko: Discriminants and generic spectra of nonselfadjoint Hill's operators, Advances in Sov. Math., AMS, 1994, **19**, 41–71.

J.-J.Sansuc, U.F.R. de Mathématiques,
Université Paris 7 Denis Diderot, 2 Place Jussieu F-75251, Paris, France

V.Tkachenko, Department of Mathematics and Computer Science,
Ben-Gurion University of the Negev, Beer–Sheva 84105, Israel

AMS classification # 34B30, 34L05

Operator Theory
Advances and Applications, Vol. 98

A LOWER CONFIDENCE LIMIT FOR THE MULTIPLE CORRELATION COEFFICIENT

I.M. SLIVNYAK

We introduce a notion of the best monotonic lower confidence limit for unknown parameter of distribution. Its existence is proved for the multiple correlation coefficient in the case of the normal law. An effective method of computing such a limit is described.

1 Introduction

In his last years, I.M.Glazman displayed an interest in the foundations of the modern mathematical statistics, in particular, in problems of testing statistical hypotheses and estimating parameters of distributions. Being already grievously ill, he organized a seminar for members of his department in the Institute for Low Temperature Physics and Engineering to study these topics. The present article has been written to commemorate that short-lived seminar.

All necessary information on the multiple correlation factor and its sample variant is contained in Sections 1 and 2. The notion of the best monotone lower confidence (BMLC) limit for unknown distribution parameter is introduced in Section 3. We give here a sufficient condition for the existence of such a limit. Using it the existence of the (BMLC)-limit for the multiple coefficient of correlation is established in the case of normal distribution, and an algorithm for its construction is described.

1. Let us consider a probabilistic space $(\mathbb{R}^p, B, \mu)$ where B is the σ-algebra of Borel sets, μ is a probability measure with a density $f(x)$, $x = (x_i)_1^n \in \mathbb{R}^p$, with respect to the Lebesgue measure. From the point of view of probability theory, x is a random vector with the density function $f(x)$, and any real B-measurable function $g(x)$ is a random variable. Let us introduce the functional Hilbert space H with the scalar product

$$(g_1, g_2) = \int\limits_{\mathbb{R}^p} g_1(x) g_2(x) f(x) dx. \tag{1}$$

We replace every element $g \in H$ by its projection g^0 on the hyperplane orthogonal to the function $\equiv 1$, and use the notation g instead of g^0.

The projection of the function x_1 on the closed linear span of x_2, ..., x_p is called the *linear regression* x_1 *related to the set* $\xi = \{x_2, \ldots, x_p\}$. We denote it by $x_{1,\xi}$.

Let V be a non-degenerate $(p \times p)$-matrix with entries $v_{ij} = (x_i, x_j)$. Then

$$x_{1,\xi} = -\sum_{j=2}^{p} \frac{V^{1j}}{V^{11}} x_j,$$

where V^{ij} is the co-factor of the entry v_{ij} in V. The *multiple coefficient of correlation between* x_1 *and* ξ is

$$R = \frac{(x_1, x_{1,\xi})}{\|x_1\| \cdot \|x_{1,\xi}\|} \tag{2}$$

In Section 3 the results related to R are obtained for the normal distribution

$$f(x) = (2\pi)^{-\frac{p}{2}} \mid V \mid^{-\frac{1}{2}} \exp(-\frac{1}{2} x' V^{-1} x). \tag{3}$$

2. As usual in statistics, the density $f(x)$ is not known completely. In the case (3), it depends on $\frac{1}{2}p(p+1)$ unknown parameters v_{ij}. Therefore the expressions containing the scalar products of form (1) are, generally speaking, beyond the reach of calculation. This difficulty can be overcome by the sample method which prescribes to pass from a single test to a series of independent tests, i.e., from the initial probabilistic space to its n-th Cartesian product. In this scheme the random vector $x \in \mathbb{R}^p$ has to be replaced by a set of independent commonly distributed vectors $x^{(1)}, ..., x^{(n)}$, and then any random variable $g(x)$ is replaced by the vector

$$\tilde{g} = (g(x^{(i)}))_1^n.$$

We fix the values $x^{(i)}$, $i = 1, ..., n$, and define the Hilbert space H_n of vectors $\tilde{g}$ with the scalar product

$$(\tilde{g}_1, \tilde{g}_2)_n = \frac{1}{n} \sum_{i=1}^{n} g_1(x^{(i)}) g_2(x^{(i)}) \tag{4}$$

which is a sample analogue of (1). Let every element $\tilde{g} \in H_n$ be replaced by its projection $\tilde{g}^0$ on a $(n-1)$-dimensional hyperplane which is orthogonal to $(1, ..., 1)$ with $\tilde{g}$ and n written instead of $\tilde{g}^0, n-1$.

Any characteristic θ of the measure μ has its sample analogue $\tilde{\theta}$. In particular, replacement of (x_i, x_j) by $(\tilde{x}_i, \tilde{x}_j)_n$ in (2) yields the multiple sample coefficient of correlation $\tilde{R}$. It should be noted that due to the law of large numbers, $\tilde{\theta} \approx \theta$ when $n >> 1$.

To obtain some probability statements on $\tilde{\theta}$, we have to consider $(x^{(1)}, ..., x^{(n)})$ as a set of random vectors with the distribution law of direct product of measures μ in H_n. However, there remains the problem of unknown parameters of the law μ on which the distribution of the random variable $\tilde{\theta}$ depends. In the case (3) the elements v_{ij} of the matrix V are such parameters. Fortunately, for many important characteristics θ the parameters v_{ij} enter into the distributions only via the variable θ itself. It is true, in particular, for R. It was shown by R.A.Fisher [1, Section 27.30] that in the case (3) the distribution of the variable $\tilde{R}$ contains just one unknown parameter R.

3. In some problems of the regression analysis it is desirable to obtain a sufficiently reliable estimate of R from below. To this end, let us use the notion of the lower confidence (LC)-limit of the parameter of a distribution.

Let τ be a random variable whose distribution depends on unknown parameter θ. Any function $\underline{\theta} = \psi(\tau)$ such that the probability P_θ of the random event $\{\psi(\tau) < \theta\}$ at any value θ satisfies the condition

$$P_\theta \{\psi(\tau) < \theta\} \geq \gamma, \tag{5}$$

is called *the* (LC)-*limit of the parameter* θ *with the confidence coefficient* γ. From now on the value γ is fixed.

Let us describe a method to construct $\underline{\theta}$. Let $\theta = \psi(t)$ be a function, and let G be a domain in the plane (t, θ) located above its graph. Suppose that for any θ the set Δ_θ of values t obtained by dissection G by the horizontal straight line on the height θ, satisfies the condition

$$P_\theta \{\tau \in \Delta_\theta\} \geq \gamma. \tag{6}$$

Then

$$P_\theta \{\psi(\tau) < \theta\} = P_\theta \{\tau \in \Delta_\theta\} \geq \gamma,$$

i.e., the function $\theta = \psi(t)$ is a required (LC)-limit $\underline{\theta}$. Obviously, a domain G can be chosen by infinitely many ways.

Let us impose an additional restriction on $\underline{\theta}$ which is natural in view of our problem.

For the sake of simplicity we assume that $\theta \in [0,1]$, $\tau \in [0,1]$, and the distribution function of the random variable τ,

$$F_\theta(t) = P_\theta \{\tau < t\} \tag{7}$$

is continuous in t for any θ. If we consider τ as an approximation to the true value θ (for example, $\tau = \tilde{\theta}$), then it is desirable that the (LC)-limit $\underline{\theta}(\tau)$ retains the character of the behavior of τ. More exactly, suppose that $\underline{\theta}(t)$ is a continuous nondecreasing function on the interval $[0,1]$, strictly increasing on a smaller interval $[t_0, t_1]$. Moreover, let

$$\underline{\theta}(t) = 0 \quad (t \leq t_0), \quad \underline{\theta}(t) = 1 \quad (t \geq t_1). \tag{8}$$

Let us call such a (LC)-limit *monotone.*

Among such monotone (LC)-limits for θ, there can exist the most exact one which is placed not lower than any other limit at every value τ. We will call it *the best monotone* (LC)-*limit* (BMLC-*limit).* Let us determine a sufficient conditions for the existence of such a limit.

The monotonicity of the (LC)-limit $\theta(\tau)$ means that for any θ

$$\Delta_\theta = \{t : 0 \leq t \leq \underline{\theta}^{-1}(\theta)\},$$

where $\underline{\theta}^{-1}(\theta)$ is the inverse function of $\underline{\theta}(t)$ on $[t_0, t_1]$.

A monotone (LC)-limit $\underline{\theta}(\tau)$ is the best if, for any θ, the interval Δ_θ is the smallest one. If it is the case, then the equality

$$P_\theta \{\tau \in \Delta_\theta\} = F_\theta(\underline{\theta}^{-1}(\theta)) = \gamma,$$

must be valid for all θ, where F_θ is defined by (7). Thus, the function $t = \underline{\theta}^{-1}(\theta)$ is a solution of the equation

$$F_\theta(t) = \gamma. \tag{9}$$

In particular, $t_0 = \underline{\theta}^{-1}(0)$, $t_1 = \underline{\theta}^{-1}(1)$ in (8).

It remains to ensure the growth of these functions. It holds if the distribution of random variable τ is shifted to the right when θ increases, i.e., for any θ the inequality

$$\frac{\partial F_\theta(t)}{\partial \theta} < 0, \quad 0 < t < 1 \tag{10}$$

holds.

Let us apply the obtained results to determine the (BMLC)-limit $R = \underline{R}(\tilde{R})$. It is convenient to take θ and τ equal to R^2 and $\tilde{R}^2$. At first, let us check (10). The distribution function $F_\theta(t) = P_{R^2}\{\tilde{R}^2 < t\}$ of the random variable $\tilde{R}^2$ was found by R.A.Fisher. However it is expressed by a slowly convergent series. On the other hand, according to Fisher's remark, if $(n-p)$ is even, then this series can be explicitly written in the form

$$F_\theta(t) = \left(\frac{1-\theta}{1-\theta t}\right)^a t^{a-1} \sum_{k=0}^{m-1} \frac{\Gamma(a+k)}{k!} \left(\frac{1-t}{1-\theta t}\theta\right)^k \varphi_{m-k-1}\left(\frac{1-t}{t}\right), \quad 0 < t < 1, \tag{11}$$

where $a = \frac{n-1}{2}$, $m = \frac{n-p}{2} \geq 1$,

$$\varphi_s(z) = \sum_{i=0}^{s} \frac{z^i}{i!\Gamma(a-i)}.$$

Passing in (11) from θ to the increasing function

$$v = \frac{(1-t)\theta}{1-\theta t},$$

we obtain

$$F_\theta(t) = t^{a-1}(1-v)^a \sum_{k=0}^{m-1} \frac{\Gamma(a+k)}{k!} v^k \varphi_{m-k-1}\left(\frac{1-t}{t}\right) = \psi(t,v).$$

It can be directly verified that $\frac{\partial}{\partial v}\psi(t,v) < 0$. Hence condition (10) is fulfilled and the (BMLC)-limit $\underline{R}(\tilde{R})$ exists.

Now let us introduce the variable

$$u = \left(\frac{1-\theta t}{1-\theta}\right)^a.$$

If t is fixed, then equation (9) can be represented in the form $\Phi(u) = u$, where

$$\Phi(u) = \frac{t^{a-1}}{\gamma} \sum_{k=1}^{m-1} \frac{\Gamma(a+k)}{k!} (1-u^{-\frac{1}{a}})^k \varphi_{m-k-1}\left(\frac{1-t}{t}\right).$$

By virtue of properties $\Phi(1) > 1$, $\Phi'(u) > 0$, $\Phi(\infty) < \infty$, the iteration process

$$u_1 = 1, \quad u_i = \Phi(u_{i-1})$$

converges. It is an effective mean to solve (9) as some calculations show.

In conclusion, it should be noted that the condition "$(n-p)$ is even" is not essential, since, as a matter of fact, $n >> p$, and it is always possible to neglect one of n observations.

References

1. M.G.Kendall and A.Stuart, The advanced theory of statistics, **2**, London, 1962.

P.O.B. 621, 44845
Immanuel, Israel

AMS classification 46S50

Operator Theory
Advances and Applications, Vol. 98

ANALYSIS IN CLASSES OF DISCONTINUOUS FUNCTIONS AND PARTIAL DIFFERENTIAL EQUATIONS

A.I.VOLPERT

Review of old and new results in analysis in classes of functions whose generalized derivatives are measures and its applications to partial differential equations and continuum mechanics is given.

Introduction

This paper consists of two parts. The first (Sections 1.1-1.5) is devoted to the analysis in the space of functions whose generalized derivatives are measures. The results obtained here are applied in the second part (Sections 2.1-2.4) to the study of partial differential equations.

One of the fundamental questions studied in the first part is the differentiation of discontinuous functions. To obtain a generalization of the usual differentiation formulas to discontinuous functions we must be able to multiply generalized derivatives of discontinuous functions by discontinuous functions. This is possible if the generalized derivatives are measures. However, the formal possibility of multiplication does not remove the fundamental difficulty. The question consists in assigning values of functions on sets on which the singular parts of measures are concentrated. To answer this question it has been necessary to introduce a new concept of superposition which is called functional superposition. It turns out that the formulas for the generalized differentiation of the usual superposition are not valid if the usual superposition is used on the right-hand sides. However, these formulas do hold if the functional superposition appears in the right-hand sides (see Section 1.4).

The second of fundamental questions of the analysis is connected with the integral formulas. In Section 1.5 we give one of the most general versions of Gauss-Green formulas which have proved to be useful in many applications.

This analysis is based on the results on the structure of BV-functions (functions whose generalized derivatives are measures) and BVsets (sets whose characteristic functions are BV-functions). It requires a more subtle topology that the usual topology in R^n. This is the topology where the inner points are points of density with respect to Lebesgue measure and the limit is the approximate limit.

In this topology the following results hold up to the sets which are negligible in the analysis in question. The BV-set E has a generalized normal at points of its boundary. The BV-function has the limit at points of the boundary from inside and outside of E which we call the inward and outward trace. It gives the possibility to obtain the generalization of Gauss-Green formulas mentioned. Moreover, the BV-function has two-sided limits at any point. So the functional superposition may be defined and the formulas for differentiation may be obtained.

The exact formulation of these and other results will be given in what follows (part 1). For complete proofs of some of them see [V1], [V2], [VH1] if other references are not mentioned.

We shall mention some applications of these results (see Part 2). They proved to be useful in the study of discontinuous solutions of hyperbolic conservation laws and degenerate quasi-linear parabolic equations. As to the general first-order hyperbolic systems, the analysis turned out to be essential in the very definition of the solution. It is an interesting problem to construct a general theory of such systems which, in particular, will generalize the known theory of conservation laws. The sequential approach to the solution of hyperbolic systems is helpful in constructing nonconservative numerical schemes.

The analysis under review was also applied to some questions of continuum mechanics. First of them is the axiomatic approach which has been developing in response to the 6 Hilbert problem. A new version of axioms is constructed which embraces discontinuous solutions. Among other questions there are those concerning nonconservative forms of equations. It became possible to write equations for entropy and other variables which do not admit conservation laws.

There are also applications to parabolic and elliptic equations.

1 Analysis in classes of discontinuous functions

1.1 The Space BV

1.1.1. The Space BV

The space $BV(\mathcal{G})$ is the space of functions $u(x)$ defined and summable on the open set $\mathcal{G} \subseteq R^n$, whose first generalized derivatives are measures.

The latter is to be understood in the following sense: there exists a vector-valued Borel measure μ such that

$$\int \nabla\varphi u dx = -\int \varphi\mu(dx) \tag{1}$$

for all $\varphi \in C_1^0(\mathcal{G})$ (differentiable functions of compact support contained in $\mathcal{G}$). We shall denote $\mu = \nabla u$.

Using Riesz's theorem on functionals in spaces of continuous functions we obtain the following criterion: $u \in BV$ iff there exists a constant K such that

$$\left| \int \nabla\varphi u dx \right| \leq K \sup_{x\in\mathcal{G}} |\varphi(x)|$$

for all $C_1^0(\mathcal{G})$.

$BV(\mathcal{G})$ is a Banach space with the norm

$$\|u\| = \int_{\mathcal{G}} | u(x) | \, dx + \int_{\mathcal{G}} | \nabla u | \, dx,$$

where $| \nabla u |$ is the total variation of the measure ∇u.

For simplicity, we shall sometimes assume that $\mathcal{G} = R^n$. Moreover, since we shall be interested in the behavior of a function in a finite portion of the space (not at infinity) it will be sufficient to assume that the functions considered are finitary. In this case we shall write $u \in BV$ and call u a BV-function.

1.1.2. Functions of Bounded Variation

We say that $u(x)$ $(x = (x^1, \dots, x^n))$ has a bounded Tonelli variation if for any $i = 1, \dots, n$ the function $u(x)$ considered as a function of x^i has essential (up to the set of one-dimensional measure zero) variation which is summable with respect to other variables: $x^1, \dots, x^{i-1}, x^{i+1}, \dots, x^n$.

Theorem 1 ([F1], [K]). *$u \in BV$ iff $u(x)$ has bounded Tonelli variation.*

1.1.3. Integral Condition

Theorem 2. *$u \in BV$ iff there exists a constant K such that for any vector $h \in R^n$ the inequality*

$$\int | u(x+h) - u(x) | \, dx \leq K \, | h |$$

holds, where $| h |$ is the Euclidean norm of h.

1.1.4. BV-Sets

If $E \subseteq R^n$ and its characteristic functions $\chi_E \in BV$, then we shall call E a BV-set or a set with finite perimeter [DG1] and number

$$P(E) = | \nabla \chi_E | (R^n)$$

will be called its perimeter.

1.1.5. Total Variation of the Gradient

Denote by E_t the set of those $x \in R^n$ for which $u(x) > t$.

Theorem 3. ([FR]). *If $u \in BV$, then for almost all t the set E_t has finite perimeter $P(E_t)$ and the equality*

$$| \nabla u | (R^n) = \int_{-\infty}^{\infty} P(E_t) dt$$

holds.

1.2 Regular Points

1.2.1. Approximate limit

If E is a set in R^n, then its point of rarefaction is a point x_0 for which

$$\lim_{r\to 0} \frac{| E \cap K(x_0, r) |}{| K(x_0, r) |} = 0,$$

where $||$ denotes the (outer) Lebesgue measure, $K(x_0, r)$ is the ball $| x - x_0 | < r$ in R^n. A point of density of E is a point of rarefaction of its complement CE. The topology where the inner point is a point of density we shall call the d-topology (cf. [GW]).

Suppose that $u(x)$ is a function defined on a set $A \subseteq R^n$ and x_0 is not a point of rarefaction for A. Then $l_A u(x_0)$ will denote the *approximate limit* of the function $u(x)$ at the point x_0 with respect to the set A. This means, by definition, that for any $\varepsilon > 0$ the point x_0 is a point of rarefaction of the set $\{x : | u(x) - l_A u(x_0) | > \varepsilon, \ x \in A\}$. In what follows we shall always have in mind a *finite limit*. For a vector-valued function we deal with its components.

Proposition 1.1. *Let $f(u)$ be a function that is defined in a neighborhood of some point $u_0 \in R^p$ and continuous at the point u_0. Further, let $u(x)$ be a vector-valued function in a neighborhood of the point $x_0 \in R^n$ and suppose the approximate limit $l_A u(x_0) = u_0$ exists. Then the equality*

$$l_A f(u(x_0)) = f(u_0)$$

holds, where the indicated approximate limit exists.

1.2.2. Regular Points

If the set A coincides with the whole space R^n or with a half-space, we introduce a special notation for the approximate limit $l_A u(x_0)$. Namely, $lu(x_0) = l_A u(x_0)$ when $A = R^n$, $l_a u(x_0) = l_A u(x_0)$ when A is the half-space $(x - x_0, a) > 0$.

Definition 1.1. *The point x_0 is said to be regular for the vector-valued function $u(x)$ if there exists a unit vector a such that $l_a u(x_0)$ and $l_{-a} u(x_0)$ exist.*

Definition 1.2. *The point x_0 is called a point of jump for $u(x)$ if it is regular and $l_a u(x_0) \neq l_{-a} u(x_0)$.*

Proposition 1.2. *If x_0 is a regular point for $u(x)$ and a is the vector in Definition 1.1, then*

1. *when $l_a u(x_0) = l_{-a} u(x_0)$, the approximate limit $lu(x_0)$ exists, and $l_b u(x_0) = lu(x_0)$ exists for any unit vector b.*
2. *when $l_a u(x_0) \neq l_{-a} u(x_0)$, the vector a is uniquely determined (except for sign).*

Definition 1.3. *If $\Gamma(u)$ is the set of points of jump for $u(x)$ and $x_0 \in \Gamma(u)$, then the vector a in the Definition 1.1 is called the normal to $\Gamma(u)$ at the point x_0.*

1.2.3. Essential Boundary

Let E be a measurable set in R^n, and E_* its set of points of density. By definition, *essential boundary* $\partial^* e$ of E is the boundary of E_* in d-topology. In other words, $\partial^* E$ is a set of points which are not points of density and points of rarefaction of E.

Definition 1.4. *A vector a is called an (inward) normal* [F2] *to the set E at the point x_0 if*

$$\lim_{r\to 0} \frac{| K_a(x_0,r) \bigcap E |}{| K_a(x_0,r) |} = 1, \quad \lim_{r\to 0} \frac{| K_{-a}(x_0,r) \bigcap E |}{| K_a(x_0,r) |} = 0,$$

where $K_a(x_0,r)$ is the hemisphere $| x - x_0 | < r$, $(x - x_0, a) > 0$.

Proposition 1.3. *A necessary and sufficient condition for a point x_0 to belong to the essential boundary of E and that the normal exists at x_0 is that x_0 be a point of jump of the characteristic function of E. In this case the definitions* 1.3 *and* 1.4 *of the normal are equivalent.*

1.3 The Structure of BV-Sets and BV-Functions

1.3.1. BV-Sets

Theorem 4. ([DG2], [V1]) *If E is a BV-set, then*

1. *Its essential boundary $\partial^* E$ has finite $(n-1)$-dimensional Hausdorff measure H_{n-1}.*

2. *The normal exists at all points of $\partial^* E$ up to the set of H_{n-1}-measure zero. In this case the equality*

$$P(E) = H_{n-1}(\partial^* E)$$

holds.

1.3.2. Sets of class Γ

A set of class Γ is a Borel set which may be covered by a finite or countable system of smooth manifolds to within a set of H_{n-1}-measure zero.

Theorem 5. ([DG2], [V1]). *The essential boundary of a BV-set is a set of class Γ and the normals to $\partial^* E$ coincide with the normals to the covering manifolds H_{n-1}-everywhere.*

1.3.3. Cross-Sections

Let a be a unit vector in R^n, Π_a the plane $(x,a) = 0$, $l_a(x')$ the straight line parallel to the vector a and passing through the point $x' \in \Pi_a$.

Theorem 6. *Let E be a BV-set and a an arbitrary unit vector in R^n. Then there exists a set F of total $(n-1)$-dimensional measure on the plane Π_a such that when $x' \in F$ the set $l_a(x') \bigcap E_*$ is the union of a finite number of (open) intervals with disjoint closures, where the union of boundary points of these intervals coincides with the set $l_a(x') \bigcap \partial^* E$.*

1.3.4. Projections

Let E be an arbitrary set in R^n, $\pi_a E$ the orthogonal projection of the set E onto the plane Π_a. We shall use the notation $\lambda_a(E) = m_{n-1}(\pi_a E)$, where m_{n-1} is the outer $(n-1)$-dimensional Lebesgue measure on the plane Π_a,

$$\lambda(E) = \sup_{|a|=1} \lambda_a(E).$$

A set E with $\lambda(E) = 0$ will be called a set of $(n-1)$-dimensional measure zero. Clearly, $\lambda(E) \leq H_{n-1}(E)$, so $\lambda(E) = 0$ if $H_{n-1}(E) = 0$. The converse is not true as follows from an example by D.E.Menshov.

Theorem 7. *If E is a BV-set, then*

$$\lambda(E_*) = \lambda(E_* \bigcup \partial^* E) = \lambda(\partial^* E) \leq P(E).$$

It is proved that $\lambda(E) = 0$ implies $\partial u/\partial x^i(E) = 0$ for all $u \in BV$.

1.3.5. BV-Functions

Now we formulate the main theorem on the structure of BV-functions.

Theorem 8. *Let $u(x)$ be a vector-valued function, $u \in BV(\mathcal{G})$. Let E be the set of regular points in $\mathcal{G}$ of $u(x)$. Then $\lambda(\mathcal{G} \setminus E) = 0$ and the set of jump points of $u(x)$ is a set of class Γ.*

1.3.6. The Mean Value

Let us introduce an averaging kernel in the usual way and denote by $\tilde{u}(x)$ the mean value of a function $u(x)$, i.e. the limit as the averaging radius tends to zero (for details, see [V1]).

Theorem 9. *Let $u(x)$ be a function that is defined and locally integrable in the neighborhood of the point x_0, where this point is regular for $u(x)$. Then the mean value $\tilde{u}(x_0)$ exists and*

$$\tilde{u}(x_0) = \sigma l_a u(x_0) + (1 - \sigma) l_{-a} u(x_0), \tag{2}$$

where a is the vector appearing in Definition 1.1, and $\sigma \in [0,1]$ is a number depending on the averaging kernel. If the averaging kernel is symmetric, then $\sigma = \frac{1}{2}$.

It follows from Theorem 9 that for BV-functions the mean value exists in all points up to the sets of $(n-1)$-dimensional measure zero.

We emphasize that results in this section are obtained up to the sets of $(n-1)$-dimensional measure zero and that these sets are negligible with respect to the measure $\partial u/\partial x^i$ for all $u \in BV$. Otherwise the analysis in spaces BV could not be constructed.

1.4 Differentiation

1.4.1. Statement of the Problem

Consider the usual formula for differentiation of superposition

$$\frac{\partial f(u(x))}{\partial x^i} = \sum_{k=1}^{p} f_k(u(x))\frac{\partial u^k}{\partial x^i}, \tag{3}$$

where $f(u)$ is a function of p variables $u = (u^1, \dots, u^p)$, $u(x)$ is a vector-valued function of n variables $x = (x^1, \dots, x^n)$,

$$f_k(u(x)) = f'_{u^k}(u(x)). \tag{4}$$

The problem is to generalize this formula for discontinuous functions $u(x)$.

For functions in BV we shall understand (3) as the equality of the values of measures. As usual, the product of a function $f_k(u(x))$ by the measure $\partial u/\partial x^k$ means the measure equal to the integral of the function with respect to the given measure.

But though the formula (3) has sense for BV-functions we have a contradiction even in simplest cases. For example, it is evident that formula

$$\frac{\partial u^m}{\partial x^i} = mu^{m-1}\frac{\partial u}{\partial x^i}$$

is wrong for $u(x)$ which are characteristic functions of sets.

To obtain the correct formula we shall introduce the notion of functional superposition.

1.4.2. Functional Superposition

Let x be a regular point for the vector-valued function $u(x) = (u^1(x), \dots, u^p(x))$. Then, by Definition 1.1 the pair of vectors $l_a u(x_0)$ and $l_{-a}u(x_0)$ exists. Suppose further that we are given a function $f(u)$ of p variables that is defined on the interval

$$u(x,\sigma) = \sigma l_a u(x) + (1-\sigma)l_{-a}u(x) \quad (0 \le \sigma \le 1) \tag{5}$$

in the space R^p and integrable over this interval. Then, by definition, the *functional superposition* is given by the equality

$$\hat{f}(u(x)) = \int_0^1 f(u(x,\sigma))d\sigma. \tag{6}$$

It is interesting to compare (5) with (2) which shows that (6) may be interpreted as averaging over all averagings. This interpretation is not essential here, but explains why the notion of the functional superposition is useful in continuum thermodynamics (see [V2]).

Clearly, $\hat{f}(u(x)) = f(u(x))$ if $u(x)$ is approximately continuous at the point x.

Theorem 10. *If $u(x)$ is a vector-valued BV-function and $\vartheta \in BV$, then $\hat{f}(u(x))$ is measurable with respect to the measure $\partial\vartheta/\partial x^i$.*

1.4.3. Formula for Differentiation of Superposition

To obtain the correct formula we must replace in (3) the usual superposition (4) by functional superposition:

$$f_k(u(x)) = \hat{f}_{u_k}(u(x)). \tag{7}$$

We now formulate the result, assuming for simplicity that the function $f(u)$ is defined on the whole space R^p and continuous together with its first derivatives.

Theorem 11. *Let $u = (u^1, \dots, u^p) \in BV(\mathcal{G})$ and let the functional superposition (7) be integrable with respect to the measure $\partial u^k/\partial x^i$ in $\mathcal{G}(k = 1, \dots, p;\ i = 1, \dots, n)$. Then $f(u(x)) \in BV(\mathcal{G})$ and the formula (3), (7) holds.*

We observe that, by virtue of Theorem 10, when $u(x)$ is a bounded vector-valued function the condition that (7) be integrable with respect to the measure $\partial u^k/\partial x^i$ is satisfied automatically. The superposition on the left-hand side in (3) can obviously be understood in the ordinary and the functional sense, since these superpositions coincide almost everywhere with respect to n-dimensional measure.

The application of the theorem to the differentiation of a product yields the following result. Let u and ϑ be BV-functions, and the symmetric mean $\overline{u}(x)(\overline{\vartheta}(x))$ of $u(\vartheta)$ be locally integrable with respect to the measure $\nabla\vartheta(\nabla u)$. Then $u\vartheta \in BV$ and

$$\nabla(u\vartheta) = \overline{u} \nabla \vartheta + \overline{\vartheta} \nabla u. \tag{8}$$

1.4.4. Generalization

Let $\hat{f}(u_+, u_-)$ be an arbitrary continuous function $R^p \times R^p \to R^p$ such that

$$f(u_+) - f(u_-) = (\hat{f}(u_+, u_-), u_+ - u_-) \tag{9}$$

for all $u_+ \in R^p$, $u_- \in R^p$, where $(,)$ denotes the inner product in R^p. In particular, we can take

$$\hat{f}(u_+, u_-) = \int_0^1 f_u'(u_+\sigma + u_-(1-\sigma))d\sigma. \tag{10}$$

Formulas (3), (7) may be written as

$$\frac{\partial f(u(x))}{\partial x^i} = \left(\hat{f}(u(x)), \frac{\partial u}{\partial x^i} \right) \tag{11}$$

with $\hat{f}$ given by (10), where $u_{\pm}(x) = l_{\pm a}u(x)$. The generalization is in the fact that we can understand formula (11) with arbitrary continuous $\hat{f}(u_+, u_-)$ satisfying (9) (see [V2]). In other words, we can use vector-valued functions which are equivalent to the functional superposition (cf. subsection 2.2.1).

For example, the generalization of formula (8) is

$$\nabla(u\vartheta) = [u_+\alpha + u_-(1-\alpha)] \nabla \vartheta + [\vartheta_-\alpha + \vartheta_+(1-\alpha)] \nabla u, \tag{12}$$

where α is an arbitrary real number. We obtain (8) for $\alpha = \frac{1}{2}$.

All these results remain true in the space BV_σ which is a generalization of the space BV (for details, see [V 2]).

1.5 Integral Formulas

1.5.1. The Trace

Let E be a measurable set, $\partial^* E$ its essential boundary, $u(x)$ a function defined in some neighborhood of the point $x_0 \in \partial^* E$.

Definition 1.5. *The inward trace* $(u^+(x_0))$ *of the function* $u(x)$ *at the point* $x_0 \in \partial^* E$ *is the approximate limit* $l_E u(x_0)$; *the outward trace* $(u_-(x_0))$ *is* $l_{CE}u(x_0)$.

Theorem 12. *Let* E *be a* BV*-set and* u *a* BV*-function. Then the inward and outward traces of the function* u *exist on* $\partial^* E$ *almost everywhere with respect to the measure* H_{n-1}.

1.5.2. Sets of Class Γ

By definition, the *jump* of a function at the regular point x_0 is the vector

$$\triangle u(x_0) = (l_a u(x_0) - l_{-a} u(x_0))a.$$

Theorem 13. *Let* S *be a set of class* Γ *and* $u \in BV$. *Then the jump* $\triangle u$ *is defined* H_{n-1}*-almost everywhere on* S *and*

$$\int_S \nabla u dx = \int_S \triangle u dH_{n-1}.$$

1.5.3. The Gauss-Green Formula

Theorem 14 ([DG2], [F3], [V1]). *Suppose that* E *is a bounded* BV*-set and* u *is a* BV*-function. Then*

$$\int_{E_*} \nabla u dx = \int_{\partial^* E} u^+ \nu dH_n, \tag{13}$$

where ν *is the outward normal.*

If $u(x)$ *is unbounded, we suppose that the integral on the right exists.*

In application it is convenient to use a combination of formula (13) with the following formula

$$\int_S \nabla u dx = \int_S (u^- - u^+)\nu dH_{n-1}, \tag{14}$$

where $S \subseteq \partial^* E$ is a Borel set.

Theorem 15. *Let* E *be a bounded* BV*-set which is open in d-topology, and* ∂E *be its boundary in this topology. If* ∂E *is of class* Γ *and* $u \in BV$*, then*

$$\int_E \nabla u dx = \int_{\partial^* E} u^+ \nu dH_{n-1} + \int_{\partial E \setminus \partial^* E} \triangle u dH_{n-1}. \tag{15}$$

If $u(x)$ *is unbounded, we suppose that the first integral on the right exists.*

2 Partial Differential Equations

2.1 Conservation Laws

2.1.1. Generalized Solutions

We consider the system of conservation laws

$$\frac{\partial u}{\partial t} + \sum_{k=1}^{n} \frac{\partial a_k(u)}{\partial x^k} = 0, \tag{16}$$

where $a_k(u)$ $(k = 1, \dots, n)$ are smooth vector-valued function of p variables $u = (u^1, \dots, u^p)$, $a_k(u) = (a_k^1(u), \dots, a_k^p(u))$. The Cauchy problem consists in finding the solution of (16) for $t > 0$ that satisfies the initial condition

$$u \mid_{t=0} = f(x). \tag{17}$$

As is well known, even for smooth initial conditions there does not exist a continuous solution in the large of the problem (16), (17). Therefore the solution must be understood in the generalized sense. In the class of bounded measurable vector-valued functions $u(x,t)$ the generalized solution of (16) is defined as a solution of the system

$$\int\int (\frac{\partial \varphi}{\partial t} u + \sum_k \frac{\partial \varphi}{\partial x^k} a_k(u)) dx\, dt = 0 \tag{18}$$

for any smooth finitary function $\varphi(x,t)$. It is also well known that in the case of discontinuous solution one must also introduce some supplementary conditions. Using physical meaning, these supplementary conditions are called entropy conditions.

2.1.2. Entropy Conditions

We begin with the scalar case of (16) $(p = 1)$. Denote $a(u) = (u, a_1(u), \dots, a_n(u))$,

$$S(u,v) = [a(u) - a(v)]\mathrm{sgn}(u - v). \tag{19}$$

Suppose first that $u \in BV(\mathcal{G})$ in a domain $\mathcal{G}$ of the space (x,t). We introduce the following *entropy condition* (see [V1]). Let $\Gamma(u) \subset \mathcal{G}$ be the set of points of jump of the function $u(x,t)$. Then H_n-almost everywhere on $\Gamma(u)$ we require that

$$(S(u_+, c), \nu) \leq (S(u_-, c)\nu), \tag{20}$$

where c is an arbitrary constant, ν is the normal to $\Gamma(u)$, $u_\pm = l_{\pm\nu}u$.

The condition (20) may be written in the following equivalent form.

Theorem 16 ([V1]). *A necessary and sufficient condition for the bounded function $u \in BV(\mathcal{G})$ to be solution of the equation (16) and satisfy the condition (20) in $\mathcal{G}$ is that*

$$\left[\frac{\partial \mid u - c \mid}{\partial t} = \sum_{k=1}^{n} \frac{\partial (a_k(u) - a_k(c))\mathrm{sgn}(u - c)}{\partial x^k}\right](E) \leq 0 \tag{21}$$

for any constant c and any bounded Borel set E whose closure belongs to $\mathcal{G}$.

It is evident that (21) is equivalent to

$$\int\int \left[\mid u - c \mid \frac{\partial \varphi}{\partial t} + \sum_{k=1}^{n} \frac{\partial \varphi}{\partial x^k}[a_k(u) - a_k(c)]\mathrm{sgn}\,(u - c)\right] dx\, dt \geq 0 \tag{22}$$

for any finitary function $\varphi(x,t) \geq 0$. It is proved that the solution of Cauchy problem satisfying the entropy condition exists and is unique (see the next subsection).

These results and the results in hydrodynamics set up a base for the following approach to the system (18) (see [FL], [G], [H], [Kr], [L], [T] and references there). A convex function $U(u)$ is called an entropy function if the following equality holds

$$(\nabla U(u), \nabla a_k(u)) = \nabla F_k(u) \quad (k = 1, \dots, n), \tag{23}$$

where $F_k(u)$ are some functions called the entropy flux. It is easy to see that in the scalar case the function (19) of u (for any constant $\vartheta = c$) provides the entropy $U(u) = \mid u - c \mid$ with the entropy flux $[a_k(u) - a_k(c)]\mathrm{sgn}\,(u - c)$. We note that the function $\mid u - c \mid$ contains an arbitrary constant c which is important, for example, in proving the uniqueness of solutions (see the next subsection).

2.1.3. Uniqueness, stability, existence

We consider first the scalar case (for details see [V1]). It is easy to show that if we have two solutions u and ϑ of (16) satisfying (20), then

$$(S(u_+, \vartheta_+), \nu) \leq (S(u_-.\vartheta_-), \nu)$$

and (21) follows with $c = \vartheta$. Applying this inequality to the characteristic cone E we get

$$\int_{|x|<r} | u(x,t) - v(x,t) | \, dx \leq \int_{|x|<r+lt} | f(x) - g(x) | \, dx, \tag{24}$$

where $f(x)$ and $g(x)$ are the initial functions, $r > 0$ is an arbitrary number and l is the slope of the characteristic cone. Stability and uniqueness of solutions follow from (24). The following existence theorem holds.

Theorem 17. *Let $f(x)$ be a bounded function belonging to the space $BV(R_0^n)$. Then there exists a generalized solution $u(x,t)$ of the problem (16), (17) in R_+^{n+1} which belongs to the space $BV(R_+^{n+1})$ and is bounded*

$$| u(x,t) | \leq \max_x | f(x) | \, .$$

Here R_0^n is the space $t = 0$ and R_+^{n+1} is the half-space $t > 0$ in R^{n+1}.

The proof of this theorem is given by the small viscosity approach considering the equation

$$\frac{\partial u}{\partial t} + \sum_{k=1}^{n} \frac{\partial a_k(u)}{\partial x^k} = \varepsilon \triangle u \quad (\varepsilon > 0)$$

with the initial condition (17).

It is interesting to make a remark about the significance of the spaces BV in this theory. It is proved that every bounded measurable generalized solution of the problem (16), (17) satisfying the entropy condition (22) belongs to BV if the initial function $f \in BV$, in particular, if it is smooth. Moreover, BV-functions have definite structure, their points of jump form a set of class Γ, so the entropy conditions may be formulated only on the sets of jumps and it is enough to obtain the uniqueness of solutions.

For systems of equations (16) we refer to the papers [Da1], [DM], [DP1], [DP2], [DP3], [G1], [Li], [S][1] devoted to the problems of uniqueness, stability, existence and asymptotic behavior of solutions, where the spaces BV or BV-analysis have been proved to be useful.

2.2 First-Order Hyperbolic Systems

Hyperbolic conservation laws are a special case of the general first-order hyperbolic systems that we consider in this section. The very definition of the nonconservative quasilinear operator requires multiplication of a discontinuous function by a measure and so uses essentially the structure of BV-functions given in Section 1.3.

[1]We can not give the complete list of references here.

2.2.1. First-Order Systems

Denote by D a domain in R^p. Two $p \times p$ matrix functions $A(u_+, u_-)$ and $B(u_+, u_-)$ defined on $D \times D(u_+ \in D,\ u_- \in D)$ and continuous are said to be equivalent iff

$$[A(u_+, u_-) - B(u_+, u_-)](u_+ - u_-) = 0.$$

We denote this equivalence by $A \approx B$. We shall consider the system

$$\sum_{n=0}^{n} A_i(u_+, u_-)\frac{\partial u}{\partial x^i} + f(u) = 0. \tag{25}$$

Here $A_i(u_+, u_-)$ $(i = 0, 1, \dots, n)$ are given $p \times p$ matrix functions defined on $D \times D$ and continuous; $f(u)$ is a continuous vector-valued function; $x = (x^0, \dots, x^n) \in \mathcal{G}$, where $\mathcal{G}$ is a domain in R^{n+1}; $u \in BV(\mathcal{G})$ and bounded with range in D. We suppose that

$$A_i(u_+, u_-) \approx A_i(u_-, u_+) \quad (i = 0, 1, \dots, n).$$

The solution of (25) is to be understood as follows:

$$\sum_{i=0}^{n} A_i(u_+(x), u_-(x))\frac{\partial u}{\partial x^i} + f(u(x)) = 0. \tag{26}$$

(local equality of measures), where

$$u_\pm(x) = l_{\pm a}u(x)$$

at each regular point x of u, a is the vector appearing in Definition 1.1.

It is proved that

$$A_i(u_+(x), u_-(x))\frac{\partial u}{\partial x^i} = B_i(u_+(x), u_-(x))\frac{\partial u}{\partial x^i} \quad (i = 0, 1, \dots, n),$$

if $A_i \approx B_i$.

Obviously, the system of conservation laws (16) is a special case of the system (26) which follows directly from the formula for differentiation of superposition (see Section 1.4).

In what follows we shall suppose that

$$\det A_0(u_+, u_-) \neq 0 \quad ((u_+ \in D,\ \ u_- \in D).$$

2.2.2. Hyperbolic Systems

Definition 2.1 *The system (25) is said to be hyperbolic in the region $D \subseteq R^p$ if there exist representatives A_i $(i = 0, 1, \dots, n)$ of equivalence classes such that for any $u_+ \in D$, $u_- \in D$ the matrix*

$$A_0^{-1}(u_+, u_-)\sum_{i=0}^{n} A_i(u_+, u_-)\omega_i \tag{27}$$

has real eigenvalues and a complete set of eigenvectors for all real ω_i.

We note that we obtain the usual definition of hyperbolicity of we set $u_+ = u_-$. So Definition 2.1 contains stronger requirements. It is proved that for systems of conservation laws (16) possessing entropy functions the corresponding system (25) is hyperbolic in the sense of Definition 2.1 (cf. [H]).

Entropy conditions and existence of solutions for some classes of nonconservative systems (25) were studied in [LF], [LL].

2.2.3. Sequential approach

From the point of view of applications and in particular of the numerical simulation it is useful to introduce the sequential approach to the definition of a solution of the system (25). A vector-valued function $u(x)$ is called a solution of the system (25) if there exists a sequence $\{u_m(x)\}$ of bounded BV-functions such that

$$u_m(x) \to u(x) \tag{28}$$

$$\sum_{i=0}^{n} A_i(u_{m+}(x), u_{m-}(x)) \frac{\partial u_m}{\partial x^i} + f(u_m) \to 0. \tag{29}$$

It should be specified additionally in what sense the convergence (28), (29) is to be understood.

In this definition we do not mean that $u \in BV$. But suppose $u \in BV$. The problem is when the solution u in this definition is a solution in the definition (26). Consider the case where the system (25) is equivalent to the system of conservation laws. It means that there exists such an invertible matrix function $C(u_+, u_-)$ that (25), being multiplied by C from the left, may be written in the form of conservation laws. In this case the both definitions coincide if (28) is the bounded convergence almost everywhere and (29) is the strong convergence of measures. We emphasize that, contrary to the systems of conservation laws, it is not right if (29) is the *weak** convergence.

2.2.4. Numerical Methods

It is known that finite difference schemes in nonconservative form converge to wrong solution even for scalar conservation laws, a wrong source-term appears (see [HL]). As shown later ([VD]), nonconservative variant of Godunov's scheme satisfies (28), (29) and so does not give a wrong source-term, but numerical experiments show that the solutions of the difference equations do not converge to the solutions of the given differential equation.It is because the convergence (29) is *weak**. In [VD] a numerical method was proposed which has realized the sequential approach with strong convergence in (29). In this case numerical experiments have shown a good approximation.

2.3 Continuum Mechanics

2.3.1. Mathematical Principles

The 6 Hilbert problem on the axiomatic approach to those physical sciences where mathematics is essential gave rise to attempts to found an axiomatic basis for continuum mechanics. Systems of axioms have been introduced (see [Tr] and references there). They deal with smooth functions and so can not cover some parts of continuum mechanics: shock waves, formation of fractures and so on. Using the analysis in classes of discontinuous functions, it became possible to do it. But the problem is not only to find a proper class of functions. For example, the Euler equation

$$\frac{dx}{dt} = u(x,t), \tag{30}$$

where u is the velocity, does not have a unique solution if u is discontinuous. In a new version of axioms n-dimensional manifolds are considered, mass is a positive Borel measure, bodies are classes of equivalent (with respect to the mass) Borel sets, the equation (30) is replaced by

$$\frac{d\chi}{dt} + (u, \nabla\chi) = 0, \tag{31}$$

where χ is the characteristic function of the moving body. A priori, the solutions of (31) are not unique, but unlike (30), the uniqueness and stability of solutions of (31) can be obtained as a consequence of the mass conservation law. Conservation laws equations for mass, momentum and energy have been derived directly for discontinuous flows.

2.3.2. Continuum Thermodynamics

(For details see [V2]). The concept of thermodynamic mean has been introduced, which is the functional superposition for any thermodynamic variable with respect to the basic variables (see Subsection 1.4.2). This is the way to choose the values of a thermodynamic variable at the points which are not points of thermodynamic equilibrium. The equations of hydrodynamics in nonconservative form contain these values (an example is given in the next subsection).

2.3.3. Nonconservative Equations

[V2]. Starting with conservation laws and using the formula for differentiation of superposition we obtain nonconservative form for equations of hydrodynamics:

$$\begin{aligned} &\tfrac{d\vartheta}{dt} - \vartheta \operatorname{div} u = 0 \\ &\tfrac{du}{dt} + \vartheta \nabla p = 0 \\ &\tfrac{de}{dt} + \vartheta p \operatorname{div} u = 0. \end{aligned} \tag{32}$$

Here ϑ, p, e are the specific volume, pressure and specific internal energy, respectively. We emphasize that though the equations (32) look like usual equations in the smooth case, they are true only if ϑ and p are understood as symmetric means (compare with the equation

$$\rho\frac{du}{dt} + \nabla p = 0,$$

where ρ is the density, which is wrong for discontinuous solutions !)

From (32) we derive the following equation for entropy S :

$$\frac{dS}{dt} + \frac{p - \hat{p}}{\hat{T}}\vartheta \operatorname{div} u = 0,$$

where T is the temperature and $\hat{}$ denotes the thermodynamic mean ($\hat{p} = p$ at continuity points).

2.3.4. Detonation

Interesting three-dimensional regimes of detonation waves propagation were observed experimentally. An attempt to describe these regimes by the bifurcation theory was made. It turned out that, unlike the known Hopf bifurcation, in this case the entire imaginary axis consists of continuum spectrum points. There are also points of discrete spectrum on this axis which are responsible for the bifurcation. In [VZ], using BV approach, the first term has been found. It describes the regimes mentioned. The problem is to develop the complete bifurcation theory.

2.4 Parabolic and Elliptic Equations

2.4.1. Degenerate Parabolic Equations

In [VH2] the Cauchy problem for the degenerate second order parabolic equation

$$\frac{\partial u}{\partial t} = \sum_{i,j=1}^{n} \frac{\partial}{\partial x^i}\left[a^{ij}(u,t,x)\frac{\partial u}{\partial x^j}\right] + \sum_{i=1}^{n}\frac{\partial b^i(u,t,x)}{\partial x^i} + c(u,t,x) \tag{33}$$

has been studied. The equation (16) is a special case of (33). The basic difficulty in studying equation (33) is caused by the mixed type of this equation with region of hyperbolicity $(a(u(t,x),t,x) = 0)$ not known beforehand (unlike the linear case) and the necessity of considering a class of functions which can have discontinuity in the region of hyperbolicity and must be somewhat smooth in the region of parabolicity.

The notion of a generalized solution of (33) of the (22) type has been introduced. Nonlocal existence theorem for solution of Cauchy problem, along with the uniqueness and stability of this solution has been proved. The solution was considered in the space of bounded BV-functions with the norm defined by the coefficients in the second order terms in (33).

2.4.2. Quasi-Linear Parabolic System

Positivity of solutions of boundary value problems in BV-domains for a second-order quasi-linear parabolic systems have been studied (see [VT1], [VT2]). Generalization of the Sobolev spaces for such domains were introduced and used.

2.4.3. Linear Elliptic Equations

The theory of the boundary value problems for linear elliptic equations well-known for domains with smooth boundary can be extended to domains with finite perimeter without requirements of smoothness of the boundary. In [VH1] investigation of boundary value problems for such domains has been done including Fredholm theorems, the eigenvalue problem, positivity of solutions.

References

[Da1] C.M. Dafermos, Admissible wave fans in nonlinear hyperbolic systems, *Arch. Rat. Mech. Anal.* **106** (1989), 243-260.

[Da2] C.M. Dafermos, Generalized characteristics in hyperbolic systems of conservation laws, *Arch. Rat. Mech. Anal.* **107** (1989), 127-155.

[DG1] E.De Giorgi, Su una teoria generale della misura $(r-1)$-dimensionale in uno spazio ad r dimensioni, *Ann. Mat. Pura Appl.* (4) **36** (1954), 191-213.

[DG2] E.De Giorgi, Nuovi teoremi relativi alle misure $(r-1)$-dimensionali in uno spazio ad r dimensioni, *Ricerche Mat.* **4** (1955), 95-113.

[DM] R.J.Di Perna and A.Maida, The validity of nonlinear geometric optics for weak solutions of conservation laws, *Comm. Math. Phys.* **98** (1985), 313-347.

[DP1] R.J. Di Perna,Decay and asymptotic behavior of solutions to nonlinear hyperbolic systems of conservation laws. *Ind. Univ. Math. J.* **24** (1975), 1047-1071.

[DP2] R.J. Di Perna, Uniqueness of solutions of hyperbolic systems of conservation laws, *Ind. Univ. Math. J.* **28** (1979), 137-188.

[DP3] R.J. Di Perna, Convergence of approximate solutions to conservation laws, *Arch. Rat. Mech. Anal.* **82** (1983), 27-70.

[FL] K.O.Friedrichs and P.D. Lax, Systems of conservation equations with a convex extension, *Proc. Nat. Acad. Sci. USA* **68** (1971), 1636-1688.

[FR] W.H.Fleming and R. Rishel, An integral formula for total gradient variation, *Arch. Math.* **11** (1960), 218-222.

[F1] H. Federer, An analytic characterization of distributions whose partial derivatives are representable by measures, *Bull. Amer. Math. Soc.* **60** (1954), 339.

[F2] H.Federer, The Gauss-Green theorem, *Trans. Amer. Math. Soc.* **58** (1945), 44-76.

[F3] H. Federer, A note on the Gauss-Green theorem, *Proc. Amer. Math. Soc.* **9** (1958), 447-451.

[G] S.K.Godunov, The problem of a generalized solutions in the theory of quasi-linear equations and in gas dynamics, *Russian Math. Surveys* **17** (1962), 145-156.

[G1] J. Glimm, Solutions in the large for nonlinear hyperbolic systems of equations, *Comm. Pure Appl. Math.* **18** (1965), 697-715.

[GW] C.Goffman and D.Waterman, Approximately continuous transformations, *Proc. Amer. Math. Soc.* **12** (1961), 116-121.

[H] A. Harten, On the symmetric form of systems of conservation laws with entropy, Journ. Comp. Physics **49** (1983), 151-164.

[HL] T.Y. Hou and Ph. Le Floch, Why nonconservative schemes converge to wrong solutions: error analysis, *Math. Comp.* **62** No.206 (1994), 497-530.

[K] K. Krickeberg, Distributionen, Funktionen bestrankter Variation and Lebesguescher Inhalt nichtparametrischer Flachen, *Ann. Math. Pura Appl.* (4) **44** (1957), 105-133.

[Kr] S.N.Kruzkov, First-order quasi-linear equations in several independent variables, *Math. USSR Sbornik* **10** (1970), 127-243.

[L] P.D. Lax, Shock waves and entropy " Contributions to Nonlinear Functional Analysis ", 603-634, Academic Press, New York, 1971.

[LF] Ph. Le Floch, Entropy weak solutions to nonlinear hyperbolic systems under nonconservative form, *Comm. Part. Dif. Eq.* **13** (1988), 669-727.

[Li] T.P.Liu, Decay to N-waves of solutions of general systems of nonlinear hyperbolic conservation laws, **C.P.A.M. 30** (1977), 585-610.

[LL] Ph. Le Floch and Tai-Ping Liu, Existence theory for nonlinear hyperbolic systems in nonconservative form, *Forum Math.* **5** No.3 (1993), 261-280.

[S] S.Schochet, Resonant nonlinear geometric optics for weak solutions of conservation laws, *J. Dif. Equat.* **113** (1994), 473-504.

[T] E.Tadmor, Entropy functions for symmetric systems of conservation laws, *Journ. Math. Anal. Appl.* **122** (1987), 355-359.

[Tr] C. Truesdell, A first course in rational continuum mechanics, 1972.

[VD] A.I. Volpert and S.I.Doronin, On numerical methods for nonconservative hyperbolic equations, Preprint, ICPh. (1992).

[VH1] A.I. Volpert and S.I. Hudjaev, Analysis in classes of discontinuous functions and equations of mathematical physics, *Martinus Nijhoff Publishers*, 1985, 678p.

[VH2] A.I.Volpert and S.I. Hudjaev, Cauchy problem for degenarate second order quasi-linear parabolic equations, *Math. USSR Sbornik* **7** (1969), 365-387.

[VT1] A.I.Volpert and R.S. Tishakova, Positive solutions of quasi-linear parabolic system of equations, Preprint, Chernogolovka, 1980, 20p. (In Russian).

[VT2] A.I.Volpert and R.S. Tishakova, Positive solutions of the second boundary value problem for quasi-linear parabolic systems, Preprint, Chernogolovka, 1981, 13p. (In Russian).

[VZ] A.I. Volpert, A.M. Zhiljaev, V.P. Filipenko, Mathematical problems of detonation waves stability, *Proceedings of 9 Symp. Comb. Explosion, Detonation* (1989), 6-9. (In Russian).

[V1] A.I. Volpert, The spaces BV and quasi-linear equations, *Math. USSR Sbornik* **2** No.2 (1967), 225-267.

[V2] A.I. Volpert, BV-Analysis and hydrodynamics, Preprint, Chernogolovka, 1991, 32p.

Department of Mathematics

Technion-Israel Institute of Technology

Haifa 32000, Israel

AMS Classification: 46E30, 35F25

Operator Theory
Advances and Applications, Vol. 98

THE BEHAVIOUR OF SOLUTIONS OF ORDINARY DIFFERENTIAL EQUATIONS IN INFINITE DOMAINS

YA. I. ZHITOMIRSKII

Under minimal restrictions on coefficients of linear ODEs with linear complex parameter λ we find asymptotics of solutions $x(t,\lambda)$ as $|\lambda| \to \infty$ and $|t| \leq q(|\lambda|)$, where $q(|\lambda|)$ is an increasing function determined by coefficients, and obtain an estimate of solutions for all $(t,\lambda) \in \mathbb{R} \times \mathbb{C}$.

1. Introduction

Many important problems of analysis, in particular, spectral theory, studying the Cauchy problem for PDE's or studying the uniqueness of integral representations, require either the exact asymptotics of solutions of parameter-dependent families of ODEs or at least estimates of such solutions (see [1-3]). The assumptions on both the coefficients of the equations and the systems are usually connected with the way the results are going to be applied (see, e.g. [4], [5]).

This paper contains two main results both obtained under minimal assumptions on the properties of coefficients of ODEs.

- We find asymptotics of solutions of the nth order ordinary differential equation

$$\sum_{k=0}^{n} p_k(t)x^{(k)}(t,\lambda) = \lambda x(t,\lambda), t \in \mathbb{R}, \lambda \in \Lambda \subset \mathbb{C}, \tag{1}$$

where $p_n(t) \equiv p \neq 0$ is a constant, $p_k(t)$, $k = 0,\ldots,n-1$, are arbitrary differentiable complex-valued functions, Λ is an unbounded set; the asymptotics of solutions is found for large values of $|\lambda|$ in the domain $|t| \leq q(|\lambda|)$ for a certain monotone function $q(x)$, $x > 0$, determined by the coefficients of (1).

- We obtain an estimate of solutions of the system

$$\frac{d\bar{x}(t,\lambda)}{dt} = P(t,\lambda)\bar{x}(t,\lambda) \tag{2}$$

for all $t \in \mathbb{R}, \lambda \in \mathbb{C}$, where $P(t,\lambda) = (P_{sk}(t,\lambda))_{s,k=1}^{n}$ is an arbitrary continuous complex-valued matrix. We also will show that our estimates for the solutions of (2) cannot be essentially improved.

2. Asymptotics of solutions of (1)

Denote by $\mu_j(t,\lambda)$, $j = 0,\ldots,n-1$, the roots of the equation

$$\sum_{k=0}^{n} p_k(t)\mu^k = \lambda \tag{3}$$

We will characterize the growth of the coefficients $p_k(t)$ of (1) by a function $p(x)$ which is introduced in the following way. Let us take an arbitrary function $\alpha(x)$ such that $\alpha(x) > 0$ as $x > 0$ and $\alpha(x) = o(1)$ as $x \to \infty$. We can find a monotone unbounded function $p(x)$ such that the following estimates hold for any $x > 0$:

$$\max_{0\le k\le n-1} \sup_{|t|\le x} |p_k(t)|^{1/(n-k)} \le \alpha(x)p(x) \tag{4}$$

$$\max_{0\le k\le n-1} \sup_{|t|\le x} |p_k'(t)(1+|t|)|^{1/(n-k)} \le \alpha(x)p(x). \tag{5}$$

Now we define a positive-valued function $q(x)$, $x > 0$ by the relation

$$p^n(q(x)) \equiv x, \ x > 0. \tag{6}$$

Obviously $q(x)$ is monotone increasing function, and $q(x) \to \infty$ as $x \to \infty$.
We fix $\delta > 0$ and denote

$$\Lambda = \{\lambda \in \mathbb{C} : |\arg\lambda - \arg p| \ge \delta > 0, \ |\arg\lambda - \arg p - \pi| \ge \delta > 0\}.$$

Theorem 1. *For arbitrary differentiable coefficients $p_k(t)$, $k = 0,\ldots,n-1$, $t \in \mathbb{R}$, equation (1) has n linearly independent solutions $x_j(t)$, $j = 0,\ldots,n-1$, such that for any $j,k \in \overline{0,n-1}$*

$$d^k x_j(t,\lambda)/dt^k = p^{-k/n}\lambda^{k/n}\exp\{2\pi ikj/n + \int_0^t \mu_j(\tau,\lambda)d\tau\}(1+o(1)) \tag{7}$$

as $|\lambda| \to \infty$, $\lambda \in \Lambda$, $|t| \le q(|\lambda|)$.

Proof. Introduce the following $(n \times n)$-matrices:

$$M(t,\lambda) \equiv \operatorname{diag}\{\mu_0(t,\lambda),\ldots,\mu_{n-1}(t,\lambda)\}, \quad \tilde{M}(t,\lambda) \equiv \left(\mu_s^k(t,\lambda)\right)_{s,k=0}^{n-1},$$

$$P(t,\lambda) \equiv \tilde{M}^{-1}(t,\lambda)d\tilde{M}(t,\lambda)/dt \equiv (P_{sk}(t,\lambda))_{s,k=0}^{n-1}.$$

The invertibility of $\tilde{M}(t,\lambda)$ for large $|\lambda|$, $\lambda \in \Lambda$, $|t| \leq q(|\lambda|)$ will be proved later on. We reduce equation (1) to the system of first order equations using the standard procedure of introducing

$$x_k(t,\lambda) = d^k x(t,\lambda)/dt^k;\ \bar{x}(t,\lambda) \equiv (x_0(t,\lambda),\ldots,x_{n-1}(t,\lambda)),$$
$$\bar{x}(t,\lambda) \equiv \tilde{M}(t,\lambda)\bar{y}(t,\lambda).$$

Equation (1) takes the following form

$$d\bar{y}(t,\lambda)/dt = M(t,\lambda)\bar{y}(t,\lambda) + P(t,\lambda)\bar{y}(t,\lambda),\ t \in \mathbf{R},\ \lambda \in \Lambda \tag{8}$$

We now transform system (8) by setting

$$\bar{y}(t,\lambda) = \bar{Z}_s(t,\lambda)\exp\{\int_0^t \mu_s(\tau,\lambda)d\tau\},\ s = \overline{0,n-1}.$$

We denote $\bar{Z}_s(t,\lambda) \equiv (Z_{0s}(t,\lambda),\ldots,Z_{n-1\,s}(t,\lambda))$ and obtain

$$dZ_{sm}(t,\lambda)/dt = [\mu_s(t,\lambda) - \mu_m(t,\lambda)]Z_{sm} + \sum_{k=0}^{n-1} P_{sk}(t,\lambda)Z_{km}(t,\lambda) \tag{9}$$

To study the coefficients of system (9) we will show, at first, that for $\lambda \in \Lambda$, $|t| \leq q(|\lambda|)$ the roots $\mu_s(t,\lambda)$ of equation (3) have the asymptotics

$$\mu_s(t,\lambda) = p^{-1/n}\lambda^{1/n}\exp\{2\pi is/n\}(1+o(1)),\ \text{as } |\lambda| \to \infty. \tag{10}$$

For this purpose we introduce $\nu_s(t,\lambda) \equiv \mu_s(t,\lambda) - p^{-1/n}\lambda^{1/n}\exp\{2\pi is/n\}$ and obtain from (3) that the functions $\nu_s(t,\lambda)$ are the roots of the equation

$$p\sum_{m=1}^{n} C_n^m \nu_s^m \lambda^{(n-m)/n} p^{(m-n)/n}\exp\{2\pi is(n-m)/n\} + \ eqno(11)$$

$$\sum_{k=0}^{n-1} p_k(t)\sum_{m=0}^{k} C_k^m \nu_s^m \lambda^{(k-m)/n} p^{(m-k)/n}\exp\{2\pi is(k-m)/n\} = 0.$$

Dividing by λ we obtain the following equation for the function $\eta_s \equiv \nu_s\lambda^{-1/n}$:

$$p\sum_{m=1}^{n} C_n^m \eta_s^m p^{(m-n)/n}\exp\{2\pi is(n-m)/n\} + \tag{12}$$

$$\sum_{k=0}^{n-1} p_k(t)\sum_{m=0}^{k} C_k^m \eta_s^m \lambda^{(k-n)/n} p^{(m-k)/n}\exp\{2\pi is(k-m)/n\} = 0.$$

For $|t| \leq q(\lambda)$ and $|\lambda| \to \infty$ (4) and (6) yield

$$|p_k(t)\lambda^{(k-n)/n}| \leq \alpha^{n-k}(q(|\lambda|))p^{n-k}(q(|\lambda|))|\lambda|^{(k-n)/n} = o(1)\ \text{as } |\lambda| \to \infty,$$

$$k = 0,\ldots,n-1.$$

Thus equation (12) has a root $\eta_s(t,\lambda) = o(1)$ as $|\lambda| \to \infty$. Then equation (11) has a root $\nu_s(t,\lambda) = o(1)\lambda^{1/n}$ which implies that (3) has a root of the form (10).

We will use (10) to derive two assertions which will be important later on:

i) $\int_{-q(|\lambda|)}^{q(|\lambda|)} |P_{km}(t,\lambda)|dt = o(1)$ as $|\lambda| \to \infty,\ k,m = 0,\dots n-1.$

ii) for sufficiently large values of $|\lambda|, \lambda \in \Lambda$, the functions $\Re(\mu_k(t,\lambda) - \mu_s(t,\lambda)),\quad k,s = 0,\dots n-1$, do not change sign on the segment $|t| \leq q(|\lambda|)$.

We begin with proving i). For $|t| \leq q(|\lambda|)$ relations (10) imply

$$\det \tilde{M}(t,\lambda) = \lambda^{(n-1)/2} p^{(1-n)/2} \left[\det(\exp(2\pi i sk/n))_{s,k=0}^{n-1} + o(1)\right] \text{ as } |\lambda| \to \infty,$$

which proves that $\det \tilde{M}(t,\lambda) \neq 0$ and also yields the following relation for the elements of the matrix $\tilde{M}^{-1} \equiv (\mu^0_{sk})_{s,k=0}^{n-1}$:

$$\mu^0_{sk} = \lambda^{-k/n}(\eta_{sk} + o(1)) \text{ as } |\lambda| \to \infty$$

for some constants $\eta_{sk},\ k,s = 0,\dots n-1$.

We now will study the behaviour of functions $d\mu_s(t,\lambda)/dt$ as $|\lambda| \to \infty,\ |t| \leq q(|\lambda|)$. Let us show that

$$|\mu_s'(t,\lambda)| = o(1)|\lambda|^{1/n} q^{-1}(|\lambda|),\ o(1) \to 0 \text{ as } |\lambda| \to \infty \tag{13}$$

In fact, (3) implies

$$\left(\sum_{k=0}^{n} p_k(t) k \mu_s^{k-1}\right) d\mu_s(t,\lambda)/dt = -\sum_{k=0}^{n-1} p_k'(t)\mu_s^k. \tag{14}$$

Using (4) and taking into account (10) we obtain

$$\left|\sum_{k=0}^{n} p_k(t) k \mu_s^{k-1}\right| \geq n|p|\,|\mu_s|^{n-1} - \sum_{k=1}^{n-1} k|p_k(t)|\,|\mu_s|^{k-1} \geq$$

$$2^{1-n}|\lambda|^{(n-1)/n} n|p|^{1/n} + o(1)|\lambda|^{(n-1)/n} \geq C_0|\lambda|^{(n-1)/n},\ C_0 > 0,$$

whereas (5) and (10) imply

$$\left|\sum_{k=0}^{n-1} p_k'(t)\mu_s^k\right| \leq \frac{o(1)|\lambda|}{1+q(|\lambda|)},\ |\lambda| \to \infty,\ |t| \leq q(|\lambda|).$$

Now (13) follows from (14) and the last two estimates.

The identity (14) leads us to the following formulas for the elements of the matrix $d\tilde{M}(t,\lambda)/dt \equiv (\mu^1_{ms})_{m,s=0}^{n-1} \equiv (m\mu_s^{m-1}\mu_s')_{m,s=0}^{n-1}$:

$$\mu^1_{ms} = o(1)|\lambda|^{m/n} q^{-1}(|\lambda|).$$

Therefore for $|t| \leq q(|\lambda|)$ we obtain $|P_{km}(t,\lambda)| = |\sum_{s=0}^{n-1} \mu^0_{ks}\mu^1_{sm}| = o(1)q(|\lambda|)$, and the assertion i) follows.

To prove ii) we note that (10) yields

$$\Re(\mu_k(t,\lambda) - \mu_s(t,\lambda)) = \tag{15}$$
$$[1+o(1)]\Re[p^{-1/n}\lambda^{1/n}(\exp\{2\pi ik/n\} - \exp\{2\pi is/n\})].$$

Since

$$\{\arg(\exp\{2\pi ik/n\} - \exp\{2\pi is/n\}),\ k,s = 0,\dots,n-1\} =$$
$$\{\arg\exp((\pi/2 + m\pi/n)i),\ m = 0,\dots,2n-1\},$$

and $\lambda \in \Lambda$, we obtain ii) from (15).

We now return to system (9) and replace it for sufficiently large $\lambda \in \Lambda$ by the system of integral equations

$$Z_{ss}(t,\lambda) = 1 + \int_{-q(|\lambda|)}^{t} \sum_{m=0}^{n-1} P_{sm}(\tau) Z_{ms}(\tau,\lambda)d\tau \tag{16}$$

$$Z_{ks}(t,\lambda) = \tag{17}$$
$$= \int_{-q(|\lambda|)}^{t} \exp\{\int_{\tau}^{t} [\mu_k(\theta,\lambda) - \mu_s(\theta,\lambda)]d\theta\} \sum_{m=0}^{n-1} P_{km}(\tau) Z_{ms}(\tau,\lambda)d\tau$$

for $k \neq s$ and $\Re(\mu_k - \mu_s) \leq 0,\ |t| \leq q(|\lambda|)$;

$$Z_{ks}(t,\lambda) = \tag{18}$$
$$= -\int_{t}^{q(|\lambda|)} \exp\{\int_{t}^{\tau} [\mu_s(\theta,\lambda) - \mu_k(\theta,\lambda)]d\theta\} \sum_{m=0}^{n-1} P_{km}(\tau) Z_{ms}(\tau,\lambda)d\tau$$

for $k \neq s$ and $\Re(\mu_k - \mu_s) \geq 0,\ |t| \leq q(|\lambda|)$;

We will solve system (16) - (18) by the method of successive approximations by setting $Z_{ks}(t,\lambda) = Z_{ks0}(t,\lambda) + \sum_{\ell=0}^{\infty}(Z_{ks,\ell+1}(t,\lambda) - Z_{ks\ell}(t,\lambda))$, where $Z_{ks0}(t,\lambda) = \delta_{ks}$ is the Kronecker symbol,

$$Z_{ks,\ell+1}(t,\lambda) = \tag{19}$$
$$= \int_{-q(|\lambda|)}^{t} \exp\{\int_{\tau}^{t} [\mu_k(\theta,\lambda) - \mu_s(\theta,\lambda)]d\theta\} \sum_{m=0}^{n-1} P_{sm}(\tau) Z_{ms\ell}(\tau,\lambda)d\tau$$

for $k \neq s$ and $\Re(\mu_k - \mu_s) \leq 0,\ |t| \leq q(|\lambda|)$;

$$Z_{ks,\ell+1}(t,\lambda) = \tag{20}$$
$$= -\int_{t}^{q(|\lambda|)} \exp\{\int_{t}^{\tau} [\mu_s(\theta,\lambda) - \mu_k(\theta,\lambda)]d\theta\} \sum_{m=0}^{n-1} P_{km}(\tau) Z_{ms\ell}(\tau,\lambda)d\tau$$

for $k \neq s$ and $\Re(\mu_k - \mu_s) \geq 0,\ |t| \leq q(|\lambda|)$;

$$Z_{kk,\ell+1}(t,\lambda) = 1 + \int_{-q(|\lambda|)}^{t} \sum_{m=0}^{n-1} P_{km}(\tau) Z_{mk\ell}(\tau,\lambda)d\tau. \tag{21}$$

If $h(|\lambda|) = \max_{0\le k\le n-1} \int_{-q(|\lambda|)}^{q(|\lambda|)} \sum_{m=0}^{n-1} |P_{km}(t)|dt$, then already proved assertion i) implies

$$h(|\lambda|) = o(1) \text{ as } |\lambda| \to \infty.$$

Formulae (19)-(21) for all $k, s = 0, \ldots, n-1$, $\ell \ge 1$ and $|t| \le q(|\lambda|)$ yield

$$|Z_{ks,\ell+1}(t,\lambda) - Z_{ks\ell}(t,\lambda)| \le$$

$$\int_{-q(|\lambda|)}^{q(|\lambda|)} \sum_{m=0}^{n-1} |P_{sm}(t)|dt \max_{0\le k,s\le n-1} |Z_{ks\ell}(t,\lambda) - Z_{ks,\ell-1}(t,\lambda)|.$$

Therefore, for $\ell \ge 1$, we have

$$\max_{\substack{k,s=\overline{0,n-1}\\ |t|\le q(|\lambda|)}} |Z_{ks,\ell+1}(t,\lambda) - Z_{ks\ell}(t,\lambda)| \le \tag{22}$$

$$h(|\lambda|) \max_{\substack{k,s=\overline{0,n-1}\\ |t|\le q(|\lambda|)}} |Z_{ks\ell}(t,\lambda) - Z_{ks,\ell-1}(t,\lambda)|$$

It is obvious that $|Z_{ks1}(t,\lambda) - Z_{ks0}(t,\lambda)| \le h(|\lambda|)$. Thus (22) implies

$$\sum_{\ell=0}^{\infty} |Z_{ks,\ell+1}(t,\lambda) - Z_{ks\ell}(t,\lambda)| \le h(|\lambda|)(1 - h(|\lambda|))^{-1},$$

$$k, s = 0, \ldots, n-1; \ |t| \le q(|\lambda|),$$

and hence system (16)-(18) has a solution

$$Z_{ks}(t,\lambda) = \begin{cases} 1 + o(1), & k = s = \overline{0, n-1}, \\ o(1), & 0 \le k \ne s \le n-1, \end{cases}$$

where $|\lambda| \to \infty$ and $|t| \le q(|\lambda|)$. Now we see that system (8) has a solution of the form $\bar{y}(t,\lambda) \equiv \bar{y}_s(t,\lambda) \equiv (y_{s0}(t,\lambda), \ldots, y_{s,n-1}(t,\lambda)) \exp \int_0^t \mu_s(\tau,\lambda)d\tau$, where

$$s = \overline{0, n-1},\ y_{sk}(t,\lambda) = \delta_{sk} + o(1), \text{ as } |\lambda| \to \infty,\ |t| \le q(|\lambda|),\ s, k = \overline{0, n-1}.$$

Finally we find the solution of equation (1) such that

$$(x_s(t,\lambda), dx_s(t,\lambda)/dt, \ldots, d^{n-1}x_s(t,\lambda)/dt^{n-1}) = \tilde{M}(t,\lambda)\bar{y}_s(t,\lambda) =$$
$$= \tilde{M}(t,\lambda)(y_{s0}(t,\lambda), \ldots, y_{s,n-1}(t,\lambda)) \exp\{\int_0^t \mu_s(\tau,\lambda)d\tau\},$$

which together with (10) implies (7). **QED**

3. The estimate of solutions of systems of equations

Let us now study a more general case of system (2) of differential equations with a complex parameter.

We set $Q(T,\Lambda) = (Q_{sk}(T,\Lambda))_{s,k=1}^n$ where $Q_{sk}(T,\Lambda) = \max\limits_{\substack{|t|\le T\\|\lambda|\le\Lambda}} |P_{sk}(t,\lambda)|$;

$$\det(\gamma I - Q(T,\Lambda)) \equiv \gamma^n + \sum_{\ell=0}^{n-1} \gamma^\ell Q_\ell(T,\Lambda);\ R(T,\Lambda) = \max_{0\le\ell\le n-1} |Q_\ell(T,\Lambda)|^{1/(n-\ell)};$$

and denote by $\Phi(t,\lambda)$ the fundamental matrix of system (2) normalized by the condition $\Phi(0,\lambda) = I$.

Theorem 2. *There exist constants $C > 0$ and $C_1 > 0$ such that*

$$\|\Phi(t,\lambda)\| \le C(1 + \|Q(|t|,|\lambda|)\||t|)^{n-1} \exp\{C_1|t|R(|t|,|\lambda|)\} \tag{23}$$

for all $t \in \mathbb{R}$, $\lambda \in \mathbb{C}$.

Proof. Let us write system (2) in the form

$$\bar{x}(t,\lambda) - \bar{x}(0,\lambda) = \int_0^t P(\tau,\lambda)\bar{x}(\tau,\lambda)d\tau,$$

which implies the following relation for $\Phi(t,\lambda)$:

$$\Phi(t,\lambda) = I + \sum_{m=0}^{\infty} \int_0^t \int_0^{t_0} \cdots \int_0^{t_{m-1}} \prod_{\ell=0}^{m} P(t_\ell,\lambda) dt_m \ldots dt_0 \equiv$$

$$I + \sum_{m=0}^{\infty} (\Phi_{msk})_{s,k=1}^n,\ m \ge 0.$$

Since

$$\left| \left[\prod_{\ell=0}^{m} P(t_\ell,\lambda) \right]_{sk} \right| \le [Q^{m+1}(|t|,|\lambda|)]_{sk},\ s,k = \overline{1,n},$$

then

$$|\Phi_{msk}(t,\lambda)| \le \frac{|t|^{m+1}}{(m+1)!} [Q^{m+1}(|t|,|\lambda|)]_{sk},$$

and it follows

$$|\Phi_{sk}(t,\lambda)| \le I_{sk} + \sum_{m=0}^{\infty} \frac{|t|^{m+1}}{(m+1)!} [Q^{m+1}(|t|,|\lambda|)]_{sk} = (\exp\{|t|Q(|t|,|\lambda|)\})_{sk}.$$

Using the estimate of the norm of $\exp(At)$ for any square matrix A given in [6] we obtain

$$\|\Phi(t,\lambda)\| \le \|\exp\{|t|Q(|t|,|\lambda|)\} \le$$
$$C\exp\{|t| \max_{1\le i\le n} \Re\gamma_i\}(1 + |t|\|Q(|t|,|\lambda|)\|)^{n-1}, \tag{24}$$

where γ_i, $i = 1,\ldots,n$, are the roots of equation

$$\det(\gamma I - Q(T,\Lambda)) \equiv \gamma^n + \sum_{\ell=0}^{n-1} \gamma^\ell Q_\ell(T,\Lambda) = 0. \tag{25}$$

We will show that the roots $\gamma_j(|t|,|\lambda|)$ of equation (25) are bounded by

$$|\gamma_j(|t|,|\lambda|)| \le C_1 R(|t|,|\lambda|),\ j=\overline{1,n}, t\in\mathbb{R}, \lambda\in\mathbb{C},\ C_1>0, \tag{26}$$

which together with (24) will imply (23).

Assume, to obtain the contradiction, that (26) is not true. Then for at least one root $\gamma_j(|t|,|\lambda|)$ there exists a sequence $(t_k,\lambda_k)\in\mathbb{R}\times\mathbb{C}$ such that

$$|\gamma_j(|t_k|,|\lambda_k|)| \ge kR(|t_k|,|\lambda_k|), \tag{27}$$

with, obviously, $|t_k|+|\lambda_k|\to\infty$ as $k\to\infty$. Since by (25) we have

$$1+\sum_{\ell=0}^{n-1}\gamma_j^{\ell-n}(|t_k|,|\lambda_k|)Q_\ell(|t_k|,|\lambda_k|)\equiv 0,\ k=1,2,\dots,$$

then for some $\ell, 0\le \ell\le n-1$, there exists a subsequence (t_{k_s},λ_{k_s}) such that

$$|\gamma_j^{\ell-n}(|t_{k_s}|,|\lambda_{k_s}|)|Q_\ell(|t_{k_s}|,|\lambda_{k_s}|)\ge C_2>0,$$

By (27) we obtain

$$|Q_\ell(|t_{k_s}|,|\lambda_{k_s}|)|\ge C_2|\gamma_j(|t_{k_s}|,|\lambda_{k_s}|)|^{n-\ell}\ge C_2k_s^{n-\ell}R^{n-\ell}(|t_{k_s}|,|\lambda_{k_s}|).$$

The last inequality is impossible by the definition of $R(t,\lambda)$. **QED**

Now we will show that estimate (23) cannot be improved. More specifically, we will prove that for every function $h(|t|,|\lambda|)=o(|t|R(|t|,|\lambda|))$ as $|t|+|\lambda|\to\infty$ and any $n\ge 1$ we can find a system defined by equation (2) for which (23) with $|t|R(|t|,|\lambda|)$ replaced by $h(|t|,|\lambda|)$ is not true with any choice of constants $C>0$, $C_1>0$. We obtain such a system by writing equation (1) in the form of system (2) and using Theorem 1. Consider the equation

$$d^n x(t,\lambda)/dt^n=(it^{2s}+\lambda)x(t,\lambda),\ s\in\mathbb{N}.$$

For this equation

$$\frac{d^k x(t,\lambda)}{dt^k}=\exp\{2\pi ik/n\}\lambda^{k/n}\exp\{\int_0^t(\lambda+i\tau^{2s})^{1/n}d\tau\}(1+o(1)), \tag{28}$$

as $|\lambda|\to\infty$, $|\arg\lambda|\ge\delta>0$, $|\arg\lambda-\pi|\ge\delta>0$, t belongs to the corresponding domain $|t|\le q(|\lambda|)$ and $(\lambda+i\tau^{2s})^{1/n}$ is one of the n values of the complex-valued function $w=z^{1/n}$ with $z=\lambda+i\tau^{2s}$. We now set $\lambda=i|\lambda|$, $|\lambda|\to\infty$ and fix $t\ne 0$. Asymptotic formula (28) implies that for a fixed $t\ne 0$ the following estimate holds with $\lambda=i\alpha, \alpha\in\mathbb{R}, \alpha\to\infty$

$$\|\Phi(t,\lambda)\|=|\lambda|^{(n-1)/n}\exp\{\int_0^t(|\lambda|+\tau^{2s})^{1/n}d\tau C_0\}(1+o(1))\ge \tag{29}$$

$$C_0'\exp\{C_0''|\lambda|^{1/n}\},\ C_0'>0,\ C_0''>0.$$

On the other hand, replacing the function $|t|R(|t|,|\lambda|)\equiv|t|(|t|^{2m}+|\lambda|)^{1/n}$ in estimate (23) by the function $h(|t|)=o(1)|t|R(|t|,|\lambda|)$ we obtain

$$\|\Phi(t,\lambda)\|\le C'|\lambda|^{n-1}\exp\{o(1)|\lambda|^{1/n}\},$$

which contradicts (29).

References

[1] Naimark, M. A. : Linear differential operators, New York, F. Ungar Pub. Co. 1967.

[2] Zolotaryov, G. N.: Uniqueness theorems for a class of integral representations, Matem. Sbornik, v.78(120), 1969, 408.

[3] Zhitomirskii, Ya. I.: Uniqueness classes for solutions of the Cauchy problem, Dokl. Akad. Nauk SSSR V. 172,1967, No.6= Soviet Math. Dokl. Vol.8,1967, No.1, 259.

[4] Zhitomirskii, Ya. I.: Uniqueness classes for the solution of the Cauchy problem for linear equations with rapidly growing coefficients, Dokl. Akad. Nauk SSSR, V. 173, 1967, No. 1= Soviet Math. Dokl. Vol. 8,1967, No.2, 317.

[5] Zhitomirskii, Ya. I.: On the asymptotics of solutions of systems of linear equations in expanding domains, Diff. Uravn., V.XII, N 8, 1976, 1427.

[6] Gel'fand, I. M. and Shilov, G. E.: Theory of differential equations Series title: Generalized functions, v. 3., New York, Academic Press, 1967, 307p.

Nehemia 10,
32294 Haifa, Israel

AMS Classification: 34A30 (34C11)

Titles previously published in the series

OPERATOR THEORY: ADVANCES AND APPLICATIONS

BIRKHÄUSER VERLAG

Edited by
I. Gohberg,
School of Mathematical Sciences, Tel-Aviv University, Ramat Aviv, Israel

This series is devoted to the publication of current research in operator theory, with particular emphasis on applications to classical analysis and the theory of integral equations, as well as to numerical analysis, mathematical physics and mathematical methods in electrical engineering.

47. **I. Gohberg** (Ed.): Extension and Interpolation of Linear Operators and Matrix Functions, 1990, (3-7643-2530-5)
48. **L. de Branges, I. Gohberg, J. Rovnyak** (Eds.): Topics in Operator Theory. Ernst D. Hellinger Memorial Volume, 1990, (3-7643-2532-1)
49. **I. Gohberg, S. Goldberg, M.A. Kaashoek:** Classes of Linear Operators, Volume I, 1990, (3-7643-2531-3)
50. **H. Bart, I. Gohberg, M.A. Kaashoek** (Eds.): Topics in Matrix and Operator Theory, 1991, (3-7643-2570-4)
51. **W. Greenberg, J. Polewczak** (Eds.): Modern Mathematical Methods in Transport Theory, 1991, (3-7643-2571-2)
52. **S. Prössdorf, B. Silbermann:** Numerical Analysis for Integral and Related Operator Equations, 1991, (3-7643-2620-4)
53. **I. Gohberg, N. Krupnik:** One-Dimensional Linear Singular Integral Equations, Volume I, Introduction, 1992, (3-7643-2584-4)
54. **I. Gohberg, N. Krupnik:** One-Dimensional Linear Singular Integral Equations, Volume II, General Theory and Applications, 1992, (3-7643-2796-0)
55. **R.R. Akhmerov, M.I. Kamenskii, A.S. Potapov, A.E. Rodkina, B.N. Sadovskii:** Measures of Noncompactness and Condensing Operators, 1992, (3-7643-2716-2)
56. **I. Gohberg** (Ed.): Time-Variant Systems and Interpolation, 1992, (3-7643-2738-3)
57. **M. Demuth, B. Gramsch, B.W. Schulze** (Eds.): Operator Calculus and Spectral Theory, 1992, (3-7643-2792-8)
58. **I. Gohberg** (Ed.): Continuous and Discrete Fourier Transforms, Extension Problems and Wiener-Hopf Equations, 1992, (3-7643-2809-6)
59. **T. Ando, I. Gohberg** (Eds.): Operator Theory and Complex Analysis, 1992, (3-7643-2824-X)
60. **P.A. Kuchment:** Floquet Theory for Partial Differential Equations, 1993, (3-7643-2901-7)

61. **A. Gheondea, D. Timotin, F.-H. Vasilescu** (Eds.): Operator Extensions, Interpolation of Functions and Related Topics, 1993, (3-7643-2902-5)
62. **T. Furuta, I. Gohberg, T. Nakazi** (Eds.): Contributions to Operator Theory and its Applications. The Tsuyoshi Ando Anniversary Volume, 1993, (3-7643-2928-9)
63. **I. Gohberg, S. Goldberg, M.A. Kaashoek:** Classes of Linear Operators, Volume 2, 1993, (3-7643-2944-0)
64. **I. Gohberg** (Ed.): New Aspects in Interpolation and Completion Theories, 1993, (3-7643-2948-3)
65. **M.M. Djrbashian:** Harmonic Analysis and Boundary Value Problems in the Complex Domain, 1993, (3-7643-2855-X)
66. **V. Khatskevich, D. Shoiykhet:** Differentiable Operators and Nonlinear Equations, 1993, (3-7643-2929-7)
67. **N.V. Govorov †:** Riemann's Boundary Problem with Infinite Index, 1994, (3-7643-2999-8)
68. **A. Halanay, V. Ionescu:** Time-Varying Discrete Linear Systems Input-Output Operators. Riccati Equations. Disturbance Attenuation, 1994, (3-7643-5012-1)
69. **A. Ashyralyev, P.E. Sobolevskii:** Well-Posedness of Parabolic Difference Equations, 1994, (3-7643-5024-5)
70. **M. Demuth, P. Exner, G. Neidhardt, V. Zagrebnov** (Eds): Mathematical Results in Quantum Mechanics. International Conference in Blossin (Germany), May 17–21, 1993, 1994, (3-7643-5025-3)
71. **E.L. Basor, I. Gohberg** (Eds): Toeplitz Operators and Related Topics. The Harold Widom Anniversary Volume. Workshop on Toeplitz and Wiener-Hopf Operators, Santa Cruz, California, September 20–22, 1992, 1994 (3-7643-5068-7)
72. **I. Gohberg, L.A. Sakhnovich** (Eds): Matrix and Operator Valued Functions. The Vladimir Petrovich Potapov Memorial Volume, (3-7643-5091-1)
73. **A. Feintuch, I. Gohberg** (Eds): Nonselfadjoint Operators and Related Topics. Workshop on Operator Theory and Its Applications, Beersheva, February 24–28, 1994, (3-7643-5097-0)
74. **R. Hagen, S. Roch, B. Silbermann:** Spectral Theory of Approximation Methods for Convolution Equations, 1994, (3-7643-5112-8)
75. **C.B. Huijsmans, M.A. Kaashoek, B. de Pagter**: Operator Theory in Function Spaces and Banach Lattices. The A.C. Zaanen Anniversary Volume, 1994 (ISBN 3-7643-5146-2)
76. **A.M. Krasnosellskii**: Asymptotics of Nonlinearities and Operator Equations, 1995, (ISBN 3-7643-5175-6)
77. **J. Lindenstrauss, V.D. Milman** (Eds): Geometric Aspects of Functional Analysis Israel Seminar GAFA 1992–94, 1995, (ISBN 3-7643-5207-8)
78. **M. Demuth, B.-W. Schulze** (Eds): Partial Differential Operators and Mathematical Physics, 1995, (ISBN 3-7643-5208-6)

79. **I. Gohberg, M.A. Kaashoek, F. van Schagen**: Partially Specified Matrices and Operators: Classification, Completion, Applications, 1995, (ISBN 3-7643-5259-0)
80. **I. Gohberg, H. Langer** (Eds): Operator Theory and Boundary Eigenvalue Problems. International Workshop in Vienna, July 27–30, 1993, 1995, (ISBN 3-7643-5275-2)
81. **H. Upmeier**: Toeplitz Operators and Index Theory in Several Complex Variables, 1996, (ISBN 3-7643-5282-5)
82. **T. Constantinescu**: Schur Parameters, Factorization and Dilation Problems, 1996, (ISBN 3-7643-5285-X)
83. **A.B. Antonevich**: Linear Functional Equations. Operator Approach, 1995, (ISBN 3-7643-2931-9)
84. **L.A. Sakhnovich**: Integral Equations with Difference Kernels on Finite Intervals, 1996, (ISBN 3-7643-5267-1)

85/ **Y.M. Berezansky, G.F. Us, Z.G. Sheftel**: Functional Analysis, Vol. I + Vol. II, 1996,
86. Vol. I (ISBN 3-7643-5344-9), Vol. II (3-7643-5345-7)

87. **I. Gohberg, P. Lancaster, P.N. Shivakumar** (Eds): Recent Developments in Operator Theory and Its Applications. International Conference in Winnipeg, October 2–6, 1994, 1996, (ISBN 3-7643-5414-5)
88. **J. van Neerven** (Ed.): The Asymptotic Behaviour of Semigroups of Linear Operators, 1996, (ISBN 3-7643-5455-0)
89. **Y. Egorov, V. Kondratiev**: On Spectral Theory of Elliptic Operators, 1996, (ISBN 3-7643-5390-2)
90. **A. Böttcher, I. Gohberg** (Eds): Singular Integral Operators and Related Topics. Joint German-Israeli Workshop, Tel Aviv, March 1–10, 1995, 1996, (ISBN 3-7643-5466-6)
91. **A.L. Skubachevskii**: Elliptic Functional Differential Equations and Applications, 1997, (ISBN 3-7643-5404-6)
92. **A.Ya. Shklyar**: Complete Second Order Linear Differential Equations in Hilbert Spaces, 1997, (ISBN 3-7643-5377-5)
93. **Y. Egorov, B.-W. Schulze**: Pseudo-Differential Operators, Singularities, Applications, 1997, (ISBN 3-7643-5484-4)
94. **M.I. Kadets, V.M. Kadets**: Series in Banach Spaces. Conditional and Unconditional Convergence, 1997, (ISBN 3-7643-5401-1)
95. **H. Dym, V. Katsnelson, B. Fritzsche, B. Kirstein** (Eds): Topics in Interpolation Theory, 1997, (ISBN 3-7643-5723-1)
96. **D. Alpay, A. Dijksma, H. de Snoo**: Schur Functions, Operator Colligations, and Reproducing Kernel Pontryagin Spaces, 1997, (ISBN 3-7643-5763-0)
97. **M.L. Gorbachuk / V.I. Gorbachuk**: M.G. Krein's Lectures on Entire Operators, 1997, (ISBN 3-7643-5704-5)